Quantification of Building Seismic Performance Factors

FEMA P695 / June 2009

Quantification of Building Seismic Performance Factors

Prepared by

APPLIED TECHNOLOGY COUNCIL
201 Redwood Shores Parkway, Suite 240
Redwood City, California 94065
www.ATCouncil.org

Prepared for

FEDERAL EMERGENCY MANAGEMENT AGENCY
Michael Mahoney, Project Officer
Robert D. Hanson, Technical Monitor
Washington, D.C.

PROJECT MANAGEMENT COMMITTEE
Charles Kircher (Project Technical Director)
Michael Constantinou
Gregory Deierlein
James R. Harris
Jon A. Heintz (Project Manager)
William T. Holmes (Project Tech. Monitor)
John Hooper
Allan R. Porush
Christopher Rojahn (Project Executive)

WORKING GROUPS
Jason Chou
Jiannis Christovasilis
Kelly Cobeen
Stephen Cranford
Brian Dean
Andre Filiatrault
Kevin Haas
Curt Haselton

WORKING GROUPS (CONT'D)
Helmut Krawinkler
Abbie Liel
Jiro Takagi
Assawin Wanitkorkul
Farzin Zareian

PROJECT REVIEW PANEL
Maryann T. Phipps (Chair)
Amr Elnashai
S.K. Ghosh
Ramon Gilsanz*
Ronald O. Hamburger
Jack Hayes
Richard E. Klingner
Philip Line
Bonnie E. Manley
Andrei M. Reinhorn
Rafael Sabelli

*ATC Board Representative

Notice

Foreword

The Federal Emergency Management Agency (FEMA) has the goal of reducing the ever-increasing cost that disasters inflict on our country. Preventing losses before they happen by designing and building to withstand anticipated forces from these hazards is one of the key components of mitigation, and is the only truly effective way of reducing the cost of these disasters.

As part of its responsibilities under the National Earthquake Hazards Reduction Program (NEHRP), and in accordance with the National Earthquake Hazards Reduction Act of 1977 (PL 94-125) as amended, FEMA is charged with supporting mitigation activities necessary to improve technical quality in the field of earthquake engineering. The primary method of addressing this charge has been supporting the investigation of seismic and related multi-hazard technical issues as they are identified by FEMA, the development and publication of technical design and construction guidance products, the dissemination of these products, and support of training and related outreach efforts. These voluntary resource guidance products present criteria for the design, construction, upgrade, and function of buildings subject to earthquake ground motions in order to minimize the hazard to life for all buildings and increase the expected performance of critical and higher occupancy structures.

The linear design procedure contained in modern building codes is based on the concept of converting the complicated nonlinear dynamic behavior of a building structure under seismic loading to an equivalent linear problem. The design process starts with the selection of a basic seismic force resisting system for the structure. The code specifies a series of prescriptive requirements for structures based on each such system. These prescriptive requirements regulate configuration, size, materials of construction, detailing, and minimum required strength and stiffness. These seismic design performance requirements are controlled through the assignment of a series of system response coefficients (R, C_d, Ω_0), which represent the material properties and design detailing of the selected system. Based on the linear dynamic response characteristics of the structure and these response coefficients, design lateral forces are distributed to the building's various structural elements using linear analysis techniques and the resulting member forces and structural deflections are calculated. Members are then proportioned to have adequate capacity to resist the calculated forces in combination with other prescribed loads to ensure that calculated displacements do not exceed maximum specified values.

As the codes have improved over the last several decades in how they address seismic design, one of the results was an expansion of code-approved seismic force resisting systems, with many individual systems classified by the type of detailing used. For each increment in detailing, response coefficients were assigned in the code, based largely on judgment and qualitative comparison with the known response capabilities of other systems. The result is that today's code includes more than 80 individual structural systems, each with individual system response coefficients somewhat arbitrarily assigned. Many of these recently defined structural systems have never been subjected to significant level of earthquake ground shaking and the potential response characteristics and ability to meet the design performance objectives is untested and unknown.

What was needed was a standard procedural methodology where the inelastic response characteristics and performance of typical structures designed to a set of structural system provisions could be quantified and the adequacy of the structural system provisions to meet the design performance objectives verified. Such a methodology would need to directly account for the potential variations in structure configuration of structures designed to a set of provisions, the variation in ground motion to which these structures may be subjected and available laboratory data on the behavioral characteristics of structural elements.

The objective of this publication was to develop a procedure to establish consistent and rational building system performance and response parameters (R, C_d, Ω_0) for the linear design methods traditionally used in current building codes. The primary application of the procedure is for the evaluation of structural systems for new construction with equivalent earthquake performance. The primary design performance objective was taken to minimize the risk of structural collapse under the seismic load of maximum considered earthquake as specified in the current *NEHRP Recommended Provisions for New Buildings and Other Structures* (FEMA 450). Although the R factor is the factor of most concern, displacements and material detailing to achieve the implied design ductilities were also included.

It is anticipated that this methodology will ultimately be used by the nation's model building codes and standards to set minimum acceptable design criteria for standard code-approved systems, and to provide guidance in the selection of appropriate design criteria for other systems when linear design methods are applied. This publication will also provide a basis for future evaluation of the current tabulation of and limitations on code-approved structural systems for adequacy to achieve the inherent seismic performance objectives. This material could then potentially be used to modify or eliminate those systems or requirements that can not reliably meet these objectives.

FEMA wishes to express its sincere gratitude to Charlie Kircher, Project Technical Director, and to the members of the Project Team for their efforts in the development of this recommended methodology. The Project Management Committee consisted of Michael Constantinou, Greg Deierlein, Jim Harris, John Hooper, and Allan Porush. They in turn guided the Project Working Groups, which included Andre Filiatrault, Helmut Krawinkler, Kelly Cobeen, Curt Haselton, Abbie Liel, Jiannis Christovasilis, Jason Chou, Stephen Cranford, Brian Dean, Kevin Haas, Jiro Takagi, Assawin Wanitkorkul, and Farzin Zareian. The Project Review Panel consisted of Maryann Phipps (Chair), Amr Elnashai, S.K. Ghosh, Ramon Gilsanz, Ron Hamburger, Jack Hayes, Rich Klingner, Phil Line, Bonnie Manley, Andrei Reinhorn, and Rafael Sabelli, and they provided technical advice and consultation over the duration of the work. The names and affiliations of all who contributed to this report are provided in the list of Project Participants.

Without their dedication and hard work, this project would not have been possible. The American public who live, work and play in buildings in seismic areas are all in their debt.

Federal Emergency Management Agency

Preface

In September 2004 the Applied Technology Council (ATC) was awarded a "Seismic and Multi-Hazard Technical Guidance Development and Support" contract (HSFEHQ-04-D-0641) by the Federal Emergency Management Agency (FEMA) to conduct a variety of tasks, including one entitled "Quantification of Building System Performance and Response Parameters" (ATC-63 Project). The purpose of this project was to establish and document a recommended methodology for reliably quantifying building system performance and response parameters for use in seismic design. These factors include the response modification coefficient (R factor), the system overstrength factor (Ω_0), and the deflection amplification factor (C_d), collectively referred to as "seismic performance factors."

Seismic performance factors are used to estimate strength and deformation demands on systems that are designed using linear methods of analysis, but are responding in the nonlinear range. Their values are fundamentally critical in the specification of seismic loading. R factors were initially introduced in the ATC-3-06 report, *Tentative Provisions for the Development of Seismic Regulations for Buildings*, published in 1978, and subsequently replaced by the *NEHRP Recommended Provisions for Seismic Regulations for New Buildings and Other Structures*, published by FEMA. Original R factors were based on judgment or on qualitative comparisons with the known response capabilities of seismic-force-resisting systems in use at the time. Since then, the number of systems addressed in current seismic codes and standards has increased substantially, and their ability to meet intended seismic performance objectives is largely unknown.

The recommended methodology described in this report is based on a review of relevant research on nonlinear response and collapse simulation, benchmarking studies of selected structural systems, and evaluations of additional structural systems to verify the technical soundness and applicability of the approach. Technical review and comment at critical developmental stages was provided by a panel of experts, which included representatives from the steel, concrete, masonry and wood material industry groups. A workshop of invited experts and other interested stakeholders was convened to receive feedback on the recommended methodology, and input from this group was instrumental in shaping the final product.

ATC is indebted to the leadership of Charlie Kircher, Project Technical Director, and to the members of the ATC-63 Project Team for their efforts in the development of this recommended methodology. The Project Management Committee, consisting of Michael Constantinou, Greg Deierlein, Jim Harris, John Hooper, and Allan Porush monitored and guided the technical efforts of the Project Working Groups, which included Andre Filiatrault, Helmut Krawinkler, Kelly Cobeen, Curt Haselton, Abbie Liel, Jiannis Christovasilis, Jason Chou, Stephen Cranford, Brian Dean, Kevin Haas, Jiro Takagi, Assawin Wanitkorkul, and Farzin Zareian. The Project Review Panel, consisting of Maryann Phipps (Chair), Amr Elnashai, S.K. Ghosh, Ramon Gilsanz, Ron Hamburger, Jack Hayes, Rich Klingner, Phil Line, Bonnie Manley, Andrei Reinhorn, and Rafael Sabelli provided technical advice and consultation over the duration of the work. The names and affiliations of all who contributed to this report are provided in the list of Project Participants.

ATC also gratefully acknowledges Michael Mahoney (FEMA Project Officer), Robert Hanson (FEMA Technical Monitor), and William Holmes (ATC Project Technical Monitor) for their input and guidance in the preparation of this report, Peter N. Mork and Ayse Hortacsu for ATC report production services, and Ramon Gilsanz as ATC Board Contact.

Jon A. Heintz
ATC Director of Projects

Christopher Rojahn
ATC Executive Director

Executive Summary

This report describes a recommended methodology for reliably quantifying building system performance and response parameters for use in seismic design. The recommended methodology (referred to herein as the Methodology) provides a rational basis for establishing global seismic performance factors (SPFs), including the response modification coefficient (R factor), the system overstrength factor (Ω_0), and deflection amplification factor (C_d), of new seismic-force-resisting systems proposed for inclusion in model building codes.

The purpose of this Methodology is to provide a rational basis for determining building seismic performance factors that, when properly implemented in the seismic design process, will result in *equivalent safety against collapse in an earthquake, comparable to the inherent safety against collapse intended by current seismic codes, for buildings with different seismic-force-resisting systems.*

As developed, the following key principles outline the scope and basis of the Methodology:

- It is applicable to new building structural systems.

- It is compatible with the *NEHRP Recommended Provisions for Seismic Regulations for New Buildings and Other Structures* (FEMA, 2004a) and ASCE/SEI 7, *Minimum Design Loads for Buildings and Other Structures*, (ASCE, 2006a).

- It is consistent with a basic life safety performance objective inherent in current seismic codes and standards.

- Earthquake hazard is based on Maximum Considered Earthquake ground motions.

- Concepts are consistent with seismic performance factor definitions in current seismic codes and standards.

- Safety is expressed in terms of a collapse margin ratio.

- Performance is quantified through nonlinear collapse simulation on a set of archetype models.

- Uncertainty is explicitly considered in the collapse performance evaluation.

The Methodology is intended to apply broadly to all buildings, recognizing that this objective may not be fully achieved for certain seismic environments and building configurations. Likewise, the Methodology has incorporated certain simplifying assumptions deemed appropriate for reliable evaluation of seismic performance. Key assumptions and potential limitations of the Methodology are presented and summarized.

In the development of the Methodology, selected seismic-force-resisting systems were evaluated to illustrate the application of the Methodology and verify its methods. Results of these studies provide insight into the collapse performance of buildings and appropriate values of seismic performance factors. Observations and conclusions in terms of generic findings applicable to all systems, and specific findings for certain types of seismic-force-resisting systems are presented. These findings should be considered generally representative, but not necessarily indicative of all possible trends, given limitations in the number and types of systems evaluated.

The Methodology is recommended for use with model building codes and resource documents to set minimum acceptable design criteria for standard code-approved seismic-force-resisting systems, and to provide guidance in the selection of appropriate design criteria for other systems when linear design methods are applied. It also provides a basis for evaluation of current code-approved systems for their ability to achieve intended seismic performance objectives. It is possible that results of future work based on this Methodology could be used to modify or eliminate those systems or requirements that cannot reliably meet these objectives.

Table of Contents

List of Figures

List of Tables

Chapter 1

Introduction

This report describes a recommended methodology for reliably quantifying building system performance and response parameters for use in seismic design. The recommended methodology (referred to herein as the Methodology) provides a rational basis for establishing global seismic performance factors (SPFs), including the response modification coefficient (R factor), the system overstrength factor (Ω_0), and deflection amplification factor (C_d), of new seismic-force-resisting systems proposed for inclusion in model building codes.

1.1   Background and Purpose

The Applied Technology Council (ATC) was commissioned by the Federal Emergency Management Agency (FEMA) under the ATC-63 Project to develop a methodology for quantitatively determining global seismic performance factors for use in seismic design.

Seismic performance factors are used in current building codes and standards to estimate strength and deformation demands on seismic-force-resisting systems that are designed using linear methods of analysis, but are responding in the nonlinear range. R factors were initially introduced in the ATC-3-06 report, *Tentative Provisions for the Development of Seismic Regulations for Buildings* (ATC, 1978), and their values have become fundamentally critical in the specification of design seismic loading.

Since then, the number of structural systems addressed in seismic codes has increased dramatically. The 2003 edition of the National Earthquake Hazards Reduction Program (NEHRP) *Recommended Provisions for Seismic Regulations for New Buildings and Other Structures* (*NEHRP Recommended Provisions*), (FEMA, 2004a), includes more than 75 individual systems, each having a somewhat arbitrarily assigned R factor.

Original R factors were based largely on judgment and qualitative comparisons with the known response capabilities of relatively few seismic-force-resisting systems in widespread use at the time. Many recently defined seismic-force-resisting systems have never been subjected to any significant level of earthquake ground shaking. As a result, the seismic response characteristics of many systems, and their ability to meet seismic design performance objectives, are both untested and unknown.

As new systems continue to be introduced during each code update cycle, uncertainty in the seismic performance capability of the new building stock continues to grow, and the need to quantify the seismic performance delivered by current seismic design regulations becomes more urgent. Advances in performance-based seismic design tools and technologies has resulted in the ability to use nonlinear collapse simulation techniques to link seismic performance factors to system performance capabilities on a probabilistic basis.

The purpose of this Methodology is to provide a rational basis for determining building system performance and response parameters that, when properly implemented in the seismic design process, will result in *equivalent safety against collapse in an earthquake, comparable to the inherent safety against collapse intended by current seismic codes, for buildings with different seismic-force-resisting systems.*

The Methodology is recommended for use with model building codes and resource documents to set minimum acceptable design criteria for standard code-approved seismic-force-resisting systems, and to provide guidance in the selection of appropriate design criteria for other systems when linear design methods are applied. It also provides a basis for evaluation of current code-approved systems for their ability to achieve intended seismic performance objectives. It is possible that results of future work based on this Methodology could be used to modify or eliminate those systems or requirements that cannot reliably meet these objectives.

1.2　Scope and Basis of the Methodology

The following key principles outline the scope and basis of the Methodology.

1.2.1　Applicable to New Building Structural Systems

The Methodology applies to the determination of seismic performance factors appropriate for the design of seismic-force-resisting systems in new building structures. While the Methodology is conceptually applicable (with some limitations) to design of non-building structures, and to retrofit of seismic-force-resisting systems in existing buildings, such systems were not explicitly considered. The Methodology is not intended to apply to the design of nonstructural systems.

1.2.2　Compatible with the NEHRP Recommended Provisions and ASCE/SEI 7

The Methodology is based on, and intended for use with, applicable design criteria and requirements of the most current editions of the *NEHRP*

Recommended Provisions for Seismic Regulations for New Buildings and Other Structures (NEHRP Recommended Provisions), (FEMA, 2004a), and the seismic provisions of ASCE/SEI 7-05, *Minimum Design Loads for Buildings and Other Structures*, (ASCE, 2006a). The Building Seismic Safety Council has adopted ASCE/SEI 7-05 as the "starting point" for the development of its 2009 and future editions of the *NEHRP Recommended Provisions*. At this time, ASCE/SEI 7-05 is the most current, published source of seismic regulations for model building codes in the United States.[1]

ASCE/SEI 7-05 provides the basis for ground motion criteria and "generic" structural design requirements applicable to currently accepted and future (proposed) seismic-force-resisting systems. ASCE/SEI 7-05 provisions include detailing requirements for currently approved systems that may also apply to new systems. By reference, other standards, such as ACI 318, *Building Code Requirements for Structural Concrete* (ACI, 2005), AISC/ANSI 341, *Seismic Provisions for Structural Steel Buildings* (AISC, 2005), ACI 530/ASCE 5/TMS 402, *Building Code Requirements for Masonry Structures* (ACI, 2002b), and ANSI/AF&PA, *National Design Specification for Wood Construction* (ANSI/AF&PA, 2005) apply to currently approved systems, and may also apply to new systems.

The Methodology requires the seismic-force-resisting system of interest to comply with all applicable design requirements in ASCE/SEI 7-05, including limits on system irregularity, drift, and height, except when such requirements are specifically excluded and explicitly evaluated in the application of the Methodology. For new (proposed) systems, the Methodology requires identification and use of applicable structural design and detailing requirements in ASCE/SEI 7-05, and development and use of new requirements as necessary to adequately describe system limitations and ensure predictable seismic behavior of components. The latest edition of the *NEHRP Recommended Provisions*, containing modifications and commentary to ASCE/SEI 7-05, may be a possible source for additional design requirements.

1.2.3 Consistent with the Life Safety Performance Objective

The Methodology is consistent with the primary "life safety" performance objective of seismic regulations in model building codes. As stated in the *Part 2: Commentary* to the *NEHRP Recommended Provisions for Seismic*

---

[1] This chapter and other sections of this document refer to ASCE/SEI 7-05 for design criteria and requirements to illustrate the Methodology, and to define the values of certain parameters used for performance evaluation. The Methodology is intended to be generally applicable, and such references should not be construed as limiting the Methodology to this edition of ASCE/SEI 7.

Regulations for New Buildings and Other Structures (Commentary to the *NEHRP Recommended Provisions*), (FEMA, 2004b), "the *Provisions* provides the minimum criteria considered prudent for protection of life safety in structures subject to earthquakes."

Design for performance other than life safety was not explicitly considered in the development of the Methodology. Accordingly, the Methodology does not address special performance or functionality objectives of ASCE/SEI 7-05 for Occupancy III and IV structures.

1.2.4 Based on Acceptably Low Probability of Structural Collapse

The Methodology achieves the primary life safety performance objective by requiring an acceptably low probability of collapse of the seismic-force-resisting system when subjected to Maximum Considered Earthquake (MCE) ground motions.

In general, life safety risk (i.e., probability of death or life-threatening injury) is difficult to calculate accurately due to uncertainty in casualty rates given collapse, and even greater uncertainty in assessing the effects of falling hazards in the absence of collapse. Collapse of a structure can lead to very different numbers of fatalities depending on variations in construction or occupancy, such as structural system type and the number of building occupants. Rather than attempting to quantify uniform protection of "life safety", the Methodology provides approximate uniform protection against collapse of the structural system.

Collapse includes both partial and global instability of the seismic-force-resisting system, but does not include local failure of components not governed by global seismic performance factors, such as localized out-of-plane failure of wall anchorage and potential life-threatening failure of non-structural systems.

Similarly, the Methodology does not explicitly address components that are not included in the seismic-force-resisting system (e.g., gravity system components and nonstructural components). It assumes that deformation compatibility and related requirements of ASCE/SEI 7-05 adequately protect such components against premature failure. Components that are not designated as part of the seismic-force-resisting system are not controlled by seismic-force-resisting system design requirements. Accordingly, they are not considered in evaluating the overall resistance to collapse.

1.2.5 Earthquake Hazard based on MCE Ground Motions

The Methodology evaluates collapse under Maximum Considered
Earthquake (MCE) ground motions for various geographic regions of
seismicity, as defined by the coefficients and mapped acceleration parameters
of the general procedure of ASCE/SEI 7-05, which is based on the maps and
procedures contained in the *NEHRP Recommended Provisions*.

While seismic performance factors apply to the design response spectrum,
taken as two-thirds of the MCE spectrum, code-defined MCE ground
motions are considered the appropriate basis for evaluating structural
collapse. As noted in the Commentary to the *NEHRP Recommended
Provisions*, "if a structure experiences a level of ground motion 1.5 times the
design level, the structure should have a low likelihood of collapse."

1.2.6 Concepts Consistent with Current Seismic Performance Factor Definitions

The Methodology remains true to the definitions of seismic performance
factors given in ASCE/SEI 7-05, and the underlying nonlinear static analysis
(pushover) concepts described in the Commentary to the *NEHRP
Recommended Provisions*. Values of the response modification coefficient,
R factor, the system overstrength factor, Ω_O , and the deflection amplification
factor, C_d, for currently approved seismic-force-resisting systems are
specified in Table 12.2-1 of ASCE/SEI 7-05. Section 4.2 of the Commentary
to the *NEHRP Recommended Provisions* provides background information
on seismic performance factors.

Figures 1-1 and 1-2 are used to explain and illustrate seismic performance
factors, and how they are used in the Methodology. Parameters are defined
in terms of equations, which in all cases are dimensionless ratios of force,
acceleration or displacement. However, in attempting to utilize the figures to
clarify and to illustrate the meanings of these ratios, graphical license is taken
in two ways. First, seismic performance factors are depicted in the figures as
incremental differences between two related parameters, rather than as ratios
of the parameters. Second, as a consequence of being depicted as
incremental differences, seismic performance factors are shown on plots with
units, when, in fact, they are dimensionless.

Figure 1-1, an adaptation of Figures C4.2-1 and C4.2-3 from the
Commentary to the *NEHRP Recommended Provisions*, defines seismic
performance factors in terms of the global inelastic response (idealized
pushover curve) of the seismic-force-resisting system. In this figure, the

horizontal axis is lateral displacement (i.e., roof drift) and the vertical axis is lateral force at the base of the system (i.e., base shear).

In Figure 1-1, the term V_E represents the force level that would be developed in the seismic-force-resisting system, if the system remained entirely linearly elastic for design earthquake ground motions. The term V_{max} represents the actual, maximum strength of the fully-yielded system, and the term V is the seismic base shear required for design. As defined in Equation 1-1, the R factor is the ratio of the force level that would be developed in the system for design earthquake ground motions (if the system remained entirely linearly elastic) to the base shear prescribed for design:

$$R = \frac{V_E}{V} \tag{1-1}$$

and, as defined in Equation 1-2, the Ω_O factor is the ratio of the maximum strength of the fully-yielded system to the design base shear:

$$\Omega_O = \frac{V_{max}}{V} \tag{1-2}$$

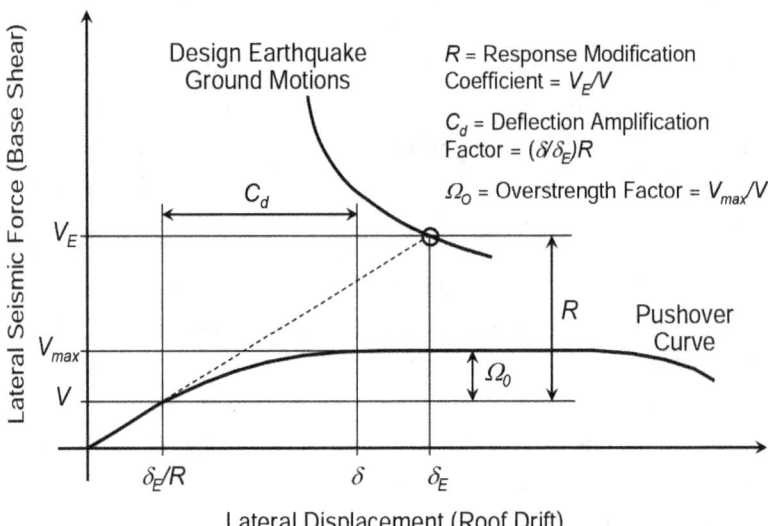

Figure 1-1   Illustration of seismic performance factors (R, Ω_O, and C_d) as defined in the Commentary to the *NEHRP Recommended Provisions* (FEMA, 2004b).

In Figure 1-1, the term δ_E/R represents roof drift of the seismic-force-resisting system corresponding to design base shear, V, assuming that the system remains essentially elastic for this level of force, and the term δ represents the assumed roof drift of the yielded system corresponding to design earthquake ground motions. As illustrated in the figure and defined by Equation 1-3, the C_d factor is some fraction of the R factor (typically less than 1.0):

$$C_d = \frac{\delta}{\delta_E} R \qquad (1-3)$$

The Methodology develops seismic performance factors consistent with the concepts and definitions described above. Figure 1-2 illustrates the seismic performance factors defined by the Methodology and their relationship to MCE ground motions.

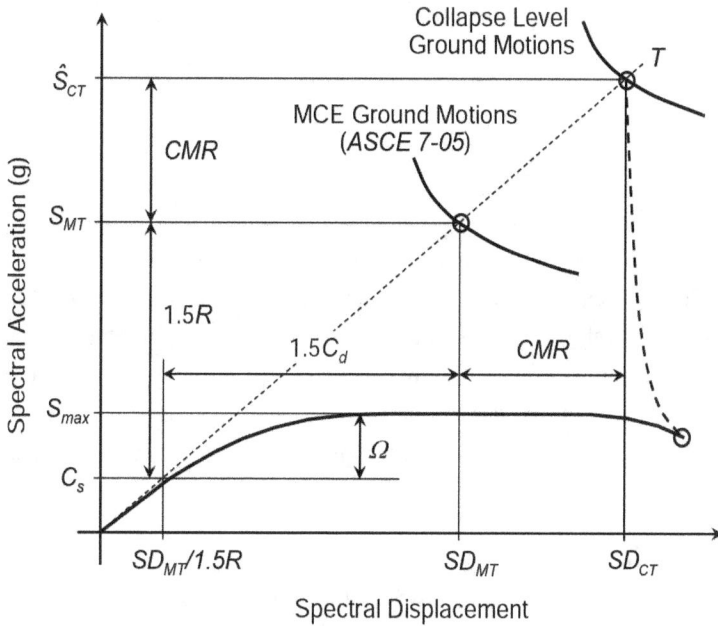

Figure 1-2   Illustration of seismic performance factors (R, Ω, and C_d) as defined by the Methodology.

Figure 1-2 parallels the "pushover" concept shown in Figure 1-1 using spectral coordinates rather than lateral force (base shear) and lateral displacement (roof drift) coordinates. Conversion to spectral coordinates is based on the assumption that 100% of the effective seismic weight of the structure, W, participates in fundamental mode at period, T, consistent with Equation 12.8-1 of ASCE/SEI 7-05:

$$V = C_s W \qquad (1-4)$$

In Figure 1-2, the term S_{MT} is the Maximum Considered Earthquake (MCE) spectral acceleration at the period of the system, T, the term S_{max} represents the maximum strength of the fully-yielded system (normalized by the effective seismic weight, W, of the structure), and the term C_s is the seismic response coefficient. As defined in Equation 1-5, the ratio of the MCE spectral acceleration to the seismic response coefficient, which is the design-level acceleration, is equal to 1.5 times the R factor:

$$1.5R = \frac{S_{MT}}{C_s} \qquad (1\text{-}5)$$

The 1.5 factor in Equation 1-5 accounts for the definition of design earthquake ground motions in ASCE/SEI 7-05, which is two-thirds of MCE ground motions.

In Figure 1-2, the overstrength parameter, Ω, is defined as the ratio of the maximum strength of the fully-yielded system, S_{max} (normalized by W), to the seismic response coefficient, C_s:

$$\Omega = \frac{S_{max}}{C_s} \qquad (1\text{-}6)$$

The Methodology calculates the overstrength parameter, Ω, based on nonlinear static (pushover) analysis. Calculated values of overstrength, Ω, are different from the overstrength factor, Ω_O, of ASCE/SEI 7-05, which is specified for design of non-ductile elements. In general, different designs of the same system will have different calculated values of overstrength, and the parameter, Ω, will vary. The single value of Ω that is considered to be most appropriate for use in design of the system of interest, is the value ultimately selected for Ω_O.

In Figure 1-2, inelastic system displacement at the MCE level is defined as $1.5C_d$ times the displacement corresponding to the design seismic response coefficient, C_s, and set equal to the MCE elastic system displacement, SD_{MT} (based on the "Newmark rule"), effectively redefining the C_d factor to be equal to the R factor:

$$C_d = R \qquad (1\text{-}7)$$

The equal displacement assumption is reasonable for most conventional systems with effective damping approximately equal to the nominal 5% level used to define response spectral acceleration and displacement. Systems with substantially higher (or lower) levels of damping would have significantly smaller (or larger) displacements than those with 5%-damped elastic response. As one example, systems with viscous dampers have significantly higher damping than 5%. For such systems, the response modification methods of Chapter 18 of ASCE/SEI 7-05 are used to determine an appropriate value of the C_d factor, as a fraction of the R factor.

1.2.7 Safety Expressed in Terms of Collapse Margin Ratio

The Methodology defines collapse level ground motions as the intensity that would result in median collapse of the seismic-force-resisting system. Median collapse occurs when one-half of the structures exposed to this intensity of ground motion would have some form of life-threatening collapse. As shown in Figure 1-2, collapse level ground motions are higher than MCE ground motions. As such, MCE ground motions would result in a comparatively smaller probability of collapse. As defined in Equation 1-8, the collapse margin ratio, CMR, is the ratio of the median 5%-damped spectral acceleration of the collapse level ground motions, $\hat{S}_{CT}$ (or corresponding displacement, SD_{CT}), to the 5%-damped spectral acceleration of the MCE ground motions, S_{MT} (or corresponding displacement, SD_{MT}), at the fundamental period of the seismic-force-resisting system:

$$CMR = \frac{\hat{S}_{CT}}{S_{MT}} = \frac{SD_{CT}}{SD_{MT}} \qquad (1\text{-}8)$$

In one sense, the collapse margin ratio, CMR, could be thought of as the amount S_{MT} must be increased to achieve building collapse by 50% of the ground motions. Collapse of the seismic-force-resisting system, and hence CMR, is influenced by many factors, including ground motion variability and uncertainty in design, analysis, and construction of the structure. These factors are considered collectively in a collapse fragility curve that describes the probability of collapse of the seismic-force-resisting system as a function of the intensity of ground motion.

1.2.8 Performance Quantified Through Nonlinear Collapse Simulation on a Set of Archetype Models

The Methodology determines the response modification coefficient, R factor, and evaluates the system over-strength factor, Ω, using nonlinear models of seismic-force-resisting system "archetypes." Archetypes capture the essence and variability of the performance characteristics of the system of interest. The Methodology requires nonlinear analysis of a sufficient number of archetype models, with parametric variations in design parameters, to broadly represent the system of interest.

The Methodology requires archetype models to meet the applicable design requirements of ASCE/SEI 7-05 and related standards, and additional criteria developed for the system of interest. Archetype design assumes a trial value of the R factor to determine the seismic response coefficient, C_s. The Methodology requires detailed modeling of nonlinear behavior of archetypes, based on representative test data sufficient to capture collapse failure modes.

Collapse failure modes that cannot be explicitly modeled are evaluated using appropriate limits on the controlling response parameter.

1.2.9 Uncertainty Considered in Performance Evaluation

The Methodology defines acceptable values of the collapse margin ratio in terms of an acceptably low probability of collapse for MCE ground motions, given uncertainty in the collapse fragility. Systems that have more robust design requirements, more comprehensive test data, and more detailed nonlinear analysis models, have less collapse uncertainty, and can achieve the same level of life safety with smaller collapse margin ratios.

Calculated values of collapse margin ratio are compared with acceptable values that reflect collapse uncertainty. If the calculated collapse margin is large enough to meet the performance objective (i.e., an acceptably small probability of collapse at the MCE), then the trial value of the R factor used in the archetype design is acceptable. If not, a new (lower) trial value of the R factor must be re-evaluated using the Methodology, or other limitations on the system of interest (e.g., height restrictions in the design requirements) must be considered.

1.3 Content and Organization

This report is written and organized to facilitate potential use and adoption by the *NEHRP Recommended Provisions*. Chapter 2 provides an overview of the Methodology, introducing the basic theory and concepts that are described in more detail in the chapters that follow.

Chapters 3 through 7 step through the elements of the Methodology, including required system information, structure archetype development, nonlinear modeling, criteria for collapse assessment, nonlinear analysis, and evaluation of seismic performance factors.

Chapter 8 defines documentation and peer review requirements, and describes recommended qualifications, expertise, and responsibilities for personnel involved with implementing the Methodology in the development and review of a proposed system.

Chapter 9 provides example applications intended to assist users in implementing the Methodology, and to validate the technical approach. Example systems include special and ordinary reinforced concrete moment frame systems, and wood light-frame systems.

Chapter 10 includes supporting studies on non-simulated collapse failure modes for steel moment frame systems, and on dynamic response characteristics, performance properties, and collapse failure modes unique to seismically-isolated structures.

Chapter 11 provides summary conclusions, recommendations, and limitations on the use of the Methodology.

Appendices A through F provide background information supporting the development of the Methodology, and expanded guidance on key aspects of the Methodology.

A glossary of definitions and list of symbols used throughout this report, along with a list of references, are provided at the end of this report.

Chapter 2
Overview of Methodology

This chapter outlines the general framework of the Methodology and describes the overall process. It introduces the key elements of the Methodology, including required system information, development of structural system archetypes, archetype models, nonlinear analysis of archetypes, performance evaluation, and documentation and peer review requirements. These elements are specified in more detail in the chapters that follow.

2.1   General Framework

The Methodology consists of a framework for establishing seismic performance factors (SPFs) that involves the development of detailed system design information and probabilistic assessment of collapse risk. It utilizes nonlinear analysis techniques, and explicitly considers uncertainties in ground motion, modeling, design, and test data. The technical approach is a combination of traditional code concepts, advanced nonlinear dynamic analyses, and risk-based assessment techniques.

Reliable analysis requires valid ground motions and representative nonlinear models of the seismic-force-resisting system. Development of representative models requires both detailed design information and comprehensive nonlinear test data on structural components and assemblies that make up the system of interest. Figure 2-1 illustrates the key elements of the Methodology.

The Methodology includes fully defined characterizations of ground motion and methods of analysis that are generically applicable to all seismic-force-resisting systems. Design information and test data will be different for each system, and may not yet exist for new systems. The Methodology includes requirements for defining the type of design information and test data that are needed for developing representative analytical models of the seismic-force-resisting system of interest.

Rather than establishing minimum requirements for design information and test data, the use of better quality information is encouraged by rewarding systems that have "done their homework." Systems that are based on well-defined design requirements and comprehensive test data will have inherently less uncertainty in their seismic performance. Such systems will need a lower margin against collapse to achieve an equivalent level of safety, as compared to systems with less robust data.

Due to the complexity of nonlinear dynamic analysis, the difficulty in modeling inelastic behavior, and the need to verify the adequacy and quality of design information and test data, the Methodology requires independent peer review of the entire process.

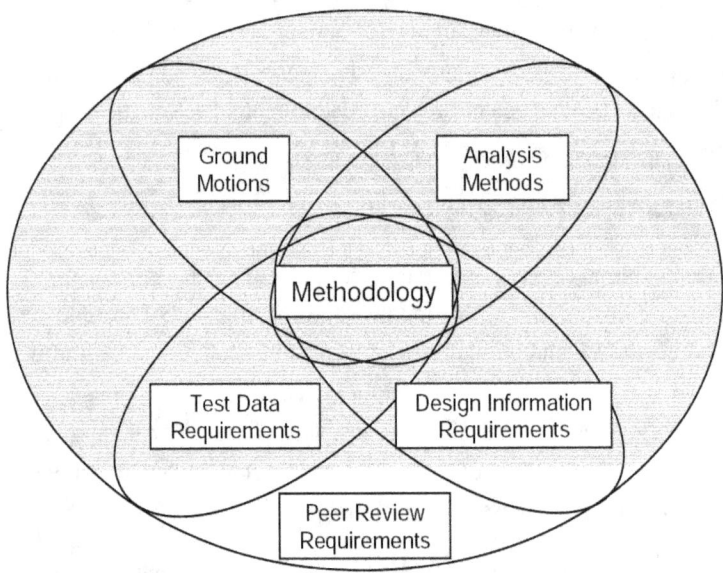

Figure 2-1   Key elements of the Methodology.

2.2   Description of Process

The steps comprising the Methodology are shown in Figure 2-2. These steps outline a process for developing system design information with enough detail and specificity to identify the permissible range of application for the proposed system, adequately simulate nonlinear response, and reliably assess the collapse risk over the proposed range of applications. Each step is linked to a corresponding chapter in this report, and described in the sections that follow.

2.3   Develop System Concept

The process begins with the development of a well-defined concept for the seismic-force-resisting system, including type of construction materials, system configuration, inelastic dissipation mechanisms, and intended range of application.

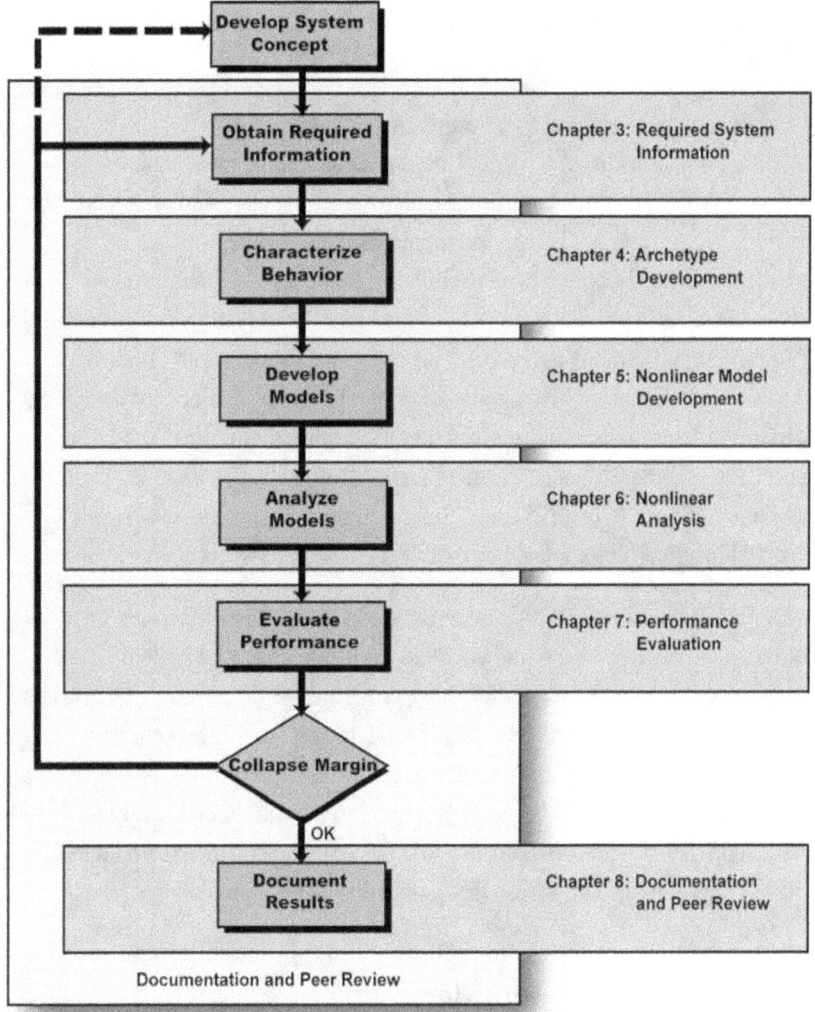

Figure 2-2   Process for quantitatively establishing and documenting seismic performance factors (SPFs).

The amount of documentation necessary to describe the system and characterize system components will vary, depending on the novelty and uniqueness of the proposed system relative to other well-established structural systems.

2.4   Obtain Required Information

Required system information is specified in Chapter 3. Required information includes detailed design requirements and results from material, component, and system testing. Design requirements include the rules that engineers will use to proportion and detail structural components of the system, and limits in the application of the system. Test results include information on component material properties, force-deformation behavior, and nonlinear response.

Comprehensive design provisions are developed within the context of the seismic provisions of ASCE/SEI 7-05 and other applicable standards. The provisions should address all significant aspects of the design and detailing of the seismic-force-resisting system and its components. Important exceptions and deviations from established building code requirements should be clearly stipulated. Design provisions should address criteria for determining minimum strength and ensuring inelastic deformation capacity through a combination of system design requirements, component design and detailing requirements, and project-specific testing requirements. Design provisions should also specify the seismic performance factors (R, Ω_0, C_d) and other criteria (e.g., drift limits, height limits, and seismic usage restrictions) that are proposed as part of the design basis for the new system.

Test data are necessary for characterizing the strength, stiffness and ductility of the materials, members, and connections of the proposed system. Test data are also necessary for establishing properties of the nonlinear analysis models used to assess collapse risk. Test data and other substantiating evidence should be acquired as the basis of the design provisions and for calibrating analysis models. Design requirements should be documented with supporting evidence to ensure sufficient strength, stiffness and ductility of the proposed system, across the intended range of application of the system.

2.5   Characterize Behavior

System behavior is characterized through the use of structural system archetypes. The concept of an archetype is described in Chapter 4. Establishment of archetypes begins with identifying the range of features and behavioral characteristic that describe the bounds of the proposed seismic-force-resisting system.

Archetypes provide a systematic means for characterizing permissible configurations and other significant features of the proposed system. Like building code provisions, archetypical systems are intended to represent typical applications of a seismic-force-resisting system, recognizing that it is practically impossible to envision or attempt to quantify performance of all possible applications. They should, however, reflect the degree of irregularity permitted within standard building code provisions.

The challenge in defining and assessing structural system archetypes is in narrowing the range of parameters and attributes to the fewest and simplest possible, while still being reasonably representative of the variations that would be permitted in actual structures. In addition to ground motion intensity (Seismic Design Category), the following characteristics are considered in defining structural system archetypes: (1) building height; (2) fundamental period; (3) structural framing configurations; (4) framing bay sizes or wall lengths; (5) magnitude of gravity loads; and (6) member and connection design and detailing requirements. Structural system archetypes are assembled into bins called performance groups, which reflect major divisions, or changes in behavior, within the archetype design space. The collapse safety of the proposed system is then evaluated for each performance group.

In the collapse assessment process, only framing components that are specifically designated as part of the seismic-force-resisting system are included in the archetypes. While it is recognized that other portions of the building (e.g., components of the gravity system or certain nonstructural components) can significantly affect collapse behavior, such components, which are not controlled by seismic-force-resisting system design requirements, cannot be relied upon for reducing collapse risk.

2.6   Develop Models

Development of structural models for collapse assessment is discussed in Chapter 5. Structural system archetypes provide the basis for preparing a finite number of trial designs and developing a corresponding number of idealized nonlinear models that sufficiently represent the range of intended applications for a proposed system. Index archetype models are developed to provide the most basic (generic) idealization of an archetypical configuration, while still capturing significant behavioral modes and key design features of the proposed seismic-force-resisting system.

Designs consider the range of seismic criteria for each applicable Seismic Design Category, variations in gravity loads, and other distinguishing features including alternative geometric configurations, varying heights, and different tributary areas that impact seismic design or system performance.

To the extent possible, nonlinear models include explicit simulation of all significant deterioration mechanisms that could lead to structural collapse. Recognizing that it is not always possible (or practical) to simulate all possible collapse modes, the Methodology includes provisions for assessing the effects of behaviors that are not explicitly simulated in the model, but could trigger collapse.

Nonlinear models must account for the seismic mass that is stabilized by the seismic-force-resisting system, including the destabilizing P-delta effects associated with the seismic mass. In most cases, elements are idealized with phenomenological models to simulate complicated component behavior. In some cases, however, two-dimensional or three-dimensional continuum finite element models may be required to properly characterize behavior. Models are calibrated using material, component, or assembly test data and other substantiating evidence to verify their ability to simulate expected nonlinear behavior.

2.7   Analyze Models

Collapse assessment is performed using both nonlinear static (pushover) and nonlinear dynamic (response history) analysis procedures described in Chapter 6. Nonlinear static analyses are used to help validate the behavior of nonlinear models and to provide statistical data on system overstrength and ductility capacity. Nonlinear dynamic analyses are used to assess median collapse capacities and collapse margin ratios.

Nonlinear response is evaluated for a set of pre-defined ground motions that are used for collapse assessment of all systems. Two sets of ground motion records are provided for nonlinear dynamic analysis. One set includes 22 ground motion record pairs from sites located greater than or equal to 10 km from fault rupture, referred to as the "Far-Field" record set. The other set includes 28 pairs of ground motions recorded at sites less than 10 km from fault rupture, referred to as the "Near-Field" record set. While both Far-Field and Near-Field record sets are provided, only the Far-Field record set is required for collapse assessment. This is done for reasons of practicality, and in recognition of the fact that there are many unresolved issues concerning the characterization of near-fault hazard and ground motion effects. The Near-Field record set is provided as supplemental information to examine issues that could arise due to near-fault directivity effects, if needed.

The record sets include records from all large-magnitude events in the Pacific Earthquake Engineering Research Center (PEER) Next-Generation Attenuation (NGA) database (PEER, 2006a). Records were selected to meet a number of sometimes conflicting objectives. To avoid event bias, no more than two of the strongest records have been taken from any one earthquake, yet the record sets have a sufficient number of motions to permit statistical evaluation of record-to-record (RTR) variability and collapse fragility. Strong ground motions were not distinguished based on either site condition or source mechanism. The Far-Field and Near-Field record sets are provided in Appendix A, along with background information on their selection.

For collapse evaluation, ground motions are systematically scaled to increasing earthquake intensities until median collapse is established. Median collapse is the ground motion intensity in which half of the records in the set cause collapse of an index archetype model. This process is similar to, but distinct from the concept of incremental dynamic analysis (IDA), as proposed by Vamvatsikos and Cornell (2002).

Figure 2-3 shows an example of IDA results for a single structure subjected to a suite of ground motions of varying intensities. In this illustration, sidesway collapse is the governing mechanism, and collapse prediction is based on lateral dynamic instability, or excessive lateral displacements. Using collapse data obtained from IDA results, a collapse fragility can be defined through a cumulative distribution function (CDF), which relates the ground motion intensity to the probability of collapse (Ibarra et al., 2002). Figure 2-4 shows an example of a cumulative distribution plot obtained by fitting a lognormal distribution to the collapse data from Figure 2-3.

While the IDA concept is useful for illustrating the collapse assessment procedure, the Methodology only requires calculation of the median collapse point, which can be calculated with fewer nonlinear analyses than would otherwise be required to calculate the full IDA curve. An abbreviated process for calculating the median collapse point is described in Chapter 6.

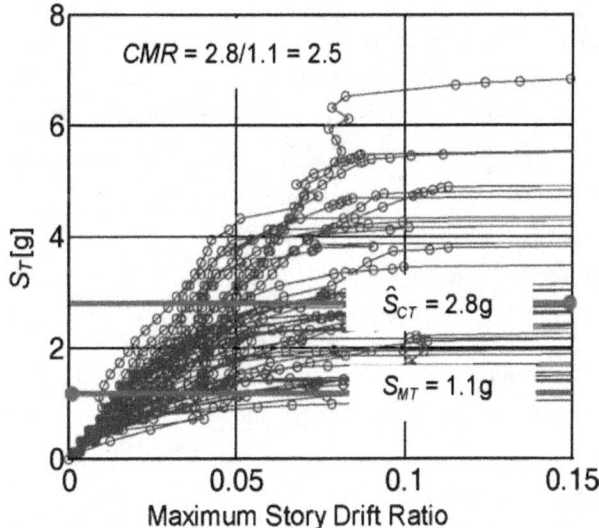

Figure 2-3   Incremental dynamic analysis response plot of spectral acceleration versus maximum story drift ratio.

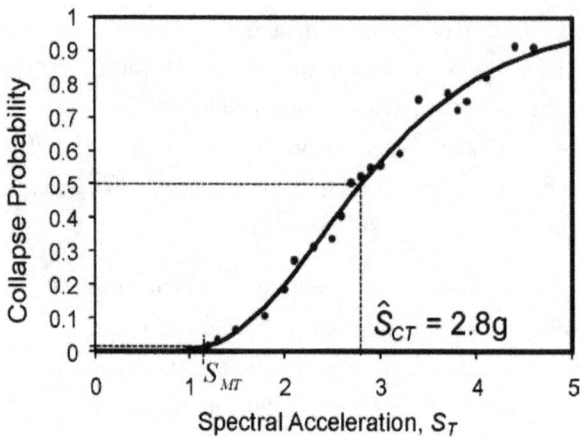

Figure 2-4   Collapse fragility curve, or cumulative distribution function.

2.8   Evaluate Performance

The performance evaluation process is described in Chapter 7. It utilizes results from nonlinear static analyses to determine an appropriate value of the system overstrength factor, Ω_O, and results from nonlinear dynamic analyses to evaluate the acceptability of a trial value of the response modification coefficient, R. The deflection amplification factor, C_d, is derived from an acceptable value of R, with consideration of the effective damping of the system of interest.

The trial value of the response modification coefficient, R, is evaluated in terms of the acceptability of a calculated collapse margin ratio, which is the ratio of the ground motion intensity that causes median collapse, to the Maximum Considered Earthquake (MCE) ground motion intensity defined by the building code. Acceptability is measured by comparing the collapse margin ratio, after some adjustment, to acceptable values that depend on the quality of information used to define the system, total system uncertainty, and established limits on acceptable probabilities of collapse.

To account for unique characteristics of extreme ground motions that lead to building collapse, the collapse margin ratio is converted to an adjusted collapse margin ratio. The adjustment is based on the shape of the spectrum of rare ground motions, and is a function of the structure ductility and period of vibration. Systems with larger ductility and longer periods benefit by larger adjustments. The background and development of this adjustment to account for the effects of spectral shape are provided in Appendix B.

Acceptable values of the collapse margin ratio are defined in terms of an acceptably low probability of collapse for MCE ground motions, considering uncertainty in collapse fragility. Systems that have more robust design requirements, more comprehensive test data, and more detailed nonlinear analysis models, have less collapse uncertainty, and can achieve the same level of life safety with smaller collapse margin ratios. The following sources of uncertainty are explicitly considered: (1) record-to-record uncertainty; (2) design requirements-related uncertainty; (3) test data-related uncertainty; and (4) modeling uncertainty.

The probability of collapse due to MCE ground motions applied to a population of archetypes is limited to 10%, on average. Each performance group is required to meet this average limit, recognizing that some individual archetypes could have collapse probabilities that exceed this value. The probability of collapse for individual archetypes is limited to 20%, or twice the average value, to evaluate acceptability of potential "outliers" within a performance group. It should be noted that these limits were selected based on judgment. Within the performance evaluation process, these values can be adjusted to reflect different values of acceptable probabilities of collapse that are deemed appropriate by governing jurisdictions or other authorities employing this Methodology to establish seismic design requirements for a proposed system.

If the adjusted collapse margin ratio is large enough to result in an acceptably small probability of collapse at the MCE, then the trial value of R is acceptable. If not, the system must be redefined by adjusting the design requirements (Chapter 3), re-characterizing behavior (Chapter 4), or redesigning with new trial values (Chapter 5), and then re-evaluated using the Methodology. In some cases, inadequate performance could require extensive revisions to the overall system concept.

2.9   Document Results

Documentation requirements are described in Chapter 8. The results of system development efforts must be thoroughly documented for review and approval by an independent peer review panel, review and approval by an authority having jurisdiction, and eventual use in design and construction.

Documentation is required at each step of the process. It should describe seismic design rules, range of applicability of the system, testing protocols and results, rationale for the selection of structural system archetypes, results of analytical investigations, evaluation of quality of information, quantification of uncertainties, results of performance evaluations, and proposed seismic performance factors.

Documentation should be of sufficient detail and clarity to allow an unfamiliar structural engineer to properly implement the design, and an unfamiliar reviewer to evaluate compliance with the design requirements. Documentation should also provide sufficient information to allow peer reviewers, code authorities, or material standard organizations to assess the viability of the proposed system and the reasonableness of the proposed seismic performance factors.

2.10   Peer Review

Peer review by an independent team of experts is a requirement of the Methodology, and should be an integral part of the process at each step. Implementation of the Methodology involves much uncertainty, judgment and potential for variation. Deciding on an appropriate level of detail to adequately characterize performance of a proposed system should be performed in collaboration with a peer review panel, on an ongoing basis, during developmental efforts.

The peer review panel is responsible for reviewing and commenting on the approach taken by the system development team, including the extent of the experimental program, testing procedures, design requirements, development of structural system archetypes, analytical approaches, extent of the nonlinear analysis investigation, and the final selection of the proposed seismic performance factors. Members of the peer review panel must be qualified to critically evaluate the development of the proposed system including testing, design, and analysis, and sufficiently independent from the system development team to provide an unbiased assessment of the developmental process.

The peer review panel, and their involvement, should be established early to clarify expectations for the collapse assessment. The peer review team is expected to exercise considerable judgment in evaluating all aspects of the process, from definition of the proposed system, to establishment of design criteria, scope of testing, and extent of analysis deemed necessary to adequately evaluate collapse safety.

Details on the required peer review process, and guidance on the selection of peer review panel members, are provided in Chapter 8.

Chapter 3
Required System Information

This chapter identifies information that is necessary for establishing seismic performance factors as part of the development, documentation, and review of a proposed seismic-force-resisting system. It describes the type of information that is required, and provides guidance on how it should be developed.

This information is used in the development of structural system archetypes in Chapter 4, and nonlinear analysis models in Chapter 5. It is subject to peer review as it is developed, and is an integral part of the reporting requirements in Chapter 8.

3.1 General

Seismic performance factors for a proposed system are established through nonlinear simulation of response to earthquake ground motion, and probabilistic assessment of collapse risk. Detailed system information is necessary for reliable prediction of structural response, and for development and validation of standardized engineering criteria that will lead to structures that perform as expected. Required system information includes:

- a comprehensive description of the proposed system, including its intended applications, physical and behavioral characteristics, and construction methods;

- a clear and complete set of design requirements and specifications for the system that provide information to quantify strength limit states, proportion and detail components, analyze predicted response, and confirm satisfactory behavior; and

- test data and other supporting evidence from an experimental investigation program to validate material properties and component behavior, calibrate nonlinear analysis models, and establish performance acceptance criteria.

The process for obtaining required system information is shown in Figure 3-1. It involves development of detailed system design requirements, acquisition of test data, and assessment of the quality of this information.

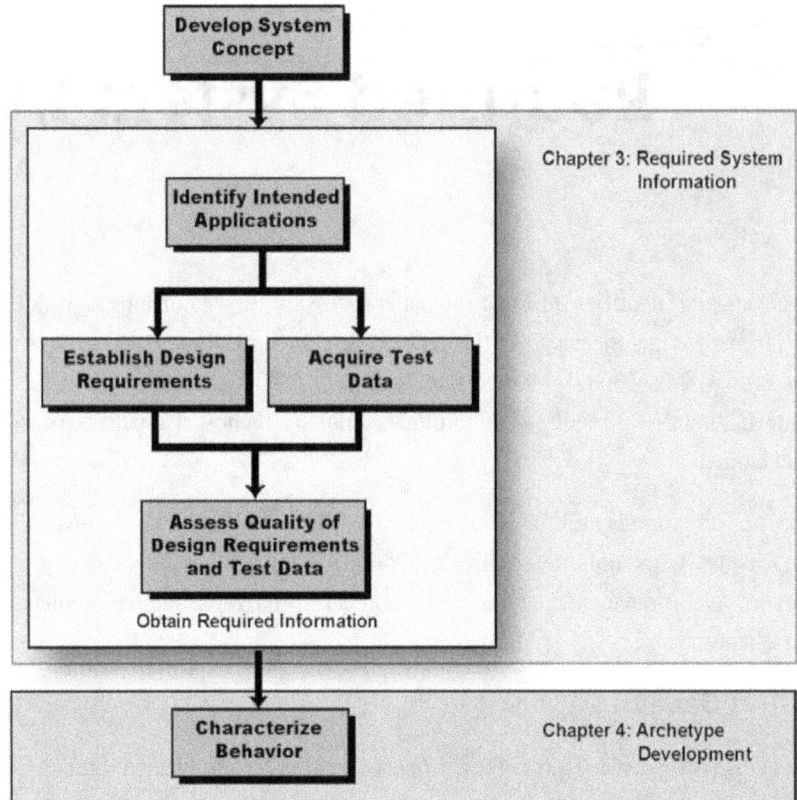

Figure 3-1   Process for obtaining required system information

Design requirements and test data are used as inputs for the development of structural system archetypes in Chapter 4. Quality ratings for design requirements and test data are used to assess total uncertainty in Chapter 7.

3.2   Intended Applications and Expected Performance

A description of the intended applications and expected performance of a proposed seismic-force-resisting system is required. This description should include: (1) the anticipated function and occupancy; (2) physical and behavioral characteristics of the system; (3) typical geometric configurations; and (4) any similarities or differences between the proposed system and current code-conforming systems. The description should also indicate how the structural system and its key components are expected to perform in an earthquake.

The following information should be used as a guide for describing the intended applications of a proposed seismic-force-resisting system:

- intended occupancies and use of facilities to be constructed using the proposed system,

- horizontal and vertical configurations (e.g., framing layout, spans, story heights, overall heights) of typical facilities to be constructed using the proposed system,

- structural gravity framing systems to be used in combination with the proposed system, including typical dead and live loads,

- geometric configurations of the proposed seismic-force-resisting system,

- expected inelastic behavior under seismic loading of varying intensity, and

- methods of construction.

3.3   Design Requirements

Design requirements establish the fundamental information that will be used to proportion and detail components, analyze the predicted response, and confirm the behavior of a proposed system. They also set boundaries in the application of the system. Design requirements are an essential input to the development of structural system archetypes in Chapter 4. Information needed to define and document system design requirements is specified in the sections that follow.

3.3.1   Basis for Design Requirements

Design requirements should be based on criteria specified in applicable sections of the latest edition of ASCE/SEI 7, *Minimum Design Loads for Buildings and Other Structures* (ASCE, 2006a), and other applicable material reference standards, such as ACI 318, *Building Code Requirements for Structural Concrete* (ACI, 2005), AISC/ANSI 341, *Seismic Provisions for Structural Steel Buildings* (AISC, 2005), ACI 530/ASCE 5/TMS 402, *Building Code Requirements for Masonry Structures* (ACI, 2002b), and ANSI/AF&PA, *National Design Specification for Wood Construction* (ANSI/AF&PA, 2005). The following statements, taken mostly from the *NEHRP Recommended Provisions for Seismic Regulations for New Buildings and Other Structures* (FEMA, 2004a), should be used as a basis for developing design requirements:

- The structure shall include complete lateral- and vertical-force-resisting systems capable of providing adequate strength, stiffness, and energy dissipation capacity to withstand the design ground motions within the prescribed limits of deformation and strength demand.

- Design ground motions shall be assumed to occur along any direction of the structure.

- Adequacy of the systems shall be demonstrated through construction of a mathematical model, and evaluation of this model for the effects of design ground motions. This evaluation shall be based on analysis in which design seismic forces are distributed and applied throughout the height of the structure in accordance with the ASCE/SEI 7-05.

- Deformations and internal forces in all members of the structure shall be determined and evaluated against acceptance criteria contained or referred to in ASCE/SEI 7-05, and as developed for the system under consideration.

- A continuous load path, or paths, shall be provided with adequate strength and stiffness to transfer all forces from the point of application to the final point of resistance.

- The foundation shall be designed to accommodate forces developed or movements imparted to the structure by design ground motions. In determining foundation design criteria, special recognition shall be given to the dynamic nature of the forces, the expected ground motions, and the design basis for strength and energy dissipation capacity of the structure.

- Design of a structure shall consider potentially adverse effects on the stability of the structure due to failure of a member, connection, or component of the seismic-force-resisting system.

3.3.2 Application Limits and Strength Limit States

The boundaries of the intended application of the proposed seismic-force-resisting system must be clearly stated, including, for example, any proposed height limitations or restrictions to certain Seismic Design Categories.

Design requirements must address material properties, components, connections, assemblies, and seismic-force-resisting system overall behavior. With generally accepted modeling criteria and good engineering judgment, design requirements should be of sufficient detail that analytical models of component behavior can be developed. They must address the details of stiffness models for members, connections, assemblies, and the overall system, recognizing that seismic performance factors will be used in the context of linear analyses and response to equivalent static forces. Where size effects are important, they must be included.

Design requirements must provide information necessary to quantify all pertinent strength limit states, including:

- tension, compression, bending, shear;

- yield, rupture, brittle fracture;

- local, member, and global instability.

Proposed systems that rely on standard structural materials, or minor modifications to existing, proven systems, can reference much of the requirements to existing standards. However, such references must be clearly justifiable and verifiable. New systems that behave outside the bounds of existing system behavior must include consideration of behavioral effects on other elements of building construction, including the gravity load system and nonstructural components.

3.3.3 Overstrength Design Criteria

It is expected that most seismic-force-resisting systems will rely on inelastic behavior somewhere within the system. Overstrength criteria should be applied to the design of components that are judged to have small inelastic deformation capacity followed by rapid deterioration in strength. This is especially important if the component is also an essential part of the gravity load system. Design requirements should be written in a manner that clearly identifies all such components so that it is not left up to the judgment of the designer to make this identification.

Design procedures utilizing linear static equivalent lateral force analyses should follow the current standard method of requiring that such components be designed for gravity loads plus Ω_0 times the seismic loads, or for the maximum forces that can be delivered to the component by other elements in the system. It is understood that the overstrength factor used for this purpose is based on judgment, and can vary by a large amount depending on system configuration. To provide adequate protection, this factor should be a high estimate of the expected ratio of maximum force to design force, particularly for systems or materials that are non-ductile or have significant variability or uncertainty in response.

3.3.4 Configuration Issues

Design requirements should comprehensively address all expected system configurations. Emphasis should be placed on criteria that protect against the occurrence of non-ductile failure modes and unintended concentration of inelastic action in limited portions of the system.

When determining the design strength of components that are affected by combined actions, such as axial-shear force interaction, consideration should be given to system configurations that might have an effect on the magnitude of combined loads. Beneficial effects of gravity loading must not be permitted in configurations that result in little or no gravity load on seismic-force-resisting components. Similarly, possible detrimental effects of induced vertical loads should be considered in configurations that generate high axial loads on seismic-force-resisting components.

Design requirements should address issues of multi-directional loading, and simultaneous in-plane and out-of-plane loading, unless the combined load effects are demonstrated to be unimportant.

3.3.5 Material Properties

Design requirements should document all material properties that will serve as reference values for design of components, as well as criteria for determining and measuring these properties. Documentation is not needed for material properties that are prescribed in existing codes and material reference standards. To the extent possible, the experimental determination of material properties should be based on testing procedures specified in ASTM standards. Material properties of interest include:

- tensile, compressive, and shear stress and strain properties,

- friction properties between parts that might possibly slide,

- bond properties at the interface of two materials, and

- other properties on which component behavior depends strongly.

In the determination of material properties, consideration should be given to the simulation of common field conditions during testing, including confinement conditions (e.g., bi-axial or tri-axial states of stress or strain), environmental effects (e.g., temperature, moisture, solar radiation), and cyclic loading. Effects of aging should be quantified, if deemed important to seismic behavior.

Design requirements should include criteria for field testing of material properties for systems in which the reference properties depend strongly on case-specific mix proportions, placement, curing or other similar aspects of construction.

3.3.6 Strength and Stiffness Requirements

Design requirements should contain comprehensive guidance for the determination of design strength and effective elastic stiffness of structural components, and assemblies of components.

- **Stiffness requirements.** Guidance on determination of the effective elastic stiffness of structural components should be provided. The effective elastic stiffness is defined as the stiffness that, if utilized in an

3: Required System Information

analytical model, will provide a good estimate of the story drift demand at the design level.

- **Component strength requirements.** The nominal strength of a component should be expressed in terms of material properties, and quantified for the range of loads, and combinations of loads, that might be experienced as the system is subjected to collapse-level ground motions.

  Uncertainty inherent in a strength design equation, as well as the severity of the consequence of failure, should be reflected in the resistance factor (ϕ-factor) associated with the strength design equation. Resistance factors calibrated for use with common gravity load combinations are recommended for use. Although these factors may not be anchored in reliability analyses for seismic load combinations, design requirements will be utilized in conjunction with linear analyses and equivalent static forces, and the use of resistance factors will have an important effect on the overall capacity of the system.

  If the strength or deformation capacity of one component is affected significantly by interaction with other components, then this interaction should be accounted for in design equations.

  If a component is subjected to a load effect, or combination of load effects, that will cause rapid deterioration in strength in the inelastic range, then this load effect, or this combination of load effects, must be clearly identified as a non-ductile mode, and should trigger overstrength design criteria.

  Design requirements should be specific about component detailing needed to ensure adequate strength during inelastic deformation.

- **Connection strength requirements.** Design requirements should be specific about design of connections. In general, connections are considered to be non-ductile. If a proposed system is based on a ductile connection, then design requirements must clearly result in connections that will have sufficient deformation capacity to avoid significant deterioration before any of the connected components reach their expected limits.

- **Sensitivity to gravity loads.** Where the strength of a member or connection is sensitive to compression or tension from gravity loads, design requirements must account for the effect of vertical ground motions. Standard factors in existing standards for additive effects ($1.2 + 0.2S_{DS}$) and counteracting effects ($0.9 - 0.2S_{DS}$) may be used only if justified through studies on structural system archetypes.

3.3.7 Approximate Fundamental Period

The design requirements should include formulas for calculation of the approximate fundamental period, T_a, when the formulas of Section 12.8.2.1 of ASCE/SEI 7-05 do not apply to the system of interest (e.g., when the approximate period parameters given in Table 12.8-2 of ASCE/SEI 7-05 for "all other structural systems" are not appropriate). The formula(s) for T_a should provide an estimate of the mean minus one sigma value of the actual first mode period of buildings in which the system is utilized.

3.4 Quality Rating for Design Requirements

Quality of information is related to uncertainty, which factors into the performance evaluation for a proposed seismic-force-resisting system. The quality of the proposed design requirements is rated in accordance with the requirements of this section, and approved by the peer review panel.

Design requirements are rated between (A) Superior and (D) Poor, as shown in Table 3-1. The selection of a quality rating considers the completeness and robustness of the design requirements, and confidence in the basis for the design equations. Quantitative values of design requirements-related collapse uncertainty are: (A) Superior, $\beta_{DR} = 0.10$; (B) Good, $\beta_{DR} = 0.20$; (C) Fair, $\beta_{DR} = 0.35$; and (D) Poor, $\beta_{DR} = 0.50$. Use of these values is described in Section 7.3.

Table 3-1   Quality Rating of Design Requirements

Completeness and Robustness	Confidence in Basis of Design Requirements		
	High	Medium	Low
High. Extensive safeguards against unanticipated failure modes. All important design and quality assurance issues are addressed.	(A) Superior $\beta_{DR} = 0.10$	(B) Good $\beta_{DR} = 0.20$	(C) Fair $\beta_{DR} = 0.35$
Medium. Reasonable safeguards against unanticipated failure modes. Most of the important design and quality assurance issues are addressed.	(B) Good $\beta_{DR} = 0.20$	(C) Fair $\beta_{DR} = 0.35$	(D) Poor $\beta_{DR} = 0.50$
Low. Questionable safeguards against unanticipated failure modes. Many important design and quality assurance issues are not addressed.	(C) Fair $\beta_{DR} = 0.35$	(D) Poor $\beta_{DR} = 0.50$	--

The highest rating of (A) Superior applies to systems that include a comprehensive set of design requirements that provide safeguards against unanticipated failure modes. Therefore, for a superior rating, there must be a high level of confidence that the design requirements produce the anticipated structural behavior. Existing code requirements for special concrete moment frames and special steel moment frames, for example, have been vetted with detailed experimental results and documented performance in earthquakes. Design and detailing provisions include capacity design requirements to safeguard against unanticipated behaviors. A set of design requirements such as these would be rated (A) Superior.

The lowest rating of (D) Poor applies to design requirements that have minimal safeguards against unanticipated failure modes, do not ensure a hierarchy of yielding and failure, and would generally be associated with systems that exhibit behavior that is difficult to predict.

3.4.1 Completeness and Robustness Characteristics

Completeness and robustness characteristics are related to how well the design requirements address issues that could potentially lead to unanticipated failure modes, as well as proper implementation of designs through fabrication, erection and final construction. Completeness and robustness characteristics are rated from high to low, as follows:

- **High.** Design requirements are extensive, well-vetted and provide extensive safeguards against unanticipated failure modes. They establish a definite hierarchy of component yielding and failure. All important issues regarding system behavior have been addressed, resulting in a high reliability in the behavior of the system. Through mature construction practices, and tightly specified quality assurance requirements, there is a high likelihood that the design provisions will be well executed through fabrication, erection and final construction.

- **Medium.** Design requirements are reasonably extensive and provide reasonable safeguards against unanticipated failure modes, leaving some limited potential for the occurrence of such modes. Design requirements establish a suggested hierarchy of component yielding and failure. While most important behavioral issues have been addressed, some have not, which somewhat reduces the reliability of the system. Quality assurance requirements are specified but do not fully address all the important aspects of fabrication, erection and final construction.

- **Low.** Design requirements provide questionable safeguards against unanticipated failure modes. Hierarchy of component yielding and failure has been only marginally addressed (if at all), and there is a likelihood of the occurrence of unanticipated failure modes. Design requirements do not address all important behaviors, resulting in marginally reliable behavior of the system. Quality assurance is lacking, written guidance is not provided, and construction practices are not well-developed for the type of system and materials.

Simplified, but conservative, design requirements by themselves are not a reason for a low Completeness and Robustness rating as long as conservatism is quantifiable in the context of unanticipated failure modes.

3.4.2 Confidence in Design Requirements

Confidence in the basis of the design requirements refers to the degree to which the prescribed material properties, strength criteria, stiffness parameters, and design equations are representative of actual behavior and will achieve the intended result. Confidence is rated from high to low, as follows:

- **High.** There is substantiating evidence (experimental data, history of use, similarity with other systems) that results in a high level of confidence that the properties, criteria, and equations provided in the design requirements will result in component designs that perform as intended.

- **Medium.** There is some substantiating evidence that results in a moderate level of confidence that the properties, criteria, and equations provided in the design requirements will result in component designs that perform as intended.

- **Low.** There is little substantiating evidence (little experimental data, no history of use, no similarity with other systems) that results in a low level of confidence that the properties, criteria, and equations provided in the design requirements will result in component designs that perform as intended.

3.5 Data from Experimental Investigation

Analytical modeling alone is not adequate for predicting nonlinear seismic response with confidence, particularly for structural systems that have not been subjected to past earthquakes. A comprehensive experimental investigation program is necessary to establish material properties, develop design criteria, calibrate and validate component models, confirm behavior, and calibrate analyses for a proposed seismic-force-resisting system. Experimental results from other testing programs can be used to supplement an experimental investigation program, but these results must come from reliable sources, and their applicability to the system under consideration must be demonstrated.

There are practical limitations on how comprehensive an experimental testing program can be. It must be understood, however, that limitations on available experimental data will affect the uncertainty and reliability of the collapse assessment of a proposed system, and will factor directly into the performance evaluation process. The scope of an experimental investigation program should be developed in consultation with the peer review panel.

3.5.1 Objectives of Testing Program

Testing is used to develop basic information so that the combination of experimental and analytical data is sufficient to achieve the following two objectives:

- Predict the seismic response of structures in the regime of interest to the establishment of seismic performance factors, which occurs when the structure, or any portion of the structure, is subjected to large seismic demands and approaches a state of lateral dynamic instability (collapse). This implies the need to model strength and stiffness properties of important components, and reliably capture structural response, from elastic behavior through the state of lateral dynamic instability, over the entire range of possible structural configurations permitted by the design requirements.

- Develop and validate standardized engineering design criteria that can be used to design structures that perform as expected, given the specified seismic performance factors.

Achievement of these objectives requires a coordinated material, component, connection, assembly, and system testing program that will provide the following information:

- **Material test data.** Data that serve as reliable reference values for the prediction of strength, stiffness, and deformation properties of structural components and connections under earthquake loading.

- **Component and connection test data.** Information needed to develop and calibrate design criteria and analytical models of cyclic load-deformation characteristics for components and connections that form an essential part of the seismic-force-resisting system.

- **Assembly and system test data.** Information needed to quantify interactions between structural components and connections that cannot be predicted by analysis with confidence.

3.5.2 General Testing Issues

In developing a comprehensive testing program, the following issues should be considered:

- **Cumulative damage effects.** Structural materials and components experience history-dependent cumulative damage during repeated cyclic loading. The loading history used in testing should be representative of the cyclic response that a material, component, connection, or assembly would experience as part of a typical structural system subjected to a severe earthquake.

- **Size effects.** Tests should be performed on full-size specimens unless it can be shown by theory and experiment that testing of reduced-scale specimens will not significantly affect behavior.

- **Strain rate effects.** If the load-deformation characteristics of the specimen are sensitive to strain rate effects, then testing should be done at strain rates commensurate with those that would be experienced in a severe earthquake.

- **Boundary conditions.** The boundary conditions of component and assembly tests should be: (1) representative of constraints that a component or assembly would experience in a typical structural system; and (2) sufficiently general so that the results can be applied to boundary conditions that might be experienced in other system configurations. Boundary conditions should not impose beneficial effects on seismic behavior that would not exist in common system configurations.

- **Load application.** Loads should be applied to test specimens in a manner that replicates the transfer of forces to the component or assembly commonly occurring in in-situ conditions.

- **Configuration and number of component/assembly test specimens.** The configuration and number of component and assembly test specimens should be such that all common failure modes that could occur in typical system configurations are represented and evaluated. Emphasis should be on the detection and evaluation of failure modes that lead to a rapid deterioration in strength (e.g., brittle failure modes).

- **Interaction between structural components.** Test configurations should consider important interactions between structural components, unless these interactions can be predicted with confidence by analysis.

- **Direction(s) of loading.** Structural components that resist seismic forces in more than one direction (e.g., concrete core walls) should be tested such that the combined load effects are adequately considered, unless these effects can be predicted with confidence by analysis.

- **In-plane and out-of-plane load effects.** Planar structural components (e.g., walls, diaphragms) should be subjected to simultaneous in-plane and out-of-plane loading, unless these effects can be superimposed with confidence by analysis.

- **Gravity load effects.** Effects of gravity loads should be considered in the experimental program, unless these effects can be superimposed on lateral load effects with confidence by analysis.

- **Statistical variability.** A sufficiently large number of tests should be performed so that statistical variations can be evaluated from the data directly, or can be deduced in combination with data from other sources.

- **Environmental conditions.** If environmental conditions during construction or service (e.g., temperature, humidity) will significantly affect behavior, then the range of conditions that could exist in practice should be simulated during testing.

- **Test specimen construction.** Specimens should be constructed in a setting that simulates commonly encountered field conditions. For example, if field conditions necessitate overhead welding then this type of welding should be applied in test specimen construction.

- **Quality of test specimen construction.** Construction of component, connection, assembly, and system specimens should match the level of quality that will be commonly implemented in the field. Special construction techniques or quality control measures should not be employed, unless they are part of the design requirements.

- **Past experience.** Laboratory testing cannot fully replace experience gained from observation of system behavior in actual use. A benefit should be given to structural systems whose performance has been documented in past earthquakes or other use.

- **Documentation of tests and test results.** Documentation of experiments should be comprehensive, and should include: (1) geometric data, test setup, and boundary conditions; (2) important details of the test specimen, including construction process and fabrication details; (3) type and location of instruments used for measurement of important response parameters; (4) material test data needed for performance evaluation; (5) a written record of all important events prior and during the test; (6) a comprehensive log of all important visual observations; and (7) a comprehensive set of digital experimental data needed for performance evaluation.

The above list should be used as a guide. There may be other issues that are equally important to a given system, but cannot be placed in a general context. Assistance in the identification of important testing issues can be obtained from references available in the literature (e.g., ACI, 2001; AISC, 2005; ASTM, 2003; ATC, 1992; Clark et al., 1997; FEMA, 2007; and ICC, 2009). Specimen fabrication, testing procedures, loading protocol, and test documentation should follow guidelines established in these references as appropriate for the system being evaluated, and as approved by the peer review panel.

Testing laboratories used to conduct an experimental investigation program should comply with national or international accreditation criteria, such as the ISO/IEC 17025 *General Requirements for the Competence of Testing and Calibration Laboratories* (ISO/IEC, 2005). Testing Laboratories that are not accredited may be used for the experimental investigation program, subject to approval by the peer review panel.

3.5.3 Material Testing Program

A material testing program is required to provide reliable stress-strain relationships for the prediction of strength, stiffness, and deformation properties of structural components and connections under the type of loading experienced during an earthquake. In addition to general testing issues, materials testing should consider the following:

- low-cycle fatigue and fracture properties,

- bi-axial and tri-axial stress conditions,

- utilization of applicable ASTM Standards,

- evaluation of variability in material properties,

- effects of aging, and

- effects of environmental conditions.

Material testing programs should be performed in accordance with all applicable ASTM Standards and other testing criteria specified in nationally accepted industry standards and specifications. Material test data available from past tests that conform to all applicable standards may be used.

3.5.4 Component, Connection, and Assembly Testing Program

A component, connection, and assembly testing program is required to provide information for the development, calibration, and validation of analytical models of cyclic load-deformation characteristics of components and connections that form an essential part of the seismic-force-resisting system, as well as development of the design criteria for such components and connections.

Components, connections, and assemblies that have low inelastic deformation capacities (commonly referred to as non-ductile elements) must be identified, and should undergo sufficient testing to validate both the design strength of such elements as well as the strength properties used in the analytical models employed in the collapse assessment.

Testing of Structural Components

Component testing serves to identify and quantify component parameters that significantly affect seismic response. Cyclic behavior is characterized by a basic hysteresis loop, which deteriorates with the number and amplitude of cycles. It is critical that a test is continued until severe strength deterioration is evident, and all important characteristics that enter design equations and analytical models have been verified experimentally. Two hysteretic responses of a structural component (in this case a steel beam and a plywood panel) are shown in Figure 3-2. In the left figure it appears that the loops stabilize at very large amplitudes, and that more (and larger) deformation cycles can be sustained. However, the possibility of fracture at large deformations is high, and once this fracture occurs, the resistance will deteriorate, rapidly approaching zero. For this reason, no credit should be given to residual strength beyond the deformation at which the test is terminated.

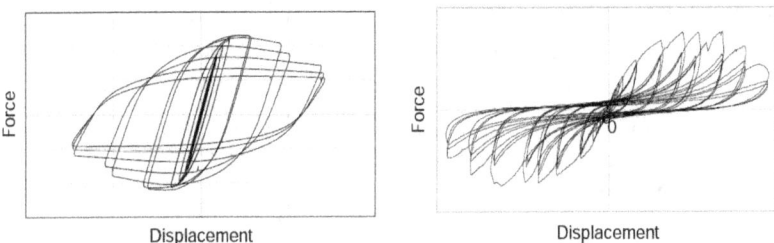

Figure 3-2   Characteristics of force-deformation response: (a) steel beam (Uang et al. 2000); (b) plywood shear wall panel (Gatto and Uang, 2002).

A sufficiently large number of tests should be performed so that important statistical variations can be evaluated from the data directly, or can be deduced in combination with data from other sources. A minimum of two tests is required for each set of primary variables in a test configuration. If rapid deterioration occurs, such as that caused by brittle fracture, a minimum of three tests should be performed. If rapid deterioration is not of concern, it is recommended that one of the tests be a monotonic loading test, which facilitates analytical modeling of the type discussed in Sections 5.4 and 5.5.

Component tests are conducted for the purpose of evaluating all force and deformation characteristics that have a significant effect on the seismic response up to the state of incipient collapse. Gravity loads should be represented, unless it can be shown that their effect is not detrimental to the seismic behavior or can be predicted with confidence from analytical models. The load application and loading history should be representative of what components will experience as part of a typical structural system subjected to a severe earthquake. Instrumentation should permit the measurement of all relevant stiffness and strength properties.

In contrast to "qualification testing", which is intended to gain approval of the use of certain components for specific applications, the main objectives of these component tests are to develop design criteria and to calibrate and validate nonlinear models that are used in collapse assessment. Types and configurations of component tests, together with the loading protocol, should be planned in conjunction with development of the nonlinear analysis models in Chapter 5.

Testing of Connections

A connection is the medium that transfers forces and deformations between adjacent components. Connections should be tested in configurations that simulate gravity load effects as well as seismic load effects, unless gravity loading results in more favorable connection behavior. Connection tests should provide all information necessary to develop connection design criteria in conformance with the latest edition of ASCE/SEI 7-05, and to permit simulation of connections in analytical models.

Testing of Assemblies

An assembly is an arrangement of structural components whose seismic behavior can be described in terms of a single response quantity, such as story drift. An assembly testing program is required if important interactions between adjacent components (or between components and connections) cannot be deduced with confidence from a combination of material, component, and connection tests in combination with analytical modeling. Unless strain-rate effects are important, assembly tests can be performed by imposing load(s) to control point(s) in a quasi-static manner, following a predetermined loading history.

3.5.5 Loading History

Structural elements have limited strength and deformation capacities. Collapse safety depends on the ability to assess these capacities with some confidence. Strength and deformation capacities depend on cumulative damage, which implies that every component has a "memory" of past damaging events, and that all past excursions (or cycles) that have contributed to its current state of health will affect future behavior. Thus, performance depends on the history of previously applied damaging cycles, and assessing the consequences of loading history requires replication of the load and deformation cycles that a component will undergo in an earthquake (or several earthquakes, if appropriate). The objective of a loading history is to achieve this in an approximate, but consistent manner.

There is no unique or best loading history, because no two earthquakes are alike, and a specimen may be part of many different structural configurations. The overriding issue is to account for cumulative damage effects through cyclic loading. The number and amplitude of cycles applied to the specimen may be derived from analytical studies in which models of representative structural systems are subjected to representative earthquake ground motions, and the response is evaluated statistically. In analytical modeling, it should be assumed that specimen resistance deteriorates to zero following the maximum amplitude executed in the test. No credit should be given to deformation capability beyond the largest deformation that a specimen experiences in a test.

Many loading protocols have been proposed in the literature, and several have been used in multi-institutional testing programs (e.g., ATC, 1992; Clark et al., 1997; Krawinkler et al., 2000), or are contained in standards or are proposed for standards (e.g., AISC, 2005; ASTM, 2003; FEMA, 2007; ICC, 2009). These protocols recommend somewhat different loading histories, but they differ more in detail than in concept. Comprehensive discussions of loading histories and their origin and objectives are presented in Filiatrault et al. (2008), Krawinkler et al. (2000), and Krawinkler (1996).

The loading protocols referenced above have been developed with a design level or maximum considered ground motion in mind, and not for the purpose of collapse evaluation. As a result the loading histories are symmetric with step-wise increasing deformation cycles (with the exception of the SAC near-fault loading protocol presented in Clark et al., 1997). While such histories are not representative of cyclic response approaching collapse, they do serve the purpose of quantifying deterioration properties for analytical modeling, particularly if a cyclic test is complemented with a monotonic loading test that provides information on the force-displacement capacity boundary (FEMA, 2009).

Loading history should be deformation-controlled, with the following two exceptions:

- Force-control may be applied for small excursions in which a component will remain essentially elastic. Force-control is encouraged for stiff specimens tested in a relatively flexible test set-up, in order to facilitate measurements and test control in the early stages of testing.

- Force-control may be necessary to test components that are an essential part of the load path, but are fully force-controlled in the in-situ condition, and have no reliable inelastic deformation capacity. One such example would be an anchor controlled by the maximum force exerted by a connected component. In such a case, the loading history should be determined based on the strength and deformation capacities of the connected component. Criteria for a force-controlled loading history are presented in FEMA 461 *Interim Testing Protocols for Determining the Seismic Performance Characteristics of Structural and Nonstructural Components* (FEMA, 2007).

3.5.6 System Testing Program

Testing of an essentially complete structural system should be performed if important response characteristics or important interactions between components and connections cannot be evaluated with good confidence by analytical models that have been calibrated through material, component, connection, or assembly tests.

System tests should be used as a validation tool for a proposed analytical model rather than as an exploratory test from which analytical models will be developed. System tests should not be used to replace any testing at the material, component, connection, or assembly level.

Dynamic System Tests

A dynamic system test should be performed if the response near collapse depends strongly on dynamic characteristics that cannot be predicted with good confidence by analytical models or from component or assembly tests.

Such a test should be performed on an earthquake simulator (shake table) utilizing realistic MCE-level (or collapse level) ground motions, unless it can be demonstrated that equivalent response evaluation can be achieved through alternative means.

Quasi-Static System Tests

Quasi-static cyclic testing consists of loads that are applied to one or more control points by means of hydraulic actuators whose displacement or load values are varied in a cyclic manner in accordance with a predetermined loading history. The loading history used in testing should be representative of the cyclic response that a material, component, connection, or assembly would experience as part of a typical structural system subjected to a severe earthquake.

A quasi-static system test should be performed if important behaviors or interactions between components and connections cannot be evaluated with good confidence by means of calibrated analytical models. Examples include interactions between horizontal and vertical components (floor diaphragms and vertical seismic-force-resisting units) and between vertical components that resist seismic forces in orthogonal directions.

3.6   Quality Rating of Test Data

Quality of test data is related to uncertainty, which factors into the performance evaluation for a proposed seismic-force-resisting system. The quality of test data obtained from an experimental investigation program is rated in accordance with the requirements of this section, and approved by the peer review panel.

Test data are rated between (A) Superior and (D) Poor, as shown in Table 3-2. This rating depends not only on the quality of the testing program, but on how well the tests address key parameters and behavioral issues. The selection of a quality rating for test data considers the completeness and robustness of the overall testing program, and confidence in the test results. Quantitative values of test data-related collapse uncertainty are: (A) Superior, $\beta_{TD} = 0.10$; (B) Good, $\beta_{TD} = 0.20$; (C) Fair, $\beta_{TD} = 0.35$; and (D) Poor, $\beta_{TD} = 0.50$. Use of these values is described in Section 7.3.

Table 3-2   Quality Rating of Test Data from an Experimental Investigation Program

Completeness and Robustness	Confidence in Test Results		
	High	**Medium**	**Low**
High. Material, component, connection, assembly, and system behavior well understood and accounted for. All, or nearly all, important testing issues addressed.	(A) Superior $\beta_{TD} = 0.10$	(B) Good $\beta_{TD} = 0.20$	(C) Fair $\beta_{TD} = 0.35$
Medium. Material, component, connection, assembly, and system behavior generally understood and accounted for. Most important testing issues addressed.	(B) Good $\beta_{TD} = 0.20$	(C) Fair $\beta_{TD} = 0.35$	(D) Poor $\beta_{TD} = 0.50$
Low. Material, component, connection, assembly, and system behavior fairly understood and accounted for. Several important testing issues not addressed.	(C) Fair $\beta_{TD} = 0.35$	(D) Poor $\beta_{TD} = 0.50$	--

3.6.1   Completeness and Robustness Characteristics

Completeness and robustness characteristics are related to: (1) the degree to which relevant testing issues have been considered in the development of the testing program; and (2) the extent to which the testing program and other documented experimental evidence quantify the necessary material, component, connection, assembly, and system properties and important behavior and failure modes. Completeness and robustness characteristics are rated from high to low, as follows:

- **High.** All, or nearly all, important general testing issues of Section 3.5.2 are addressed comprehensively in the testing program and other supporting evidence. Experimental evidence is sufficient so that all, or nearly all, important behavior aspects at all levels (from material to system) are well understood, and the results can be used to quantify all important parameters that affect design requirements and analytical modeling.

- **Medium.** Most of the important general testing issues of Section 3.5.2 are addressed adequately in the testing program and other supporting evidence. Experimental evidence is sufficient so that all, or nearly all, important behavior aspects at all levels (from material to system) are generally understood, and the results can be used to quantify or deduce most of the important parameters that significantly affect design requirements and analytical modeling.

- **Low.** Several important general testing issues of Section 3.5.2 are not addressed adequately in the testing program and other supporting evidence. Experimental evidence is sufficient so that the most important behavior aspects at all levels (from material to system) are fairly well understood, but the results are not adequate to quantify or deduce, with high confidence, many of the important parameters that significantly affect design requirements and analytical modeling.

3.6.2 Confidence in Test Results

Confidence in test results is related to the reliability and repeatability of the results obtained from the testing program, and corroboration with available results from other relevant testing programs. It includes consideration as to whether or not experimental results consistently record performance to failure for all modes of behavior (limited ductility to large ductility), and if sufficient information is provided to assess uncertainties in the design requirements (e.g., ϕ factors) and analytical models. Confidence in test results is rated from high to low, as follows:

- **High.** Reliable experimental information is produced on all important parameters that affect design requirements and analytical modeling. Comparable tests from other testing programs have produced results that are fully compatible with those from the system-specific testing program. A sufficient number of tests are performed so that statistical variations in important parameters can be assessed. Test results are fully supported by basic principles of mechanics.

- **Medium.** Moderately reliable experimental information is produced on all important parameters that affect design requirements and analytical modeling. Comparable tests from other testing programs do not contradict, but do not fully corroborate, results from the system-specific testing program. A measure of uncertainty in important parameters can be estimated from the test results. Test results are supported by basic principles of mechanics.

- **Low.** Experimental information produced on many of the important parameters that affect design requirements and analytical modeling is of limited reliability. Comparable tests from other testing programs do not support the results from the system-specific testing program. Insufficient data exists to assess uncertainty in many important parameters. Basic principles of mechanics do not support some of the results of the testing program.

Chapter 4

Archetype Development

This chapter describes the development of structural system *archetypes*, which provide a systematic means for characterizing key features and behaviors related to collapse performance of a proposed seismic-force-resisting system. It defines how archetype descriptions and performance characteristics are used to develop a set of building configurations (*index archetype configurations*) that together describe the overall range of permissible configurations (*archetype design space*) of a system, which is then separated into groups sharing common features or behavioral characteristics (*performance groups*) for assessing collapse performance. Specific structural designs (*index archetype designs*) are then developed for each configuration based on the specified design criteria, and these designs then form the basis of nonlinear analysis models (*index archetype models*) that are analyzed to assess collapse performance. Guidelines related to index archetype designs and index archetype models are presented in Chapter 5.

4.1   Development of Structural System Archetypes

Behavior of a proposed seismic-force-resisting system is investigated through the use of archetypes. An *archetype* is a prototypical representation of a seismic-force-resisting system. Archetypes are intended to reflect the range of design parameters and system attributes that are judged to be reasonable representations of the feasible design space and have a measurable impact on system response. They are used to bridge the gap between collapse performance of a single specific building and the generalized predictions of behavior needed to quantify performance for an entire class of buildings.

An *index archetype configuration* is a prototypical representation of a seismic-force-resisting system configuration that embodies key features and behaviors related to collapse performance when the system is subjected to earthquake ground motions. Given that building codes permit significant latitude with respect to system configurations within a building class, index archetype configurations are not intended to represent every conceivable configuration of the system of interest. Rather, the intent is to investigate a reasonably broad range of parameters that are permitted by the specified design requirements and represent conditions that are feasible in design and construction practice.

Collectively, the set of index archetype configurations describe the *archetype design space*, which defines the overall range of permissible configurations, structural design parameters, and other properties that define the application limits for a seismic-force-resisting system. For performance evaluation, the archetype design space is divided into *performance groups*, which are groups of index archetype configurations that share a common set of features or behavioral characteristics.

Development of structural system archetypes follows the process outlined in Figure 4-1. Using the design requirements and test data developed under Chapter 3 as inputs, the development of structural system archetypes considers both structural configuration issues and seismic behavioral effects.

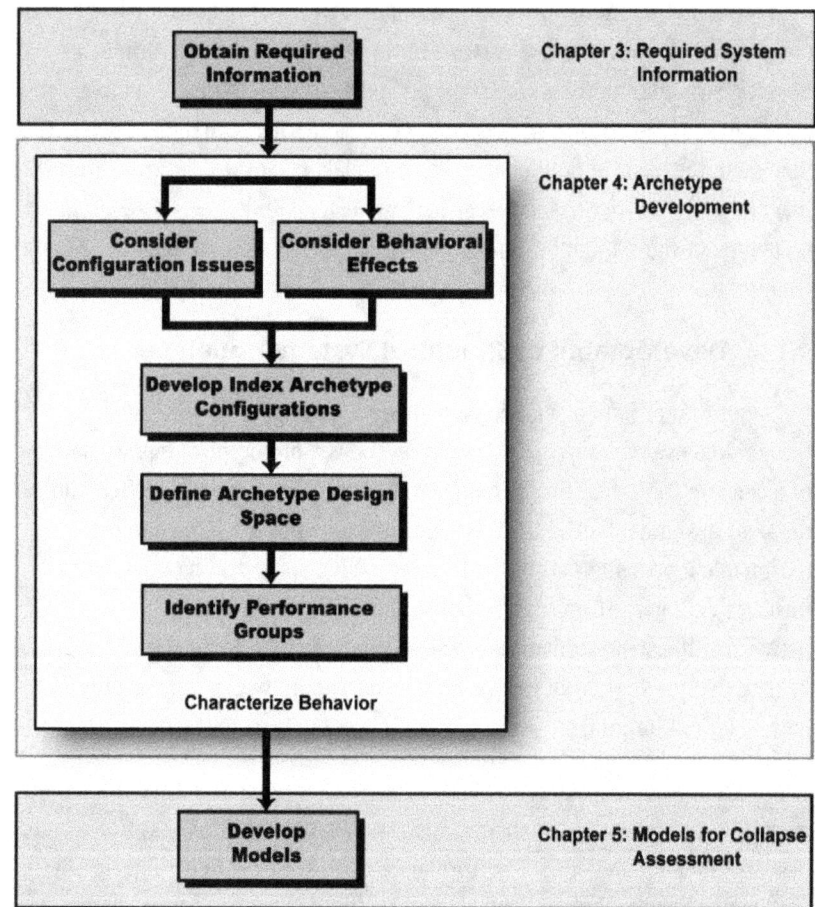

Figure 4-1   Process for development of structural system archetypes.

4.2   Index Archetype Configurations

Index archetype configurations must be sufficiently broad in scope to capture the range of situations that are feasible under the design requirements, but sufficiently limited to be practical to evaluate. The intent is to both: (1)

assess situations that will be generally representative of practice that meets minimum specified requirements for seismic design and construction of a proposed seismic-force-resisting system; and (2) assess designs that are at the limits of the range of design configurations that are allowable based on the design requirements.

While index archetype configurations are not intended to represent every conceivable combination of design parameters, the archetype configurations must encompass the full design space permitted by the design requirements. An exception to this occurs when the collapse safety trends for the assessed archetype designs show that certain configurations will not control the system performance assessment.

Index archetype configurations should not incorporate "standard" practices that may routinely exceed minimum code requirements. For example, use of one member size at multiple locations in a building is a design and construction practice that can result in member overstrengths, which exceed minimum design requirements, and should not be built into index archetype configurations.

It is expected that the set of index archetype configurations will generally include about twenty to thirty specific structural configurations, though the specific number will depend on the characteristics of the seismic-force-resisting system and the limits of the archetype design space. Where the seismic design requirements are relatively loose and cover a broad range of possible design situations, the number of required index archetype configurations may be significantly larger than for systems with more limited applications. The final selection of index archetype configurations, and their corresponding design parameters, should be reviewed and approved by the peer review panel.

The development of index archetype configurations involves the following steps:

- Identify key design variables and related physical properties based on structural configuration issues summarized in Section 4.2.1. Investigate physical properties that affect collapse performance to identify critical design variables that should be reflected in index archetype configurations.

- Establish bounds for key design variables that define the archetype design space. The design space is limited primarily by the seismic design requirements and practical constraints on design and construction.

- Identify behavioral issues and related design considerations based on behavioral effects summarized in Section 4.2.2. Investigate possible deterioration modes that could result in local and global collapse scenarios, and assess the likelihood of those scenarios.

- Develop a set of index archetype configurations based on key design variables and behavioral effects that are likely to result in local or global collapse scenarios.

4.2.1 Structural Configuration Issues

Structural configuration issues include occupancy and program influences, framing type and geometric variations, and gravity and lateral load intensities. Typical configuration design variables that can affect the behavior of a seismic-force-resisting system are summarized in Table 4-1. These structural issues should be used as a guide in establishing index archetype configurations, as follows:

- **Occupancy and Use.** Building occupancy and use can influence the structural layout, framing system, configuration, and loading intensity. Framing spans, story heights, and live loads for seismic-force-resisting systems intended for residential occupancies are usually quite different from office occupancies. Similarly, steel moment frames and associated gravity framing used for industrial occupancies can be different from those used for office or institutional buildings. Occupancies with large live load demands may have larger inherent overstrength in comparison with systems for other occupancies. Major changes in structural configuration resulting from different occupancies and use should be reflected in the index archetype configurations.

- **Elevation and Plan Configuration.** The range of elevation and plan configurations permitted by the system design requirements should be reflected in the index archetype configurations. This could include, for example: (1) the range of framing span lengths of the seismic-force-resisting system; (2) alternative configurations of steel bracing (e.g., chevron versus x-bracing); (3) variations in shear wall aspect ratios that result in flexure- versus shear-dominated behavior; (4) floor diaphragm characteristics that are addressed in the seismic system designation; and (5) the extent of gravity loading tributary to the system. The extent to which such factors are significant will vary depending on the type of seismic-force-resisting system.

- **Building Height.** The range of story heights and number of stories permitted by the system design requirements should be reflected in the index archetype configurations to the extent that these parameters affect the structural period and the localization of inelastic deformations. Since the inelastic response of structures with short periods (in the constant acceleration region of the hazard spectrum) tends to be different from structures with longer periods (in the constant velocity region of the hazard spectrum), the response of short-period and long-period structures are characterized separately. Due to significant differences in the response characteristics of very tall buildings, and limited low frequency content in the ground motions specified for collapse assessment, use of this Methodology should be limited to buildings with a fundamental period of $T \leq 0.4$ seconds (i.e., building heights of about 20 to 30 stories for moment frame systems, and 30 to 40 stories for braced frame and shear wall systems).

Table 4-1   Configuration Design Variables and Related Physical Properties

Design Variable	Related Physical Properties
Occupancy and Use	• Typical framing layout • Distribution of seismic-force-resisting system components • Gravity load intensity • Component overstrength
Elevation and Plan Configuration	• Distribution of seismic-force-resisting components • Typical framing layout • Permitted vertical (strength and stiffness) irregularities • Beam spans, number of framing bays, system regularity • Wall length, aspect ratio, plan geometry, wall coupling • Braced bay size, number of braced bays, bracing configuration • Diaphragm proportions, strength, and stiffness (or flexibility) • Ratio of seismic mass to seismic-force-resisting components • Ratio of tributary gravity load to seismic load
Building Height	• Story heights • Number of stories
Structural Component Type	• Moment frame connection types • Bracing component types • Shear wall sheathing and fastener types • Isolator properties and types
Seismic Design Category	• Design ground motion intensity • Special design/detailing requirements • Application limits
Gravity Load	• Gravity load intensity • Typical framing layout • Ratio of tributary gravity load to seismic load • Component overstrength

- **Structural Component Type.** The extent that structural component types can vary within a given seismic-force-resisting system should be reflected in index archetype configurations. Examples include different types of moment connection details (e.g., welded, bolted, or reduced-beam section), steel bracing members (e.g., HSS, pipe, or W-shape), and light-frame wood shear wall sheathing, framing, and fasteners.

- **Seismic Design Category.** Systems should be evaluated for the highest (most severe) Seismic Design Category for which they are proposed, and then verified in lower Seismic Design Categories. Index archetype configurations within a Seismic Design Category should consider spectral intensities corresponding to the maximum and minimum values for that category, associated design and detailing requirements, and any restrictions on use that are keyed to Seismic Design Category.

- **Gravity Load.** Large gravity load demands can result in overstrength with respect to seismic demands. For some components (e.g., columns in moment frames) axial load ratio can significantly impact inelastic deformation capacity. The nature, magnitude and variation of gravity loads, including structure self weight, occupancy-related superimposed dead loads, and occupancy-related live loads should be considered, and design parameters that affect tributary gravity load, such as bay sizes and building height, should be reflected in the index archetype configurations. It is anticipated that, for most systems, two levels of gravity load (high and low) would be sufficient.

An example of how configuration issues are considered in the development of index archetype configurations for a special reinforced concrete moment frame system conforming to design requirements contained in ASCE/SEI 7-05 is provided in Appendix C.

4.2.2 Seismic Behavioral Effects

Consideration of seismic behavioral effects includes identifying dominant deterioration and collapse mechanisms that are possible, and assessing the likelihood that they will occur. How a component or system behaves under seismic loading is often influenced by configuration decisions, so behavioral effects and configuration issues should be considered concurrently in the development of index archetype configurations.

Seismic collapse resistance depends on the strength, stiffness, and deformation capacity of individual structural components and the overall seismic-force-resisting system. Each of these properties can be addressed directly through system design requirements, but each are also influenced by aspects of the configuration that could change the way a system behaves across the range of an archetype design space. For example, requirements for ductile confinement of reinforced concrete columns will directly affect inelastic deformation capacity, but the magnitude of column axial load, which is influenced by elevation and plan configurations, also has a large impact.

Consideration of behavioral effects is used to help identify major changes in system behavior as the configuration varies. Once potential deterioration and collapse mechanisms are identified, they are addressed through one of the following methods: (1) by ruling out failure modes that are unlikely to occur based on system design and detailing requirements; (2) explicit simulation of failure modes through nonlinear analyses; or (3) evaluation of "non-simulated" failure modes using alternative limit state checks on demand quantities from nonlinear analyses.

Typical behavioral issues and related design considerations that can have an effect on the behavior of a seismic force-resisting system are summarized in Table 4-2.

Table 4-2   Seismic Behavioral Effects and Related Design Considerations

Behavioral Issue	Related Design Considerations
Strength	• Minimum design member forces • Calculated member forces • Capacity design requirements • Component overstrength
Stiffness	• Design member forces • Drift limits • Plan and elevation configuration • Calculated inter-story drifts • Diaphragm stiffness (or flexibility) • Foundation stiffness (or flexibility)
Inelastic deformation capacity	• Component detailing requirements • Member geometric proportions • Capacity design requirements • Calculated member forces • Redundancy of the seismic force-resisting system
Seismic Design Category	• Design ground motion intensity • Special design/detailing requirements
Inelastic system mobilization	• Building height and period • Diaphragm strength and stiffness • Permitted strength and stiffness irregularities • Capacity design requirements

These behavioral issues should be used as a guide in establishing index archetype configurations, as follows:

- **Strength.** Differences between design strength and calculated seismic demands should be reflected in the index archetype configurations. Design strength is a function of the design earthquake intensity, component detailing requirements, capacity design requirements, and overstrength resulting from gravity loads and other minimum load requirements. Calculated demands are a function of the structural configuration and gravity load (dead and live load) intensity. In cases where gravity load, minimum seismic or wind loads, other minimum loads, or stiffness considerations control, there can be significant overstrength relative to seismic design forces. Capacity design provisions control yielding by requiring strengths of certain components to be greater than would otherwise be required by minimum seismic design forces to protect them from inelastic demands.

- **Stiffness.** The elastic lateral stiffness of the seismic-force-resisting system affects the dynamic behavior, sensitivity to sidesway stability (P-delta) effects, and induced deformation demands on critical components. Stiffness is a function of the design earthquake intensity and imposed drift limits, system configuration, and relative stiffness (or flexibility) of certain key components, such as diaphragms or foundations. Index archetype configurations should take into account system types that are more sensitive to drift limits than others, and should identify configurations that probe limits on minimum stiffness within the system design requirements. Where behavior of certain elements (e.g., diaphragm or foundation flexibility), is likely to influence the performance of the system, these effects should be considered in the development of index archetype configurations. For example, in low-rise industrial structures with large floor plans and stiff seismic-force-resisting elements, flexibility of the floor and roof diaphragms could significantly alter the period of vibration of the system, and should, therefore, be considered.

- **Inelastic Deformation Capacity.** Inelastic deformation capacity of components is a function of design requirements, including detailing rules and capacity design provisions, member geometric proportions, and calculated member forces that can vary with structural configuration. Where the inelastic deformation capacity of components is influenced by the force distribution (such as differences in the level of axial forces in walls or columns), factors that influence force distribution (such as plan configuration) should be considered in the index archetype configurations. The impact of structural redundancy on the distribution of inelastic deformations should also be reflected in the index archetype configurations.

- **Seismic Design Category.** Applicable Seismic Design Categories establish the design ground motion intensities, which influence seismic-force-resisting system strength and stiffness. Seismic Design Category designations can also trigger special design and detailing requirements that will influence component inelastic deformation capacity. Index archetype configurations should reflect behavioral effects that are influenced by the Seismic Design Categories for which a system is being proposed.

- **Inelastic System Mobilization.** Inelastic system mobilization is the extent to which inelastic action is distributed throughout the seismic-force-resisting system. Inelastic system mobilization is influenced by limits on stiffness and strength irregularities and other system design requirements such as strong-column-weak-beam criteria. Design and configuration decisions can affect whether yielding is distributed vertically across many stories or tends to concentrate in just a few stories. To the extent that the diaphragm design is specified as part of the lateral system, diaphragm strength and stiffness will influence the dynamic response and distribution of forces among the seismic-force-resisting elements. Index archetype configurations should identify and test configurations that are permitted by system design requirements, and will result in a lower bound of inelastic system mobilization.

An example of how behavioral effects are considered in the development of index archetype configurations for a special reinforced concrete moment frame system conforming to design requirements contained in ASCE/SEI 7-05 is provided in Appendix D.

4.2.3   Load Path and Components Not Designated as Part of the Seismic-Force-Resisting System

The complete load path of the seismic-force-resisting system should be considered when defining index archetype configurations. Only those components that are either specified in the design requirements of the seismic-force-resisting system, or otherwise have a significant effect on the system, should be incorporated in the archetype assessment. Portions of the structure that comprise the gravity load system, but are not specifically designed to resist seismic forces, should not be included as part of the index archetype configuration. However, the seismic mass and destabilizing

P-delta effects of the gravity system should be included in the index archetype models.

Conversely, potential failure modes of the gravity system, floor diaphragms, and collector elements need not be reflected in the index archetype configurations, unless those elements are specifically defined in the design requirements for the seismic-force-resisting system. While it is recognized that failure of gravity elements may trigger collapse, their design is usually considered through separate code requirements that are common to many seismic-force-resisting systems and specific to none. If the deformation demands at collapse of a proposed seismic-force-resisting system, however, are excessive relative to those normally experienced by other systems, typical controls for deformation compatibility of gravity framing may not be adequate. In such cases, special deformation criteria for gravity system components should be included in the design requirements for the proposed seismic-force-resisting system.

4.2.4   Overstrength Due to Non-Seismic Loading

Index archetype designs should reflect the inherent overstrength that results from earthquake and gravity load design requirements in addition to any other minimum force requirements that are specified by building code provisions that will always be applicable to a structure. For example, the minimum wind load requirements specified in ASCE/SEI 7-05 (10 psf on the projected surface area exposed to wind) are required in all structures, and should be considered when evaluating potential overstrength. Similarly, minimum seismic design force requirements, which are specified generally for all seismic force resisting systems, should be incorporated in index archetype designs.

Overstrength from design loading that could exist, but is not assured in all cases, should not be considered in the development of index archetype configurations. Such an example could include hurricane wind forces, or other similar location- or use-specific non-seismic loads, that would not apply to all buildings everywhere. Only non-seismic requirements that apply universally to all structures should be considered when evaluating potential overstrength relative to earthquake loading.

4.3   Performance Groups

Index archetype configurations are assembled into performance groups (or bins) that reflect major differences in configuration, design gravity and seismic load intensity, structural period, and other factors that may significantly affect seismic behavior within the archetype design space.

Performance groups should contain multiple index archetype configurations that reflect the expected range of permissible variation in size and other key parameters defined by the archetype design space. For example, each performance group should contain index archetype configurations that cover the range of building heights (up to the limits that can be assessed using the specified ground motions) permitted by the system design requirements.

Binning of index archetype configurations into performance groups provides the basis for statistical assessment of minimum and average collapse margin ratios for performance evaluation in Chapter 7. Performance group populations should not be made larger than necessary, biased towards certain configurations, or otherwise manipulated to bias the average collapse statistics for the bin. Binning of index archetype configurations into performance groups should be reviewed and approved by the peer review panel.

4.3.1 Identification of Performance Groups

As illustrated by the generic performance group table shown in Table 4-3, performance groups should be organized to consider: (1) basic structural configuration; (2) gravity load level; (3) seismic design category; and (4) period domain. The number of basic structural configurations will vary by system (i.e., 1 through N), and variation in gravity load levels may (or may not) affect the performance of certain systems. As a minimum, assuming one basic structural configuration and no dependence on gravity loads, all systems will have at least four performance groups based on combinations of two seismic design levels and two period domains. These parameters should be used as a guide in establishing performance groups, as follows:

- **Basic Structural Configuration.** Changes in the basic structural configuration are intended to capture major variations in the seismic-force-resisting system that are permissible within the design requirements and are likely to affect the structural response. Examples of alternative configurations that could be separated into performance groups include: (1) variations in bracing configurations (e.g., X-bracing versus chevron bracing) in steel concentrically braced frames; (2) variations in framing spans and story heights in moment frames; and (3) variations in shear wall aspect ratios that may influence whether a wall responds in shear or flexural behavior.

- **Gravity Load Level.** To the extent that the gravity load intensity and distribution of gravity loads affect the response of a seismic-force-resisting system, gravity loads should be varied in the index archetype configurations and separated into different performance groups.

Variation in gravity load level is related to the intensity of the gravity load (as affected by the type of gravity framing and use of the structure) and the amount of seismic mass that is tributary to the seismic-force-resisting system in the form of directly applied gravity loads. In moment frames, tributary gravity loads can be distinguished through the use of perimeter frames versus space frame configurations. In wall systems, differences can be distinguished through the use of bearing walls (where most of the gravity loads are directly supported by the walls) versus non-bearing walls (where gravity loads are supported by other means, such as frames).

Table 4-3   Generic Performance Group Matrix

Performance Group Summary					
Group No.	Grouping Criteria				Number of Archetypes
	Basic Configuration	Design Load Level		Period Domain	
		Gravity	Seismic		
PG-1	Type 1	High	Max SDC	Short	≥ 3
PG-2				Long	≥ 3
PG-3			Min SDC	Short	≥ 3
PG-4				Long	≥ 3
PG-5		Low	Max SDC	Short	≥ 3
PG-6				Long	≥ 3
PG-7			Min SDC	Short	≥ 3
PG-8				Long	≥ 3
PG-9	Type 2	High	Max SDC	Short	≥ 3
PG-10				Long	≥ 3
PG-11			Min SDC	Short	≥ 3
PG-12				Long	≥ 3
PG-13		Low	Max SDC	Short	≥ 3
PG-14				Long	≥ 3
PG-15			Min SDC	Short	≥ 3
PG-16				Long	≥ 3
PG-17	Type N	High	Max SDC	Short	≥ 3
PG-18				Long	≥ 3
PG-19			Min SDC	Short	≥ 3
PG-20				Long	≥ 3
PG-21		Low	Max SDC	Short	≥ 3
PG-22				Long	≥ 3
PG-23			Min SDC	Short	≥ 3
PG-24				Long	≥ 3

- **Seismic Design Category.** In concept, the full range of Seismic Design Categories for which the seismic-force-resisting system will be permitted should be reflected in the index archetype configurations and separated into different performance groups. Generally, however, it should suffice to check a system for the maximum and minimum spectral intensities of the highest Seismic Design Category (SDC) in which the system will be permitted. For example, systems intended for SDC D should be designed and assessed for the maximum and minimum spectral acceleration values (SDC D_{max} and SDC D_{min}) as given in Table 5-1A and Table 5-1B of Chapter 5. Usually, designs for the maximum spectral acceleration of the highest Seismic Design Category will control the collapse performance of the system, which is an indication that assessment of performance in lower Seismic Design Categories will not be required. If, however, the minimum spectral acceleration of the highest Seismic Design Category controls performance, then the system should also be designed and assessed for the minimum spectral acceleration of the next lowest Seismic Design Category.

- **Period Domain.** Differences in fundamental period, T, between short-period and long-period systems should be reflected in the index archetype configurations and separated into different performance groups. Period domain is described in Chapter 5, and defined by the boundary between the constant acceleration and constant velocity regions of the design spectrum. Since, within a given structural system, building period varies with the building height, bins of short-period and long-period archetypes will typically be distinguished by building height (or number of stories). The range of index archetypes should generally extend from short-period configurations for one-story buildings up to long-period configurations for the tallest practical buildings within the archetype design space. For the potentially limited number of short-period building archetypes, development of index archetypes should consider variations in both the number of stories and the story height.

Each performance group should include at least three index archetypes. There is no maximum number of archetypes in each performance group, but it is expected that each group will typically have three to six index archetype configurations. This minimum requirement may be waived if it is infeasible to have three alternative designs within a specific performance group. For example, in the case of flexible moment frame systems, it may not be possible to have three distinct index archetype configurations within the short-period domain. Further guidance on the minimum number of performance groups, and index archetype configurations in each group, is provided in Chapter 7.

Chapter 5
Nonlinear Model Development

This chapter describes the development of analytical models for collapse assessment of a proposed seismic-force-resisting system. It first defines how *index archetype designs* are prepared from index archetype configurations using the proposed design requirements for the system of interest. It then outlines how *index archetype models* are developed using nonlinear component properties and limit state criteria developed and calibrated with test data. Index archetype models are based on structural system archetypes defined in Chapter 4, and are used to perform nonlinear analyses described in Chapter 6.

5.1 Development of Nonlinear Models for Collapse Simulation

Since the index archetype configurations are developed with explicit consideration of features to be investigated through nonlinear collapse simulation, the development of nonlinear models and structural system archetypes is interdependent. Nonlinear model development includes preparation of: (1) *index archetype designs*, which are index archetype configurations that have been proportioned and detailed using the design requirements for a proposed seismic-force-resisting system; and (2) *index archetype models*, which are idealized mathematical representations of index archetype designs used to simulate collapse in nonlinear static and dynamic analyses.

Development of nonlinear models for collapse simulation follows the process outlined in Figure 5-1. Using system design requirements (Chapter 3) and structural system archetypes (Chapter 4), design criteria and material test data are applied to each index archetype configuration to develop index archetype designs based on trial values of R, C_d, and Ω_o. Each design is then idealized into an index archetype model for nonlinear analysis (Chapter 6). Model parameters and collapse assessment criteria are calibrated to component, connection, assembly, and system tests. Quality ratings for index archetype models are used to assess modeling uncertainty in Chapter 7.

5.2 Index Archetype Designs

Index archetype designs for a seismic-force-resisting system are prepared by applying the proposed design requirements, substantiated by test data, to a set of index archetype configurations. Index archetype designs should include all significant design features that are likely to affect structural response and collapse behavior. All seismic-force-resisting components and connections should be designed in strict accordance with the minimum requirements of the seismic design provisions for the proposed system. Designs should also meet applicable provisions of ASCE/SEI 7-05 (or other model building code) and any other referenced standards.

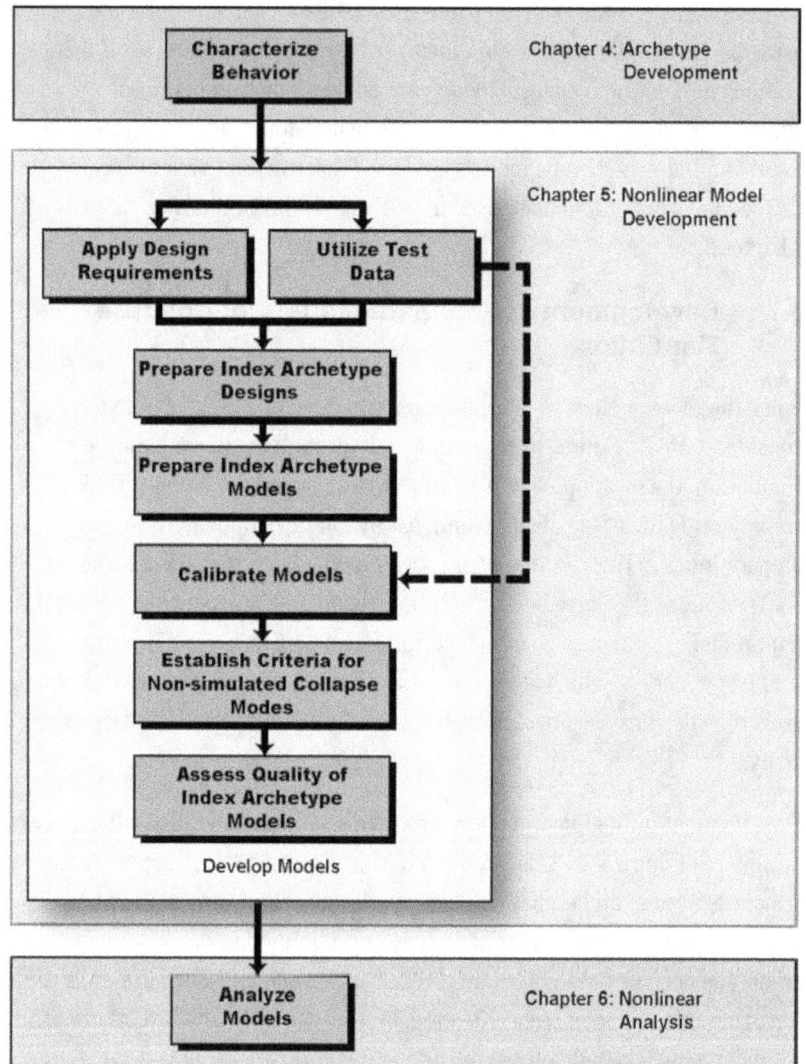

Figure 5-1   Process for development of index archetype models.

While index archetype designs are intended to interrogate the feasible range of the archetype design space, they are not intended to capture all feasible "outliers" of superior or poor seismic performance. Nor are they intended to interrogate seismic design criteria that are common across all seismic-force-resisting systems.

Explicit assessment of redundancy, structural irregularities, soil-structure interaction, importance factors, and other general seismic design criteria are not addressed in the Methodology. It is assumed that such generic design requirements of ASCE/SEI 7-05 (or other applicable building codes) are equally effective for all seismic-force-resisting systems. It is, therefore, important that index archetype designs comply with all relevant limitations in ASCE/SEI 7-05 provisions that are generic to all systems.

The following sections describe design methods, seismic criteria, design loads, load combinations and related requirements for preparing index archetype designs. They are included here to illustrate how seismic design base shears are calculated relative to the Maximum Considered Earthquake (MCE) spectrum used to assess collapse margin, and to highlight specific aspects of seismic design criteria that are relevant to the Methodology.

These criteria are primarily based on the design requirements contained within ASCE/SEI 7-05, modified as appropriate for design of index archetypes. While the specific requirements discussed here are based on ASCE/SEI 7-05, this is not meant to imply a limitation on the application of the Methodology, which is intended to be generally applicable to any set of comprehensive seismic design procedures.

5.2.1 Seismic Design Methods

In general, the Equivalent Lateral Force (ELF) method of Section 12.8, ASCE/SEI 7-05, should be used to develop index archetype designs, except as noted below:

- The Response Spectrum Analysis (RSA) method of Section 12.9, ASCE/SEI 7-05, should be used to develop archetype designs when the ELF method is not permitted by ASCE/SEI 7-05, or by the specific design requirements of the system of interest. For example, the ELF method is not permitted for the design of taller structures in Seismic Design Category D that have a fundamental period, T, greater than $3.5 T_s$ (Table 12.6-1, ASCE/SEI 7-05).

- Response history analysis (RHA) methods should be used to develop archetype designs when ELF and RSA methods are not permitted by ASCE/SEI 7-05, or by the specific design requirements of the system of interest. For example, response history methods are required for design of seismically isolated structures with certain performance characteristics (Section 17.4.2.2, ASCE/SEI 7-05).

- The RSA method (or RHA methods) may be used to develop index archetype designs when such methods are (or are expected to be) commonly used in practice in lieu of the ELF method.

5.2.2 Criteria for Seismic Design Loading

The provisions of ASCE/SEI 7-05 specify seismic loads and design criteria in terms of Seismic Design Category (SDC), which is a function of the level of design earthquake (DE) ground motions and the Occupancy Category of the structure. The Methodology is based on life safety performance and assumes all structures to be either Occupancy Category I or II (i.e., structures that do not have special functionality requirements) with a corresponding importance factor equal to unity. Seismic Design Categories for Occupancy I and II structures vary from SDC A to SDC E in regions of the lowest and highest seismicity, respectively.

The Methodology defines MCE and DE ground motions for structures in Seismic Design Categories B, C, and D. The Methodology ignores both SDC A structures, which are not subject to seismic design (other than the minimum, 1% lateral load specified by Section 11.7.2 of ASCE/SEI 7-05), and SDC E structures which are located in deterministic MCE ground motion regions near active faults.

The provisions of ASCE/SEI 7-05 define MCE demand in terms of mapped values of short-period spectral acceleration, S_S, and 1-second spectral acceleration, S_I, site coefficients, F_a and F_v, and a standard response spectrum shape. For seismic design of the structural system, ASCE/SEI 7-05 defines the DE demand as two-thirds of the MCE demand. The Methodology requires archetypical systems to be designed for DE seismic criteria, and then evaluated for collapse with respect to MCE demand.

For Seismic Design Categories B, C, and D, maximum and minimum ground motions are based on the respective upper-bound and lower-bound values of MCE and DE spectral acceleration, as given in Table 11.6-1 of ASCE/SEI 7-05, for short-period response, and in Table 11.6-2 of ASCE/SEI 7-05, for 1-second response. MCE spectral accelerations are derived from DE spectral accelerations for site coefficients corresponding to Site Class D (stiff soil) following the requirements of Section 11.4 of ASCE/SEI 7-05. For the purpose of assessing performance across all possible site classifications, the Methodology uses values based on the default Site Class D uniformly for design of all archetypes.

Tables 5-1A and 5-1B list values of maximum and minimum spectral acceleration, site coefficients, and design parameters for Seismic Design

Categories B, C, and D. Figure 5-2 shows DE response spectra for ground motions associated with these parameters, based on the standard shape of the design response spectrum shown in Figure 11.4-1 of ASCE/SEI 7-05.

Table 5-1A Summary of Mapped Values of Short-Period Spectral Acceleration, Site Coefficients and Design Parameters for Seismic Design Categories B, C, and D

| Seismic Design Category | | Maximum Considered Earthquake | | | Design |
Maximum	Minimum	S_S (g)	F_a	S_{MS} (g)	S_{DS} (g)
D		1.5	1.0	1.5	1.0
C	D	0.55	1.36	0.75	0.50
B	C	0.33	1.53	0.50	0.33
	B	0.156	1.6	0.25	0.167

Table 5-1B Summary of Mapped Values of 1-Second Spectral Acceleration, Site Coefficients and Design Parameters for Seismic Design Categories B, C, and D

| Seismic Design Category | | Maximum Considered Earthquake | | | Design |
Maximum	Minimum	S_1 (g)	F_v	S_{M1} (g)	S_{D1} (g)
D		0.60[1]	1.50	0.90	0.60
C	D	0.132	2.28	0.30	0.20
B	C	0.083	2.4	0.20	0.133
	B	0.042	2.4	0.10	0.067

1. Value of 1-second MCE spectral acceleration rounded to 0.60 g.

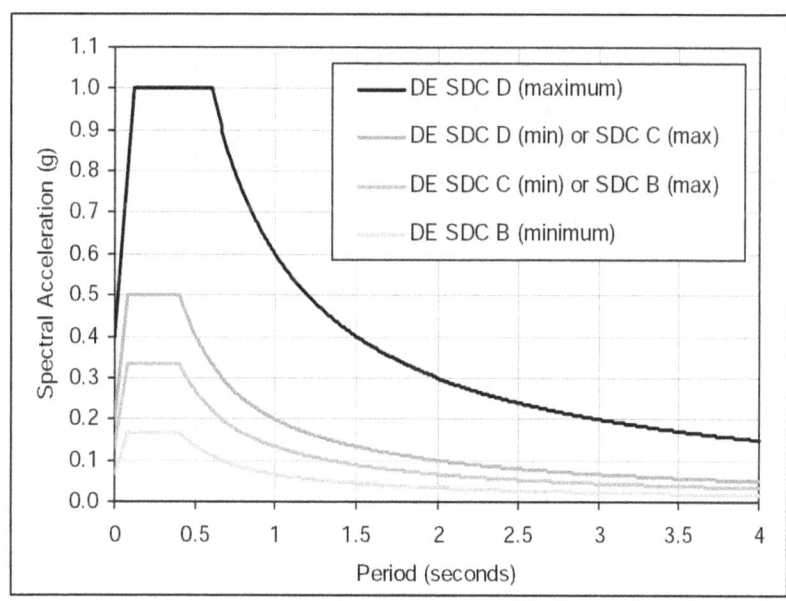

Figure 5-2   Plots of design earthquake (DE) response spectral accelerations used for design of Seismic Design Category D, C and B structure archetypes, respectively.

Maximum values of spectral acceleration for SDC D (S_s = 1.5 g and S_1 = 0.60 g) are based on the effective boundary between deterministic (near-source) and probabilistic regions of MCE ground motions, as defined in Section 21.2 of ASCE/SEI 7-05. The Methodology purposely excludes SDC D structures at deterministic (near-source) sites defined by 1-second spectral acceleration equal to or greater than 0.60 g, although Section 11.6 of ASCE/SEI 7-05 defines SDC D structures as having 1-second spectral acceleration values as high as 0.75 g. The 1-second value of MCE spectral acceleration shown in Table 5-1B, S_1 = 0.60, is rounded upward slightly for convenience, and should be taken as S_1 < 0.60g, for the purpose of evaluating minimum base shear design requirements. This is done to avoid triggering Equation 12.8-6 of ASCE-SEI 7-05 for design of index archetypes.

An internal study, documented in Appendix A, found that the collapse margin ratio (*CMR*) was somewhat smaller in systems designed using SDC E seismic criteria and evaluated using near-field ground motions, than in systems designed using SDC D seismic criteria and evaluated using far-field ground motions. By ignoring SDC E structures, the Methodology implicitly accepts a somewhat greater collapse risk for buildings located close to active faults. This is consistent with the approach in ASCE/SEI 7-05, which implicitly accepts greater risk for buildings near active faults by limiting MCE ground motions to deterministic values of seismic hazard.

5.2.3 Transition Period, T_s

The Methodology requires statistical evaluation of short-period archetypes separately from long-period archetypes, and distinguishes between them on the basis of the transition period, T_s. The transition period defines the boundary between the region of constant acceleration and the region of constant-velocity of the design (or MCE) response spectrum, as illustrated in Figure 5-2. The transition period, T_s, is defined as:

$$T_s = \frac{S_{D1}}{S_{DS}} = \frac{S_{M1}}{S_{MS}} \tag{5-1}$$

where the values of S_{D1} and S_{DS} (and S_{M1} and S_{MS}) are given in Table 5-1A and Table 5-1B, respectively, for each Seismic Design Category. The value of T_s is 0.6 seconds for the upper bound of Seismic Design Category D (SDC D_{max}), 0.4 seconds for the lower bound of Seismic Design Category D (SDC D_{min}), and 0.4 seconds for the upper and lower bounds of the other Seismic Design Categories.

5.2.4 Seismic Base Shear, V

Index archetypes are designed using the seismic base shear, V, as defined by Equation 12.8-1 of ASCE/SEI 7-05:

$$V = C_s W \qquad (5\text{-}2)$$

where C_s is the seismic response coefficient and W is the effective seismic weight. This base shear equation is used as the basis of the applied forces when the Equivalent Lateral Force procedure is used for design, and to scale design values in accordance with Section 12.9.4 of ASCE/SEI 7-05 when the Response Spectrum Analysis procedure is used for design. The seismic coefficient, C_s, is defined for short-period archetypes ($T \le T_s$) as:

$$C_s = \frac{S_{DS}}{R} \qquad (5\text{-}3)$$

and for long-period archetypes ($T > T_s$) as:

$$C_s = \frac{S_{D1}}{T\,R} \ge\ 0.44\,S_{DS} \qquad (5\text{-}4)$$

where S_{D1} and S_{DS} are given in Tables 5-1A and 5-1B, respectively, R is the trial value of the response modification factor, and T is the fundamental period of the index archetype.

Equations 5-3 and 5-4 have an implied occupancy importance factor of $I=1.0$, since the index archetype designs are defined for Occupancy Categories I and II. These equations are also constrained by the minimum seismic coefficient, $C_s = 0.01$, required by Equation 12.8-5 of ASCE/SEI 7-05 and by the minimum static lateral force, $F_x = 0.01\,w_x$, required by Equation 11.7-1 of ASCE/SEI 7-05 for design of all structures. Note that Equation 12.8-6 of ASCE-SEI 7-05 does not apply since the Methodology defines 1-second MCE spectral acceleration as $S_1 < 0.60$ g, in all cases.

ASCE/SEI 7-05 reduces the value of C_s for very long period structures that have a fundamental period, $T > T_L$, where T_L is the transition period between the constant velocity and constant displacement response domains. Values of T_L range from 4 seconds to 16 seconds, as defined in Section 11.4.5 of ASCE/SEI 7-05. Reduced values of C_s for very long period structures do not apply, since the Methodology limits index archetype designs to configurations with fundamental periods less than 4 seconds (due to possible limitations on the low frequency content of ground motion records used for the nonlinear dynamic analysis).

5.2.5 Fundamental Period, T

The fundamental period, T, is used within the Methodology in two ways. First, it is used in establishing the design base shear through Equation 5-4. Second, it is used in defining the ground motion spectral intensity to establish the collapse margin ratio (CMR) in nonlinear dynamic analysis procedures (Chapter 6). In both cases, the Methodology defines the fundamental period, T, as:

$$T = C_u T_a = C_u \ C_t \ h_n^x \geq 0.25 \text{ seconds} \qquad (5\text{-}5)$$

where h_n is the building height, the values of the coefficient, C_u, are given in Table 12.8-1 of ASCE/SEI 7-05, and values of period parameters C_t and x are given in Table 12.8-2 of ASCE/SEI 7-05. The approximate fundamental period, T_a, is based on regression analysis of actual building data, and represents lower-bound (mean minus one standard deviation) values of building period (Chopra et al., 1998). The value of the coefficient, C_u, ranges from 1.4 in high seismic regions to 1.7 in low seismic regions, and the product, $C_u T_a$, approximates the average value of building period. Alternative formulas for estimating the approximate fundamental period of masonry or concrete shear wall structures are given in Section 12.8.2.1 of ASCE/SEI 7-05, and may be used in lieu of T_a in Equation 5-5. Alternative formulas for the approximate fundamental period of proposed seismic-force-resisting systems with different dynamic characteristics should be specified as part of the design requirements for such systems.

Use of Equation 5-5 provides a consistent basis for determining the building period for the purpose of calculating the design base shear and evaluating the MCE spectral intensity at collapse. Based on scatter in ground motion spectra at small periods, the fundamental period, T, as calculated by Equation 5-5 (or other alterative formulas) includes a lower limit of 0.25 seconds for the purpose of evaluating the CMR.

5.2.6 Loads and Load Combinations

Index archetype designs should be prepared considering gravity and seismic loading, in accordance with the seismic load effects and load combinations of Section 12.4 of ASCE/SEI 7-05 and guidance provided in this section. Basic seismic load combinations for strength design (ignoring snow load, S, and foundation load, H) are:

$$\left(1.2 + 0.2 S_{DS}\right) D + Q_E + L \qquad (5\text{-}6a)$$

$$\left(0.9 - 0.2 S_{DS}\right) D + Q_E \qquad (5\text{-}6b)$$

where D includes the structural self weight and superimposed dead loads, L is the live load (including appropriate live load reduction factors), and Q_E is the effect of horizontal seismic forces resulting from the base shear, V. The basic load combinations, defined above, purposely do not include the redundancy factor, ρ, which is conservatively assumed to be 1.0 in all cases. Seismic loads, and hence capacity of index archetype designs, should not be increased for a possible lack of redundancy that may not exist in all applications of the proposed system.

Where the seismic load effect with overstrength is required, basic load combinations for strength design (ignoring snow load, S, and foundation load, H) are:

$$(1.2 + 0.2 S_{DS}) D + \Omega_O Q_E + L \qquad (5\text{-}7a)$$

$$(0.9 - 0.2 S_{DS}) D + \Omega_O Q_E \qquad (5\text{-}7b)$$

where Ω_o is the overstrength factor. Snow load, S, and foundations loads, H, are not required for design of index archetypes. Snow load is an environmental load that varies independently of seismic intensity, and is not considered to be a primary factor affecting seismic performance. Foundation loads, H, are not typical and when present, generally apply to design of structural components that do not affect performance of the seismic-force-resisting system.

Wind load is also not required for design of index archetypes, since wind load does not occur (in load combinations) with earthquake load. However, index archetypes may be designed for minimum values of wind load in lieu of earthquake load when minimum values of wind load exceed earthquake load. If wind load is used for archetype design, minimum values of wind load should be based on the lowest basic wind speed in Chapter 6 of ASCE/SEI 7-05 for all regions of the United States, and consider building (plan) configurations which minimize lateral forces due to wind load.

5.2.7 Trial Values of Seismic Performance Factors

Preparation of index archetype designs requires selection of trial values of the response modification coefficient, R, displacement amplification coefficient, C_d, and overstrength factor, Ω_o. Initial values of R, C_d, and Ω_o may need to be revised, and the index archetypes redesigned, based on the outcome of the performance evaluation process in Chapter 7.

A trial value of the response modification factor, R, is required for all index archetype designs to determine seismic base shear, V, and the related effect of horizontal seismic force, Q_E. For certain archetypes, Equations 5-7a and 5-7b require a trial value of the overstrength factor, Ω_O, for design of structural components that are subject to rapid deterioration and are sensitive to overload conditions. For index archetype designs governed by drift, Section 12.8.6 of ASCE/SEI 7-05 requires a trial value of the deflection amplification factor, C_d, to determine story drift. Guidance on initial selection of trial values may be found in the acceptance criteria of Chapter 7.

5.2.8 Performance Group Design Variations

As described in Chapter 4, performance groups for each archetype configuration include design load variations based on Seismic Design Category and gravity load intensity, and design height variations that influence the fundamental period of the structure.

Maximum and Minimum Seismic Loads. Strictly speaking, index archetype designs should reflect all Seismic Design Categories for which the seismic-force-resisting system will be permitted. Typically, however, the highest SDC will have the smallest collapse margin ratios, and govern system performance. To reasonably cover the design space, the performance groupings (Chapter 4) require assessment of index archetypes that are designed for both the maximum and minimum spectral intensities of the highest SDC in which the system is allowed. For example, index archetypes for systems permitted in all SDCs must be designed for SDC D_{max} and SDC D_{min}. Similarly, index archetypes for systems permitted in SDC A or SDC B must be designed for SDC B_{max} and SDC B_{min}.

In certain cases, performance might be controlled by lower SDCs. For example, if collapse margin ratios calculated for index archetypes designed using SDC D_{max} are not consistently larger than those for index archetypes designed using SDC D_{min}, then lower SDCs might control collapse performance. In such cases, index archetypes should also be designed and assessed for minimum spectral accelerations corresponding to the next lowest SDC. If index archetypes designed for the next lower SDC are found to have even lower collapse margin ratios, then additional index archetypes designs should be prepared to confirm whether or not lower SDCs control system performance.

High and Low Gravity Loads. Where gravity loads tributary to the seismic-force-resisting system significantly influence collapse behavior, index archetype designs must be prepared for high gravity and low gravity load intensities. High and low gravity load intensities reflect differences in the self weight of alternative gravity framing components (e.g., metal deck versus concrete slabs) as well as framing configurations (e.g., space frame versus perimeter frame configurations or bearing wall versus non-bearing wall components). Where gravity load effects are significant, design dead load, D, and live load, L, intensities should be established considering different building occupancies and the range of likely gravity load intensities in the archetype design space.

Building Height. Index archetypes must be designed to populate performance groups in the short-period ($T \le T_s$) and long-period ($T > T_s$) ranges. Equation 5-5 can be used to relate building height to number of stories and story heights. Ideally, each performance group should have at least 3 index archetype designs. However, flexible systems (e.g., steel moment frames) can have relatively long fundamental periods, T, greater then T_s, potentially limiting the number of index archetype designs in short-period performance groups. Conversely, systems with building height limits may have few archetypes with fundamental periods, T, greater than T_s, limiting the number of archetypes in long-period performance groups. In such cases, performance groups are permitted to have less than three archetypes, provided that the group has at least one index archetype design for each feasible building height (combination of number of stories and story heights).

Unless restricted by height, long-period performance groups should contain index archetypes of different heights (number of stories) that have fundamental periods ranging from about T_s to about 4 seconds. Index archetypes should be designed such that the fundamental periods of the performance group are well distributed over the full range of periods. It should be noted that more than three archetypes per long-period performance group may be required to properly evaluate the performance of taller systems.

5.3   Index Archetype Models

General considerations for developing index archetype models are summarized in Table 5-2. These considerations should be used as a guide in establishing index archetype models, as follows:

- **Model Idealization.** Definition of index archetype models includes selection of the type of idealization used to represent structural behavior. At the one extreme are nonlinear continuum finite element models, which, in theory, are capable of representing the underlying structural mechanics most directly. At the other extreme are phenomenological models, which represent the overall force-deformation response through concentrated nonlinear springs. A nonlinear beam-column hinge model is an example of such a phenomenological model, in which moment-rotation behavior is related to beam-column design parameters through semi-empirical models that are calibrated to beam-column subassembly tests.

In between these two extremes are models that utilize both continuum and phenomenological representations. A "fiber-type" model of a reinforced concrete shear wall is an example of such a combined model, where flexural effects are modeled with uniaxial stress-strain behavior for reinforcing steel and concrete, and where shear behavior (or combined shear-flexural behavior) is represented through a stress-resultant (force-based) phenomenological model. Regardless of type, models must be validated against test data and other substantiating evidence to assess how accurately they capture nonlinear response and critical limit state behavior.

Table 5-2   General Considerations for Developing Index Archetype Models

Model Attributes	Considerations
Mathematical Idealization	• Continuum (physics-based) versus phenomenological elements
Plan and Elevation Configurations	• Number of moment frame bays, regularity. • Planar versus 3-D wall representations, openings, coupling beams, regularity. • Number of bracing bays, bracing configuration, regularity. • Variations to reflect diaphragm effects on stiffness and 3-D force distributions
2-D versus 3-D Component Behavior	• Prevalence of 2-D versus 3-D systems in design practice • Impact on structural response, including provisions for 3-D (out-of-plane failures) in 2-D models
2-D versus 3-D System Behavior	• Characteristics of index archetype configurations, such as diaphragm flexibility • Impact on structural response that is specific to certain structural systems

- **Elevation and Plan Configurations.** Representation of elevation and plan configurations in index archetype models will depend on both the index archetype configurations and the structural system behavior. While vertical and horizontal irregularities will certainly influence collapse, for the purpose of evaluating general design provisions, currently permissible elevation and plan irregularities in ASCE/SEI 7-05 are not addressed in index archetype models. As illustrated by examples in Chapter 9, two-dimensional, three-bay frames of regular proportions are judged sufficient to represent typical behavior of reinforced concrete moment frame systems. The extent to which this type of model will suffice for studies of other moment frame types should be established based on the specific behavioral effects of the specific moment frame system. For walls, the issue of planar versus three-dimensional response is a key consideration, as is the presence of wall openings, boundary elements, and coupling beams. For example, reinforced concrete walls with large boundary members (e.g., flanged walls) are likely to exhibit more shear-critical behavior than planar walls without boundary members. For braced frame systems, one or two bays of framing are likely to be sufficient unless the system relies on the specific interaction between two adjacent bays. Representation of alternative brace configurations is likely to be a dominant variable in collapse assessment of braced-frame systems. Where diaphragm flexibility has a significant effect on the lateral system response and performance, this flexibility should be incorporated in the index archetype model.

- **Two-Dimensional versus Three-Dimensional Component Behavior.** The need for models that simulate two-dimensional versus three-dimensional behavior will generally depend on: (1) the type of structural configurations common in the design space; and (2) the expected influence of three-dimensional effects on structural response. For most structural framing types, two-dimensional models are likely to be sufficient. However, there may be cases where three-dimensional behavior (e.g., out-of-plane torsional-flexural instability of laterally unbraced beam-columns or braces) or three-dimensional geometry (e.g., reinforced-concrete C-shaped core walls) are important to simulate. For wall systems, two-dimensional wall models may be sufficiently accurate for some system configurations (e.g., wooden shear walls, planar reinforced concrete walls) but less accurate and perhaps inappropriate for others (e.g., C-shaped and I-shaped reinforced concrete core walls).

- **Two-Dimensional versus Three-Dimensional System Behavior.** System behavior involves the interaction of multiple seismic-force-resisting components distributed spatially within a structure. Introduction of different spatial combinations, however, could lead to an intractable number of index archetype configurations and corresponding index archetype models. Building code provisions regarding plan configuration and three-dimensional effects (e.g., redundancy, accidental torsion) are usually not system specific, so in most cases, a two-dimensional system representation should be adequate. Diaphragm flexibility may require three-dimensional index archetype model configurations if important diaphragm effects cannot be suitably incorporated in two-dimensional models.

5.3.1 Index Archetype Model Idealization

Index archetype models should provide the most basic (generic) representation of an index archetype configuration that is still capable of distinguishing between significant behavioral modes and key design features of the proposed seismic-force-resisting system. Index archetype models should be developed in cooperation with the peer review panel.

The mathematical idealization of index archetype models should capture all significant nonlinear effects related to the collapse behavior of the system. This can be done through: (1) explicit simulation of failure modes through nonlinear analyses; or (2) evaluation of non-simulated[1] failure modes using alternative limit state checks on demand quantities from nonlinear analyses.

Analytical models are generally distinguished by overall topology and element type. Topology refers to two-dimensional or three-dimensional modeling configurations. The choice of topology (2-D or 3-D) is largely a function of the index archetype configurations. The choice of element type depends on structural component behavior and the nature of component degradation. Two-dimensional topologies (e.g., planar frames or walls) do not preclude the modeling of three-dimensional effects (e.g., out-of-plane instabilities). Conversely, three-dimensional topologies (e.g., space frames or C-shaped walls) do not necessarily employ element types that capture all three-dimensional behavioral effects. Thus, the modeling decisions should be made on a case-by-case basis, depending on the specific features of the structure system archetypes.

For simulating collapse, component models must capture strength and stiffness degradation under large deformations. Structural components are usually idealized as a combination of one-dimensional line-type elements (beam-columns or axial struts) and two-dimensional continuum elements (plane-stress or plate/shell finite elements). Three-dimensional continuum elements (brick finite elements) may be appropriate and necessary in some cases. Within each element type, element formulations can be further distinguished by the extent to which the underlying structural behavior is modeled explicitly or through phenomenological representations. For

---

[1] The term "non-simulated" is used to describe potential modes of collapse failure that are not explicitly captured by the index archetype model (i.e., not explicitly simulated), but is evaluated by alternative methods of analysis and included in the evaluation of collapse performance.

example, nonlinear beam-column elements can range in sophistication from fiber-type continuum elements, in which the geometry and materials in the cross section are modeled explicitly, to concentrated spring models, in which the inelastic response is idealized through uni-axial or multi-axial springs.

Provided that they are accurately calibrated to the appropriate range of design and behavioral parameters, concentrated spring models will usually be sufficient for simulating nonlinear response of columns, beams, and beam-column connections in frame systems. These models have the practical advantage of providing a straight-forward approach to characterizing strength and inelastic deformation characteristics. However, concentrated spring models generally cannot represent behavioral effects beyond those present in the underlying data. Continuum models, which generally model the physical behavior at a more fundamental level, can, if properly formulated and validated, represent a broader range of behavioral effects that do not rely as much on tests to represent the specific parameters of the index archetype designs.

Wall systems will typically require two-dimensional continuum models that can capture significant nonlinear stress and strain variations within the walls. Continuum models may include traditional two-dimensional plane stress/strain finite elements, or alternative formulations that utilize combinations of formal finite element approaches and engineering assumptions to represent the nonlinear behavior (including the effects of strength and stiffness degradation).

In the case of moment frame systems, for example, an index archetype model might consist of the two-dimensional, three-bay frame shown in Figure 5-3. This model incorporates one-dimensional line-type elements with either concentrated spring or discrete component models to simulate the nonlinear degrading response of beams, columns, beam-column connections, and panel zones. Significant frame behaviors are captured in a two-dimensional representation, and the three-bay configuration captures differences between interior and exterior columns. The additional leaning column elements capture P-delta effects of the seismic mass that is not tributary to the frame.

For shear wall systems, an index archetype model might be as simple as a cantilever element that accounts for inelastic flexure and shear behavior at the base of the wall. However, where punched shear wall geometries are included in the index archetype configurations, then the corresponding index archetype models would need to be more complicated.

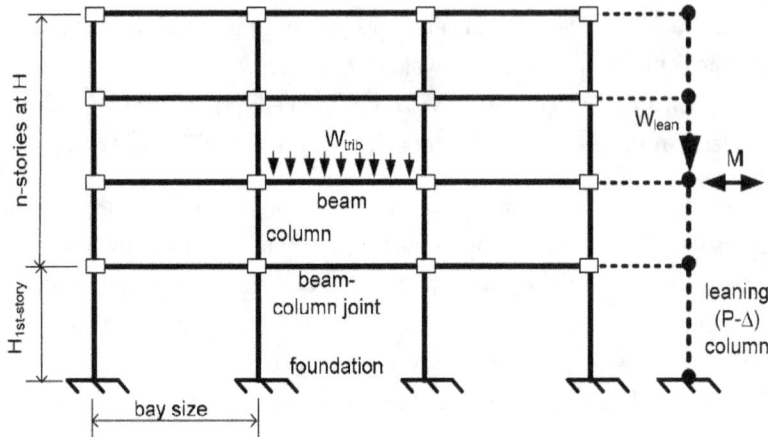

Figure 5-3   Example of index archetype model for moment resisting frame systems

5.4   Simulated Collapse Modes

To the extent possible, index archetype models should directly simulate all significant deterioration modes that contribute to collapse behavior. Typically, this is accomplished through structural component models that simulate stiffness, strength, and inelastic deformation under reverse cyclic loading. Research has demonstrated that the most significant factors influencing collapse response are the strength at yield, F_y, maximum strength (at capping point), F_c, plastic deformation capacity, δ_p, the post-capping tangent stiffness, K_{pc}, and the residual strength, F_r (Ibarra et al., 2005). These parameters can be used to define a component backbone curve, such as the one shown in Figure 5-4. Recently, such a curve has been designated a force-displacement capacity boundary (FEMA, 2009).

Cyclic deterioration, which reduces stiffness values and lowers the force-displacement capacity boundary established by the monotonic backbone curve, should also be included to the extent that it influences the collapse response in nonlinear dynamic analyses. An example of degrading hysteretic response is shown in Figure 5-5. Characterization of component backbone curves and hysteretic responses should represent the median response properties of structural components. While illustrated in an aggregate sense in Figure 5-4 and Figure 5-5, the behavior can be modeled through elements of varying degrees of sophistication using phenomenological or physics-based approaches.

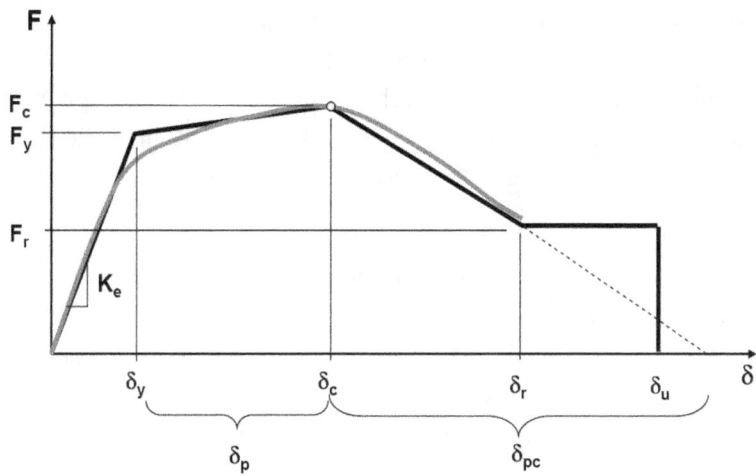

Effective yield strength and deformation (F_y and δ_y)
Effective elastic stiffness, $K_e = F_y/\delta_y$
Strength cap and associated deformation for monotonic loading (F_c and δ_c)
Pre-capping plastic deformation for monotonic loading, δ_p
Effective post-yield tangent stiffness, $K_p = (F_c\text{-}F_y)/\delta_p$
Post-capping deformation range, δ_{pc}
Effective post-capping tangent stiffness, $K_{pc} = F_c/\delta_{pc}$
Residual strength, F_r
Ultimate deformation, δ_u

Figure 5-4   Parameters of an idealized component backbone curve

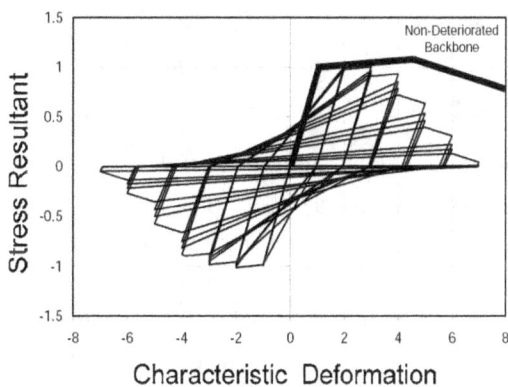

Figure 5-5   Idealized inelastic hysteretic response of structural components
                  with cyclic strength and stiffness degradation.

While of lesser importance than the definition of the maximum force and
deformation at the capping point, the initial stiffness can have a significant
effect on the ductility capacity. Element-level initial stiffness should reflect
all important contributors to deformation (e.g., flexure, bond-slip, and shear),
and should be validated against component and assembly test data. An
effective initial stiffness defined as the secant stiffness from the origin
through the point of 40% of the yield strength of the element should be

considered in phenomenological concentrated spring models. In continuum models, initial stiffness is usually modeled directly. Where results are sensitive to initial stiffness, attention should be given to effects related to initiation of cracking or yielding that may not be considered in the model, such as shrinkage cracking due to concrete curing and residual stresses due to fabrication.

Figure 5-4 and Figure 5-5 are intentionally portrayed in a generic sense, since critical response parameters will vary for each specific component and configuration. For example, in ductile reinforced concrete components (i.e., special moment frames), nonlinear response is typically associated with moment-rotation in the hinge regions where degradation occurs at large deformations through a combination of concrete crushing, confinement tie yielding/rupture, and longitudinal bar buckling. However, in less ductile reinforced concrete components (i.e., ordinary moment frames), nonlinear response may include shear failures and axial failure following shear failure. Where the seismic-force-resisting system carries significant gravity load, characteristic force and deformation quantities may need to represent vertical deformation effects as well as horizontal response effects.

The development of analytical models is case specific, and no single model is universally applicable. For many steel, reinforced concrete, and wood components, the deterioration model proposed by Ibarra et al. (2005) satisfactorily matches experimental results and analytical predictions. However, this model should be utilized for a proposed system only if it can be justified based on experimental evidence.

Referring to Figure 5-5, the backbone curve defines a boundary within which hysteresis loops are confined. The implication is that in the analytical model, the load-deformation response is not permitted to move outside this curve. Such boundaries can be based on monotonic behavior, but ideally they should be based on series of tests including monotonic loading and cyclic loading with different loading protocols (FEMA, 2009). If such boundaries are fixed in the analytical model (i.e., cyclic deterioration is not incorporated explicitly), then estimates of the backbone curve parameters should account for average cyclic deterioration, to produce a modified backbone curve. If the initial stiffness is very different from the effective elastic stiffness, then it may affect the response close to collapse, and should become part of the modeling effort.

Figure 5-6 illustrates the effect of cyclic loading relative to a backbone curve obtained from monotonic loading. In almost all cases, the plastic deformation capacity, δ_p, is reduced by cyclic loading, and in many cases it is reduced by a considerable amount from the monotonic loading case. A backbone curve is difficult to construct from a cyclic test (unless experience exists from other similar specimens) and often necessitates the execution of an additional monotonic test. If monotonic tests are not available, a curve enveloping the cyclic test (cyclic envelope) may be used as a conservative estimate of the modified backbone curve.

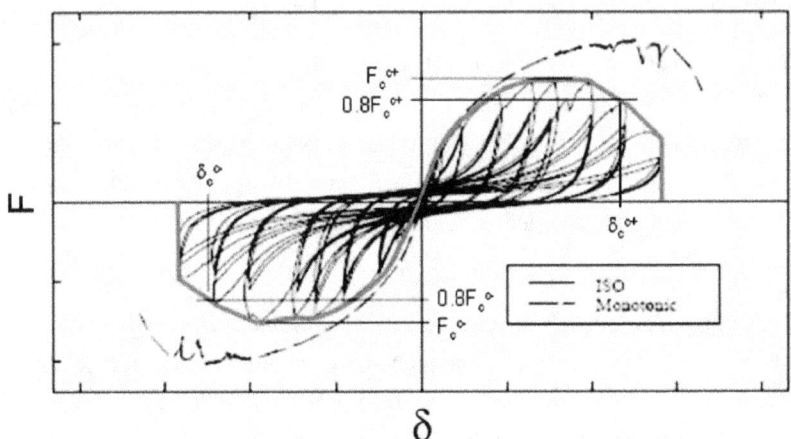

Figure 5-6   Comparison of monotonic and cyclic response, along with a cyclic envelope curve (adapted from Gatto and Uang 2002).

If the backbone curve is obtained from a monotonic test (or is deduced based on a cyclic deterioration model), then cyclic deterioration must be built into the analytical model representing component behavior. Most cyclic deterioration models are energy based (e.g., Ibarra et al., 2005; Sivaselvan and Reinhorn, 2000). Validity of the component model must be demonstrated through satisfactory matching of component, connection, or assembly test date from the experimental program.

Figure 5-6 also illustrates a simplified measure of performance, which is the deformation associated with a force value of 80% of the maximum strength measured in the test, F_c^c. In the figure, the deformation value, δ_c^c, which is obtained from the intersection of a horizontal line at $0.8F_c^c$ with the cyclic envelope, can be viewed as a conservative estimate of the ultimate deformation capacity of a component. In simplified analytical models it can be assumed that no deterioration occurs up to this value of deformation, provided that the strength of the component is assumed to drop to zero at deformations larger than this value. Both F_c^c and δ_c^c may be different in the positive and negative directions.

The monotonic backbone curve of Figures 5-4 and 5-6 is similar but distinct from the generalized force-displacement curves specified in ASCE/SEI 41,

Seismic Rehabilitation of Existing Buildings, (ASCE, 2006b). In ASCE/SEI 41-06, generalized force-displacement curves utilize cyclic envelopes that incorporate some degree of cyclic degradation and, in most cases, result in conservative estimates of median response. In this Methodology, backbone curves are intended to represent median properties of monotonic loading response, where cyclic strength and stiffness degradation are directly modeled in the analysis, and statistical variations of the component response are explicitly accounted for in the assessment process.

The type of backbone curve and cyclic hysteretic model used will also impact the amount of equivalent viscous damping used in the model. Models that have backbone curves with a large initial elastic region (which do not dissipate energy under cyclic loading) will generally use higher equivalent viscous damping than models with small initial elastic regions (which do dissipate energy under small cycles).

While component models are expected to be rigorously calibrated to test data, available data may not be comprehensive enough to fully calibrate the models. Data are often particularly scarce for evaluating the capping point and post-capping behavior that occurs at large deformations in ductile components. In such cases, test data should be augmented by engineering analysis and judgment to establish the modeling parameters.

An example of the development and calibration of nonlinear component models for reinforced concrete moment frame systems is provided in Appendix E.

5.5 Non-Simulated Collapse Modes

In cases where it is not possible (or not practical) to directly simulate all significant deterioration modes contributing to collapse behavior, non-simulated collapse modes can be indirectly evaluated using alternative limit state checks on structural response quantities measured in the analyses. Examples of possible non-simulated collapse modes might include shear failure and subsequent axial failure in reinforced concrete columns, fracture in the connections or hinge regions of steel moment frame components, or failure of tie-downs in light-frame wood shear walls. Component failures such as these may be difficult to simulate directly.

In Figure 5-7, a non-simulated collapse (NSC) limit state is shown to occur prior to the deformation at peak strength (and subsequent deterioration) that is directly simulated in the model. Non-simulated limit state checks are similar to the assessment approach of ASCE/SEI 41-06, in which component acceptance criteria are used to evaluate specific performance targets based on

5: Nonlinear Model Development

demand quantities extracted from the analyses. This approach is more of an approximation to the actual behavior of the system. It increases the uncertainty in analytical results and tends to provide conservative estimates of collapse limit states. While not ideal, it is a practical approach that provides a consistent method for evaluating the effects of deterioration and collapse mechanisms that are otherwise difficult (or impossible) to incorporate directly in the analytical model.

Compared to collapse that is simulated directly, non-simulated limit state checks will generally result in lower estimates of median collapse. Non-simulated collapse modes are usually associated with component failure modes, and a commonly applied assumption is that the first occurrence of this failure mode will lead to collapse of the structure. Collapse of an entire structure predicated on the failure of a single component can, in many cases, be overly conservative. For this reason, local failure modes should be directly simulated, if at all possible, to permit redistribution of forces to other components after a limit state has been reached. Alternatively, local limit state checks can be redefined to indirectly account for possible redistribution of forces that is known to occur prior to collapse.

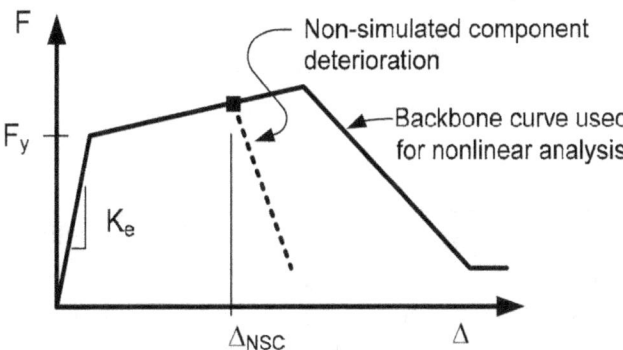

Figure 5-7   Component backbone curve showing a deterioration mode that is not directly simulated in the analysis model.

When considered in the context of incremental dynamic analyses, non-simulated component limit state checks are essentially stipulating a collapse limit prior to the point where an analysis would otherwise simulate collapse. Figure 5-8 shows a plot of the results from an incremental dynamic analysis of an index archetype model, which is subjected to a single ground motion that is scaled to increasing intensities. The point denoted SC corresponds to the collapse limit that is directly simulated in the model. The point denoted NSC represents the collapse limit as determined by applying a component limit state check on a potential collapse mode that is not directly simulated in the model. In this example, the limit state check is based on story drift, but non-simulated collapse checks could be based on any other structural response parameter measured in the analysis, such as peak force demand in an element, or peak plastic hinge rotation demand.

Limit state checks for non-simulated collapse modes should be established based on test data and other supporting evidence, and should be calibrated to represent the median value of the governing response parameter that is associated with the collapse response. When establishing limit state checks, judgment should be exercised in relating the critical condition of a component to the collapse response of the building system, since there are many cases where critical limit states for isolated components will not immediately trigger overall system collapse. Non-simulated collapse limit states should be developed in cooperation with the Methodology peer review panel.

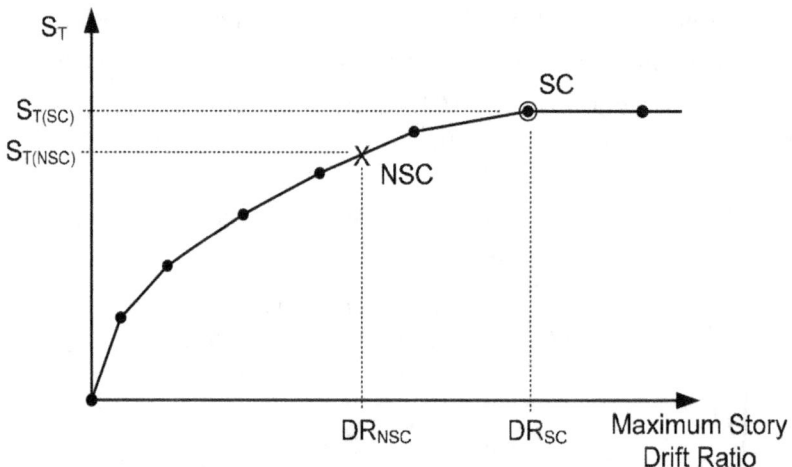

Figure 5-8   Incremental dynamic analysis results showing simulated (SC) and non-simulated (NSC) collapse modes.

5.6   Characterization of Modeling Uncertainties

In this Methodology, nonlinear analysis is used to determine the median ground motion intensity associated with collapse of a proposed seismic-force-resisting system. Index archetype models should, therefore, represent the median response of structural components that constitute the proposed system. Variability in collapse response, due to ground motion variability, modeling, and other uncertainties, is factored into the performance evaluation process in Chapter 7. When a model calibrated to median properties is used, nonlinear dynamic analysis under multiple ground motions is intended to provide a median estimate of the collapse capacity of an index archetype.

5.7   Quality Rating of Index Archetype Models

Quality of index archetype models is related to uncertainty, which factors into the performance evaluation for a proposed seismic-force-resisting system. The quality of index archetype models is rated in accordance with the requirements of this section, and approved by the peer review panel.

Index archetype models are rated between (A) Superior and (D) Poor, as shown in Table 5-3. This rating is a combined assessment of: (1) how well index archetype models represent the range of structural collapse characteristics and associated design parameters of the archetype design space; and (2) how well the analysis models capture structural collapse behavior through both direct simulation and non-simulated limit state checks. The quantitative values of modeling-related collapse uncertainty are: (A) Superior, $\beta_{MDL} = 0.10$; (B) Good, $\beta_{MDL} = 0.20$; (C) Fair, $\beta_{MDL} = 0.35$; and (D) Poor, $\beta_{MDL} = 0.50$. Use of these values is described in Section 7.3.

Table 5-3   Quality Rating of Index Archetype Models

Representation of Collapse Characteristics	Accuracy and Robustness of Models		
	High	Medium	Low
High. Index models capture the full range of the archetype design space and structural behavioral effects that contribute to collapse.	(A) Superior $\beta_{MDL} = 0.10$	(B) Good $\beta_{MDL} = 0.20$	(C) Fair $\beta_{MDL} = 0.35$
Medium. Index models are generally comprehensive and representative of the design space and behavioral effects that contribute to collapse.	(B) Good $\beta_{MDL} = 0.20$	(C) Fair $\beta_{MDL} = 0.35$	(D) Poor $\beta_{MDL} = 0.50$
Low. Significant aspects of the design space and/or collapse behavior are not captured in the index models.	(C) Fair $\beta_{MDL} = 0.35$	(D) Poor $\beta_{MDL} = 0.50$	--

The highest rating of (A) Superior applies to instances in which the index archetype models represent the complete range of structural configuration and collapse behavior, there is a high confidence in the ability of established models to simulate behavior, and the nonlinear model is of high-fidelity. The combination of low quality representation of collapse characteristics along with low quality modeling in terms of accuracy and robustness is not permitted.

An adaptation of the Methodology to assess building-specific collapse performance of an individual building is presented in Appendix F.

Differences in assigning quality ratings for an analytical model of an individual building are discussed there.

5.7.1   Representation of Collapse Characteristics

Representation of collapse characteristics refers to how completely and comprehensively the index archetype models capture the full range of design parameters and associated structural collapse behavior that is envisioned within the archetype design space. The quality of the representation is characterized as follows:

- **High.** The set of index archetype configurations and associated archetype models provides a complete and comprehensive representation of the full range of structural configurations, design parameters and behavioral characteristics that affect structural collapse. The index archetype models cover a comprehensive range of building heights, lateral system configurations, and design alternatives that are permitted by the design requirements. To the extent that 3-D component and system effects are significant, they are reflected in the index archetype models, as are other significant system effects such as diaphragm flexibility,

- **Medium.** The set of index archetype models provides a reasonably broad and complete representation of the design space. Where the complete design space is not fully represented in the set of models, there is reasonable confidence that the range of response captured by the models is indicative of the primary structural behavior characteristics that affect collapse.

- **Low.** The set of index archetype models does not capture the full range of structural configurations and collapse behavior for the system due to the combined effects of a loosely defined design space and a less than complete set of index archetype configurations. Loosely defined limits on system configurations and design parameters present a challenge in that the number of possible alternative configurations and structural design parameters are so large as to preclude systematic interrogation with a manageable number of index archetype configurations. Seismic-force-resisting systems permitted in low Seismic Design Categories that have limited requirements on design (e.g., steel ordinary moment frame systems) may fall into this category. Even for well controlled design criteria, however, representation of collapse characteristics may be low if the number and variety of index archetype configurations are not insufficient to capture the possible range in collapse behavior.

5.7.2 Accuracy and Robustness of Models

Accuracy and robustness is related to the degree to which nonlinear behaviors are directly simulated in the model, or otherwise accounted for in the assessment. Use of non-simulated collapse limit state checks will lower the accuracy and robustness of a nonlinear model. If conservatively applied, however, non-simulated collapse checks should not necessarily lower the overall quality rating of the assessment procedure. Model accuracy and robustness are characterized as follows:

- **High.** Nonlinear models directly simulate all predominate inelastic effects, from the onset of yielding through strength and stiffness degradation causing collapse. Models employ either concentrated hinges or distributed finite elements to provide spatial resolution appropriate for the proposed system. Computational solution algorithms are sufficiently robust to accurately track inelastic force redistribution, including cyclic loading and unloading, without convergence problems, up to the point of collapse.

- **Medium.** Nonlinear models capture most, but not all, nonlinear deterioration and response mechanisms leading to collapse. Models may not be sufficiently robust to track the full extent of deterioration, so that some component-based limit state checks are necessary to assess collapse.

- **Low.** Nonlinear models capture the onset of yielding and subsequent strain hardening, but do not simulate degrading response. Onset of degradation is primarily evaluated using non-simulated component limit state checks. Overall uncertainty in response quantities is increased due to inability to capture the effects of deterioration and redistribution.

Chapter 6
Nonlinear Analysis

This chapter describes nonlinear analysis procedures for collapse assessment of a proposed seismic-force-resisting system. It defines the set of input ground motions, and specifies how nonlinear static analyses and nonlinear dynamic analyses are conducted on index archetype models developed in Chapter 5. Nonlinear analyses are used to define the median collapse capacity and other parameters that are needed for performance evaluation in Chapter 7.

6.1 Nonlinear Analysis Procedures

Nonlinear analysis for collapse assessment follows the process outlined in Figure 6-1.

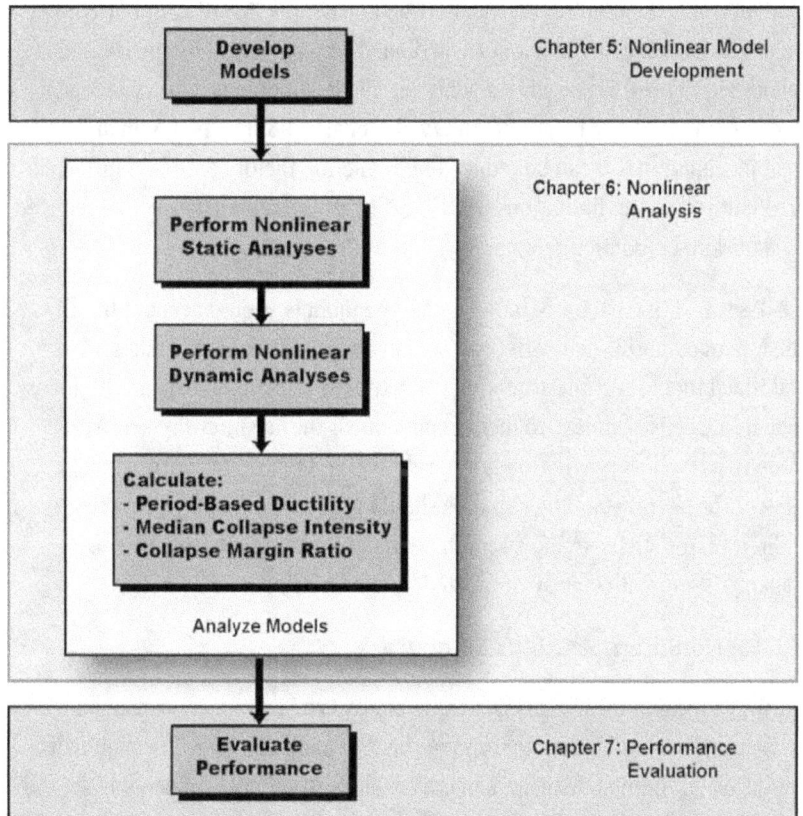

Figure 6-1   Process for performing nonlinear analyses for collapse assessment.

Nonlinear static (pushover) and dynamic (response history) analyses of all index archetype models are performed to obtain statistics for system overstrength, period elongation, and collapse capacity. Nonlinear static analyses are performed first, to help validate the model and to provide statistical data on system overstrength, Ω, and period-based ductility, μ_T. Nonlinear dynamic analyses are then performed to assess median collapse capacities, $\hat{S}_{CT}$, and collapse margin ratios, CMR. Median collapse capacity is defined as the ground motion intensity where half of the ground motions in the record set cause collapse of an index archetype model.

In all cases, modeling parameters, including the seismic mass and imposed gravity loads, should represent the median values of the structure and its components. The gravity loads for analysis are different from design gravity loads, and are given by the following load combination:

$$1.05D + 0.25L \qquad (6\text{-}1)$$

where D is the nominal dead load of the structure and the superimposed dead load, and L is the nominal live load. Load factors in Equation 6-1 are based on expected values (equivalent to median values for normally distributed random variables) reported in a study on the development of a predecessor document to ASCE/SEI 7-05 (Ellingwood et al., 1980). The nominal live load in Equation 6-1 can be reduced by reduction factors based on influence area (subject to the limitations in ASCE/SEI 7-05), but should not be reduced by additional reduction factors.

As described in Chapter 5, index archetype models should account for all seismic mass and P-delta effects associated with gravity loads that are stabilized by the seismic-force-resisting system. This includes gravity loads that are directly tributary to the components of the seismic-force-resisting system, as well as gravity loads that rely on the seismic-force-resisting system for lateral stability. Models should also account, either directly or indirectly, for stiffness and strength degradation leading to the onset of collapse along with energy dissipated by the building.

6.1.1 Nonlinear Analysis Software

Software for nonlinear analysis can be of any type that is: (1) capable of static pushover and dynamic response history analyses; and (2) capable of capturing strength and stiffness degradation in structural components at large deformations. A significant computational challenge is to accurately capture the negative post-peak response, sometimes referred to as strain-softening response, in component backbone curves. Strain-softening response leads to the need for robust iterative numerical solution strategies to minimize errors and achieve convergence at large inelastic deformations. Problems with strain-softening response have the potential for non-unique solutions and damage localization that is sensitive to numerical issues. While most modern analysis software can overcome these issues, care must be taken to investigate the sensitivity of the solution to modeling parameters and numerical aspects of the computational solution algorithms.

6.2   Input Ground Motions

Nonlinear dynamic response of index archetypes is evaluated for a set of pre-defined ground motions that are systematically scaled to increasing intensities until median collapse is established. The ratio between median collapse intensity, $\hat{S}_{CT}$, and Maximum Considered Earthquake (MCE) ground motion intensity, S_{MT}, is defined as the collapse margin ratio, CMR, which is the primary parameter used to characterize the collapse safety of the structure.

6.2.1   MCE Ground Motion Intensity

Collapse performance is evaluated relative to ground motion intensity associated with the MCE, as defined in ASCE/SEI 7-05, and related to the seismic criteria used for design of index archetypes in Chapter 5.

As described in Chapter 5, the Methodology defines DE and MCE ground motion intensities for three ranges of spectral acceleration associated with Seismic Design Categories B, C and D. Table 6-1 summarizes MCE spectral acceleration for maximum and minimum ground motions for these Seismic Design Categories, and Figure 6-1 shows MCE response spectra for the corresponding ground motion intensities.

Table 6-1   Summary of Maximum Considered Earthquake Spectral Accelerations And Transition Periods Used for Collapse Evaluation of Seismic Design Category D, C, and B Structure Archetypes, Respectively

Seismic Design Category		Maximum Considered Earthquake		Transition Period
Maximum	Minimum	S_{MS} (g)	S_{M1} (g)	T_s (sec.)
D		1.5	0.9	0.6
C	D	0.75	0.30	0.4
B	C	0.50	0.20	0.4
	B	0.25	0.10	0.4

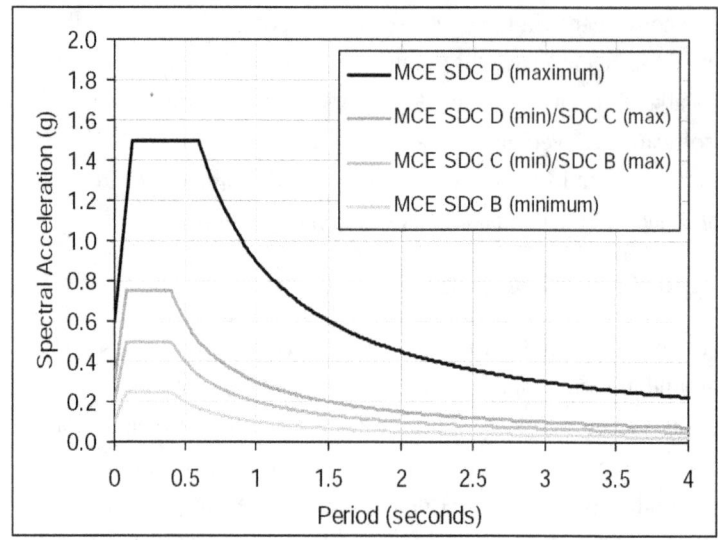

Figure 6-2   MCE response spectra required for collapse evaluation of index
             archetypes designed for Seismic Design Category (SDC) B, C,
             and D.

MCE ground motion intensity, S_{MT}, is defined for short-period archetypes
$(T \leq T_s)$ as:

$$S_{MT} = S_{MS} \qquad (6-2)$$

and for long-period archetypes $(T > T_s)$ as:

$$S_{MT} = \frac{S_{M1}}{T} \qquad (6-3)$$

where values of S_{M1} and S_{MS} are given in Table 6-1, and T is the fundamental
period of an index archetype as defined in Equation 5-5.

6.2.2   Ground Motion Record Sets

The Methodology provides two sets of ground motion records for collapse
assessment using nonlinear dynamic analysis, referred to as the Far-Field
record set and the Near-Field record set. The Far-Field record set includes
twenty-two component pairs of horizontal ground motions from sites located
greater than or equal to 10 km from fault rupture. The Near-Field record set
includes twenty-eight component pairs of horizontal ground motions
recorded at sites less than 10 km from fault rupture. The record sets do not
include the vertical component of ground motion since this direction of
earthquake shaking is not considered of primary importance for collapse
evaluation, and is not required by the Methodology for nonlinear dynamic
analysis.

The ground motion record sets each include a sufficient number of records to permit evaluation of record-to-record (RTR) variability and calculation of median collapse intensity, $\hat{S}_{CT}$. Explicit calculation of record-to-record variability, however, is not required for collapse evaluation of index archetypes. Instead, an estimate of record-to-record variability, based on previous research and developmental studies, is built into the process for calculating total system collapse uncertainty in Chapter 7. The record sets, along with selection criteria and background information on their selection, are provided in Appendix A.

The Methodology specifies use of the Far-Field record set for collapse evaluation of index archetypes designed for Seismic Design Category (SDC) B, C or D criteria (i.e., structures at sites that are located away from active faults). The Near-Field record set is provided as supplemental information, and is used in special studies of Appendix A to evaluate potential differences in the CMR for SDC E structures. Figure 6-3 shows the 44 individual response spectra (i.e., 22 records, 2 components each) of the Far-Field record set, the median response spectrum, and spectra representing one standard deviation and two-standard deviations above the median.

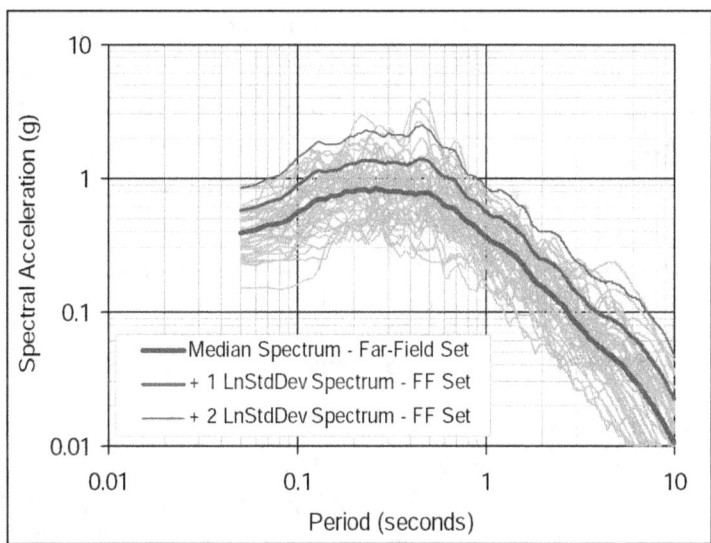

Figure 6-3   Far-Field record set response spectra.

Both ground motion record sets include strong-motion records (i.e., records with PGA > 0.2 g and PGV > 15 cm/sec) from all large-magnitude (M > 6.5) events in the Pacific Earthquake Engineering Research Center (PEER) Next-Generation Attenuation (NGA) database (PEER, 2006a). Large-magnitude events dominate collapse risk and generally have longer durations of shaking, which is important for collapse evaluation of nonlinear degrading models.

The sets include records from soft rock and stiff soil sites (predominantly Site Class C and D conditions), and from shallow crustal sources (predominantly strike-slip and thrust mechanisms). To avoid event bias, no more than two of the strongest records are taken from each earthquake.

The primary function of the Far-Field record set is to provide a fully-defined set of records for use in a consistent manner to evaluate collapse across all applicable Seismic Design Categories, located in any seismic region, and founded on any soil site classification. Actual earthquake records are used, in contrast with artificial or synthetic records, recognizing that regional variation of ground motions would not be addressed. In the United States, strong-motion records date back to the 1933 Long Beach Earthquake, with only a few records obtained from each event until the 1971 San Fernando Earthquake. Large magnitude events are rare, and few existing earthquake ground motion records are strong enough to collapse a large percentage of modern, code-compliant buildings.

Even with many instruments, existing strong motion instrumentation networks (e.g., Taiwan and California) provide coverage for only a small fraction of all regions of high seismicity. Considering the size of the earth and period of geologic time, the available sample of strong motion records from large-magnitude earthquakes is still quite limited, and potentially biased by records from more recent, relatively well-recorded events. Due to the limited number of very large earthquakes, and the frequency ranges of ground motion recording devices, the ground motion record sets are primarily intended for buildings with natural (first-mode) periods less than or equal to 4 seconds. Thus, the record set is not necessarily appropriate for tall buildings with long fundamental periods of vibration greater than 4 seconds.

6.2.3 Ground Motion Record Scaling

Ground motion records are scaled to represent a specific intensity (e.g., the collapse intensity of the index archetypes of interest). Record scaling involves two steps. First, individual records in each set are "normalized" by their respective peak ground velocities, as described in Appendix A. This step is intended to remove unwarranted variability between records due to inherent differences in event magnitude, distance to source, source type and site conditions, without eliminating overall record-to-record variability. Second, normalized ground motions are collectively scaled (or "anchored") to a specific ground motion intensity such that the median spectral acceleration of the record set matches the spectral acceleration at the fundamental period, T, of the index archetype that is being analyzed.

The first step was performed as part of the ground motion development process, so the record sets contained in Appendix A already reflect this normalization. The second step is performed as part of the nonlinear dynamic analysis procedure. This two-step scaling process parallels the ground motion scaling requirements of Section 16.1.3.2 of ASCE/SEI 7-05. Figure 6-4 shows the median spectrum of the Far-Field record set anchored to maximum and minimum MCE response spectra of Seismic Design Categories B, C and D, at a period of 1 second.

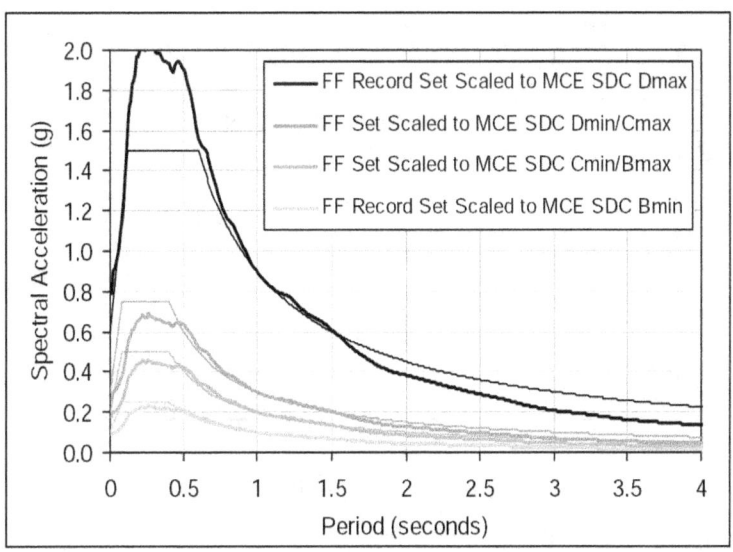

Figure 6-4   Median spectrum of the Far-Field record set anchored to maximum and minimum MCE response spectra of Seismic Design Categories B, C and D, at a period of 1 second.

6.3   Nonlinear Static (Pushover) Analyses

Nonlinear static (pushover) analyses are conducted under the factored gravity load combination of Equation 6-1 and static lateral forces. In general, pushover analysis should be performed following the nonlinear static procedure (NSP) of Section 3.3.3 of ASCE/SEI 41-06.

The vertical distribution of the lateral force, F_x, at each story level, x, should be in proportion to the fundamental mode shape of the index archetype model:

$$F_x \propto m_x \phi_{1,x} \qquad (6-4)$$

where m_x is the mass at level x; and $\phi_{1,x}$ is the ordinate of the fundamental mode at level x.

Figure 6-5 shows an idealized pushover curve and definitions of the maximum base shear capacity, V_{max} and the ultimate displacement, δ_u. V_{max} is taken as the maximum base shear strength at any point on the pushover curve, and δ_u is taken as the roof displacement at the point of 20% strength loss ($0.8\,V_{max}$).

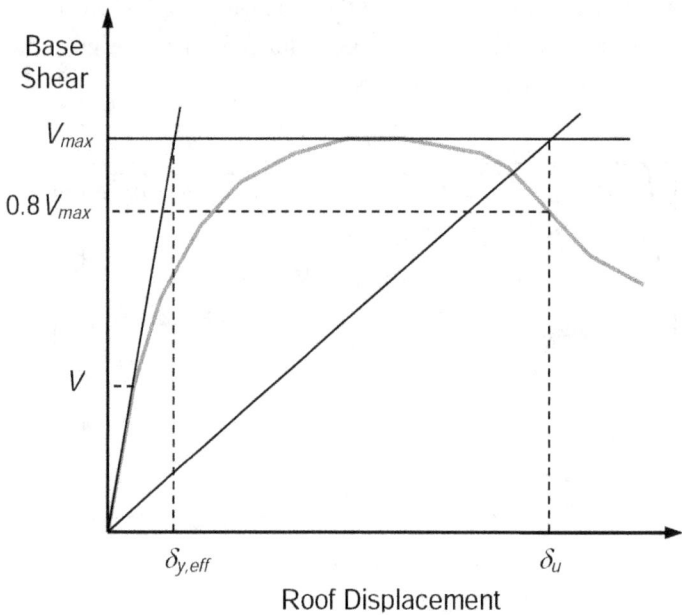

Figure 6-5   Idealized nonlinear static pushover curve

A nonlinear static pushover analysis is used to quantify V_{max} and δ_u, which are then used to compute archetype overstrength, Ω, and period-based ductility, μ_T. In order to quantify these values, the lateral loads are applied monotonically until a loss of 20% of the base shear capacity ($0.8\,V_{max}$) is achieved.

The overstrength factor for a given index archetype model, Ω, is defined as the ratio of the maximum base shear resistance, V_{max}, to the design base shear, V:

$$\Omega = \frac{V_{max}}{V} \tag{6-5}$$

The period-based ductility for a given index archetype model, μ_T, is defined as the ratio of ultimate roof drift displacement, δ_u, (defined as shown in Figure 6-5) to the effective yield roof drift displacement $\delta_{y,eff}$:

$$\mu_T = \frac{\delta_u}{\delta_{y,eff}} \tag{6-6}$$

6: Nonlinear Analysis

The effective yield roof drift displacement is as given by the formula:

$$\delta_{y,eff} = C_O \ \frac{V_{max}}{W} \left[\frac{g}{4\pi^2} \right] \ (\max(T, T_1))^2 \qquad (6\text{-}7)$$

where C_0 relates fundamental-mode (SDOF) displacement to roof displacement, V_{max}/W is the maximum base shear normalized by building weight, g is the gravity constant, T is the fundamental period (C_uT_a, defined by Equation 5-5), and T_1 is the fundamental period of the archetype model computed using eigenvalue analysis.

The coefficient C_0 is based on Equation C3-4 of ASCE/SEI 41-06, as follows:

$$C_0 = \phi_{1,r} \frac{\displaystyle\sum_1^N m_x \, \phi_{1,x}}{\displaystyle\sum_1^N m_x \phi_{1,x}^2} \qquad (6\text{-}8)$$

where m_x is the mass at level x; and $\phi_{1,x}$ ($\phi_{1,r}$) is the ordinate of the fundamental mode at level x (roof), and N is the number of levels. Additional background on period-based ductility is included in Appendix B.

Since pushover analyses are intended to verify the models and provide a conservative bound on the system overstrength factor, checks for non-simulated collapse modes are not incorporated directly. Non-simulated collapse modes should be considered when evaluating ultimate roof drift displacement, δ_u.

Where three-dimensional analyses are used, separate nonlinear static analyses should be performed to evaluate overstrength and ultimate roof drift displacement independently along the two principle axes of the index archetype model. The resulting values for overstrength and ultimate roof drift displacement are then calculated by averaging the values from each of the principle loading directions.

6.4   Nonlinear Dynamic (Response History) Analyses

Nonlinear dynamic (response history) analyses are conducted under the factored gravity load combination of Equation 6-1 and input ground motions from the Far-Field record set in Appendix A. Nonlinear dynamic analyses are used to establish the median collapse capacity, $\hat{S}_{CT}$, and collapse margin ratio, CMR, for each of the index archetype models. Ground motion intensity, S_T, is defined based on the median spectral intensity of the Far-

Field record set, measured at the fundamental period of the structure ($C_u T_a$, defined by Equation 5-5). Determination of the collapse margin ratio for each index archetype model is expected to require approximately 200 nonlinear response history analyses (approximately 5 analyses of varying intensity for each component of the 22 pairs of earthquake ground motion records).

The following sections present background on the collapse assessment methodology followed by specific guidelines for conducting nonlinear dynamic analyses to calculate the collapse parameters, $\hat{S}_{CT}$ and CMR, for index archetype models.

6.4.1   Background on Assessment of Collapse Capacity

The median collapse intensity can be visualized through the concept of incremental dynamic analysis (IDA) (Vamvatsikos and Cornell, 2002), in which individual ground motions are scaled to increasing intensities until the structure reaches a collapse point. Results from a set of an incremental dynamic analyses are illustrated in Figure 6-6, where each point in the figure corresponds to the results of one nonlinear dynamic analysis of one index archetype model subjected to one ground motion record that is scaled to one intensity level.

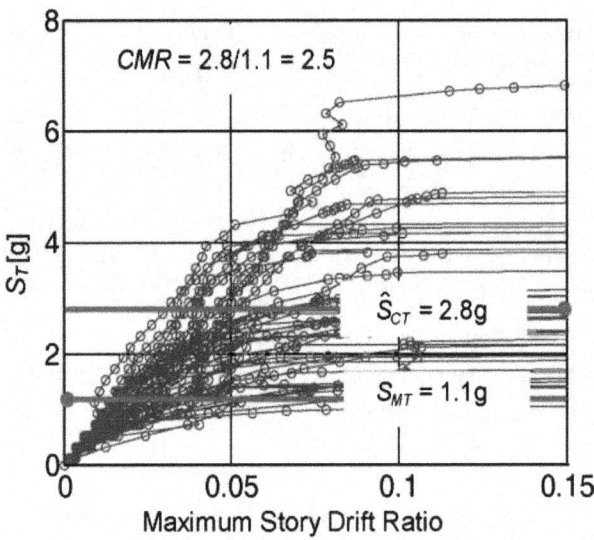

Figure 6-6   Incremental dynamic analysis response plot of spectral acceleration versus maximum story drift ratio.

In Figure 6-6, the results of each analysis are plotted in terms of the spectral intensity of the ground motion (on the vertical axis) versus maximum story drift ratio recorded in the analysis (on the horizontal axis). Each line in Figure 6-6 connects results for a given ground motion scaled to increasing spectral intensities. Differences between the lines reflect differences in the response of the same index archetype model when subjected to different ground motions with different frequency characteristics. Collapse under each ground motion is judged to occur either directly from dynamic analysis results as evidenced by excessive lateral displacements (lateral dynamic instability) or assessed indirectly through non-simulated component limit state criteria. In Figure 6-6, the median collapse capacity of $\hat{S}_{CT} = 2.8g$ is defined as the spectral intensity when half of the ground motions cause the structure to collapse.

Using collapse data from IDA results, a collapse fragility curve can be defined through a cumulative distribution function (CDF), which relates the ground motion intensity to the probability of collapse (Ibarra et al., 2002). Figure 6-7 shows an example of a cumulative distribution plot obtained by fitting a lognormal distribution through the collapse data points from Figure 6-6.

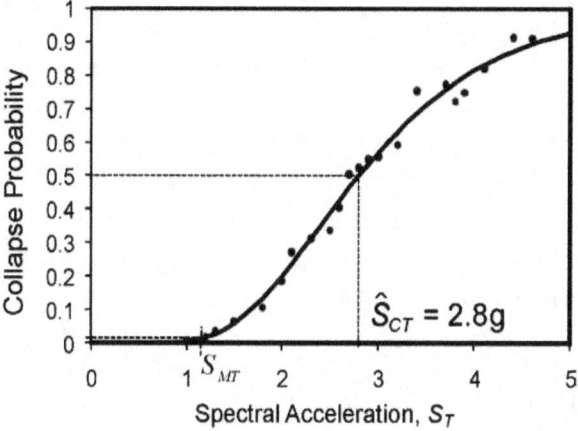

Figure 6-7   Collapse fragility curve, or cumulative distribution function.

The lognormal collapse fragility is defined by two parameters, which are the median collapse intensity, $\hat{S}_{CT}$, and the standard deviation of the natural logarithm, β_{RTR}. The median collapse capacity ($\hat{S}_{CT} = 2.8g$ in the figure) corresponds to a 50% probability of collapse. The slope of the lognormal distribution is measured by β_{RTR}, and reflects the dispersion in results due to record-to-record (*RTR*) variability (uncertainty). In this Methodology only the median collapse intensity, $\hat{S}_{CT}$, is calculated, and record-to-record variability, β_{RTR}, is set to a fixed value (i.e., $\beta_{RTR} = 0.4$ for systems with period-based ductility ≥ 3).

Values of record-to-record variability are fixed for several reasons. First, previous studies have shown that the record-to-record variability is fairly constant for different structural models and record sets. Second, more precise calculation of record-to-record variability would not significantly affect calculation of the *CMR* when combined with other sources of collapse uncertainty. Finally, record-to-record variability is fixed because accurate calculation of β_{RTR} would require collapse data from a larger number of ground motions than is necessary to calculate an accurate median collapse intensity, $\hat{S}_{CT}$.

6.4.2 Calculation of Median Collapse Capacity and CMR

While the IDA concept is useful for illustrating the collapse assessment procedure, the Methodology only requires identification of the median collapse intensity, $\hat{S}_{CT}$, which can be calculated with fewer nonlinear analyses than would otherwise be necessary for developing the full IDA curve. Referring to Figure 6-6, $\hat{S}_{CT}$ can be obtained by scaling all the records in the Far-Field record set to the MCE intensity, S_{MT}, and then increasing the intensity until just over one-half of the scaled ground motion records cause collapse. The lowest intensity at which one-half of the records cause collapse is the median collapse intensity, $\hat{S}_{CT}$. Judicious selection of earthquake intensities close to and approaching the median collapse intensity leads to a significant reduction in the number of analyses that are required. As a result, nonlinear response history analyses for median collapse assessment are computationally much less involved than the full IDA approach. While the full IDA curve is not required, a sufficient number of response points should be plotted at increasing intensities to help validate the accuracy in calculating the median collapse intensity.

The MCE intensity is obtained from the response spectrum of MCE ground motions at the fundamental period, T. In Figure 6-6, the MCE intensity, S_{MT}, is 1.1 g, taken directly from the response spectrum for SDC D_{max} in Figure 6-4. The ratio between the median collapse intensity and the MCE intensity is the collapse margin ratio, *CMR*, which is the primary parameter used to characterize the collapse safety of the structure.

$$CMR = \frac{\hat{S}_{CT}}{S_{MT}}$$
(6-9)

6.4.3 Ground Motion Record Intensity and Scaling

In the Methodology, ground motion intensities are defined in terms of the median spectral intensity of the Far-Field record set, rather than the spectral intensity of each individual record. Conceptually, this envisions the Far-Field record set as representative of a suite of records from a characteristic earthquake in which spectral intensities of individual records will exhibit

6: Nonlinear Analysis

dispersion about the median value of the set. Thus, the median collapse capacity of the structure is equal to the median capacity of the Far-Field record set at the point where half of the records in the set cause collapse of the index archetype model.

The spectral scaling intensity for the ground motion records is determined based on the median spectral acceleration of the Far-Field record set at the fundamental period, T, of the building. For purposes of scaling the spectra and calculating the corresponding MCE hazard spectra, the value of fundamental period is the same as that required for archetype design (C_uT_a, defined by Equation 5-5). The spectral acceleration of the record set at the specified period is S_T, and the intensity at the collapse point for each record is S_{CT} (i.e., S_T of the average spectra corresponding to the point when the specified record triggers collapse). The median collapse intensity for the entire record set (the median of 44 records) is $\hat{S}_{CT}$. The spectral acceleration of the MCE hazard, S_{MT}, is given in Table 6-1.

6.4.4   Energy Dissipation and Viscous Damping

To the extent that index archetype models are accurately calibrated for the loading histories encountered in the dynamic analysis, most of the structural damping will be modeled directly in the analysis through hysteretic response of the structural components. Thus, assumed viscous damping for nonlinear collapse analyses should be less than would typically be used in linear dynamic analyses. Depending on the type and characteristics of the nonlinear model, additional viscous damping may be used to simulate the portion of energy dissipation arising from both structural and nonstructural components (e.g., cladding, partitions) that is not otherwise incorporated in the model. When used, viscous damping should be consistent with the inherent damping in the structure that is not already captured by the nonlinear hysteretic response that is directly simulated in the model.

For nonlinear dynamic analyses, equivalent viscous damping is typically assumed to be in the range of 2% to 5% of critical damping for the first few vibration modes that tend to dominate the response. Care should be taken to ensure that added viscous damping does not increase beyond acceptable levels as the model yields. The appropriate amount of damping, and strategies to incorporate it in the assessment, should be confirmed with the peer review panel.

6.4.5 Guidelines for CMR Calculation using Three-Dimensional Nonlinear Dynamic Analyses

For two-dimensional analyses, all forty-four ground motion records (twenty-two pairs) are applied as independent events to calculate the median collapse intensity, $\hat{S}_{CT}$, for each index archetype model. For three-dimensional analyses, the twenty-two record pairs are applied twice to each model, once with the ground motion records oriented along one principal direction, and then again with the records rotated 90 degrees.

Because ground motions records are applied in pairs in three-dimensional nonlinear dynamic analyses, collapse behavior of each index archetype model resulting from each ground motion component is coupled. Notwithstanding other variations between the two-dimensional and three-dimensional analyses, studies have shown that the median collapse intensity resulting from three-dimensional analyses is on average about 20% less than the median collapse intensity resulting from two-dimensional analyses. Thus, the application of pairs of ground motion records in three-dimensional analyses introduces a conservative bias as compared to results from two-dimensional analyses.

To achieve parity with the two-dimensional analyses, an adjustment should be made when calculating the collapse margin ratio using three-dimensional analyses. The CMR calculated based on median collapse intensity, $\hat{S}_{CT}$, obtained from three-dimensional analyses should be multiplied by a factor of 1.2. This multiplier is applied in addition to the spectral shape factor, SSF, which is used to calculate the adjusted collapse margin ratio, $ACMR$, as part of the performance evaluation in Chapter 7.

6.4.6 Summary of Procedure for Nonlinear Dynamic Analysis

Nonlinear dynamic analysis of each index archetype model includes the following steps:

1. Using the normalized Far-Field earthquake record set in Appendix A, scale all records to an initial scale factor. A suggested initial scale factor is $S_T = 1.3S_{MT}$, which can be adjusted up or down based on the results of initial analyses.

2. Perform nonlinear response history analyses on each index archetype model using all twenty-two pairs of records in the Far-Field record set. For two-dimensional analyses, models should be analyzed separately for each ground motion component in each pair, for a total of forty-four analyses. For three-dimensional analyses, the twenty-two record pairs should be applied twice to each model, once with the ground motion

records oriented along one principal direction, and then again with the records rotated 90 degrees. Process results to check for simulated collapse (lateral dynamic instability or excessive lateral deformations signaling a sidesway collapse mechanism) or non-simulated collapse (demands that exceed certain component limit state criteria applied external to the analysis).

3. Based on results from the first set of analyses, adjust the ground motion scale factor, and perform additional analyses until collapse is detected for twenty-two of the forty-four ground motions. The median collapse intensity, $\hat{S}_{CT}$, should be taken as the collapse spectral intensity, S_T, observed for the set of analyses where twenty-two records caused collapse. To reduced the number of analyses, the median collapse intensity, $\hat{S}_{CT}$, can be conservatively estimated from the collapse spectral intensity, S_T, observed for any set of analyses where less than twenty-two records cause collapse. This may expedite the assessment for index archetype designs whose actual median collapse intensity (based on 22 records) exceeds the acceptance criteria by a significant margin.

Other strategies can be used for systematically scaling the records to determine the median collapse capacity. The most straightforward approach is to systematically scale up the entire record set, in specified increments, until collapse is detected for twenty-two of the records (or record pairs).

6.5   Documentation of Analysis Results

Nonlinear analysis results serve as the basis for performance evaluation in Chapter 7, and are subject to review by the peer review panel. As a minimum, information on model development, and data from nonlinear static and dynamic analyses, should be documented in accordance with this section.

6.5.1   Documentation of Nonlinear Models

The following information should be reported on the development of the nonlinear index archetype models:

- Description of index archetype models, including graphical representations of idealized models showing support and loading conditions and member types

- Summary of modeling parameters and substantiating test data including material strengths and stress-strain properties, component and connection strengths and deformation capacities, gravity loads and masses, and damping parameters

- Summary of criteria and substantiating test data for non-simulated collapse modes

- General documentation of analysis software

6.5.2 Data from Nonlinear Static Analyses

The following information should be reported from nonlinear static (pushover) and eigenvalue analyses of each index archetype model:

- Fundamental period of vibration, T, model period of vibration, T_1, and design base shear, V

- Distribution of lateral (pushover) loads

- Plot of base shear versus roof drift

- Fully yielded strength, V_{max}, and static overstrength factor $\Omega = V_{max}/V$

- The effective yield, $\delta_{y,eff}$, ultimate roof displacements, δ_u, and period-based ductility, μ_T

- Story drift ratios at the design base shear, the maximum load V_{max}, and $0.8\,V_{max}$ (used to gage system behavior)

6.5.3 Data from Nonlinear Dynamic Analyses

The following information should be reported from nonlinear dynamic (response history) analyses of each index archetype model:

- MCE ground motion intensity (MCE spectral acceleration), S_{MT}, and the period used to calculate this value

- Median collapse intensity, $\hat{S}_{CT}$, and collapse margin ratio, CMR

- Data used to compute the median collapse capacity, $\hat{S}_{CT}$, along with the response parameter used to identify the collapse condition (e.g., maximum story drift ratio for simulated collapse, and limit-state criteria for non-simulated collapse). Accompanying notes, plots, or narratives describing the governing mode(s) of failure

- Representative plots of hysteresis curves for selected structural components up to the collapse point

Chapter 7

Performance Evaluation

This chapter describes the process for evaluating the performance of a proposed seismic-force-resisting system, assessing the acceptability of a trial value of the response modification coefficient, R, and determining appropriate values of the system overstrength factor, Ω_O, and the deflection amplification factor, C_d.

Performance evaluation is based on the results of nonlinear static and dynamic analyses conducted in accordance with Chapter 6. It requires judgment in interpreting analytical results, assessing uncertainty, and rounding of values for design. Performance evaluation, and selection of appropriate seismic performance factors, requires the concurrence of the peer review panel.

7.1 Overview of the Performance Evaluation Process

The performance evaluation process utilizes results from nonlinear static (pushover) analyses to determine an appropriate value of the system overstrength factor, Ω_O, and results from nonlinear static and nonlinear dynamic (response history) analyses to evaluate the acceptability of a trial value of the response modification coefficient, R. The deflection amplification factor, C_d, is derived from an acceptable value of R, with consideration of the effective damping of the system of interest.

The trial value of the response modification coefficient, R, used to design index archetypes, is evaluated in terms of the acceptability of the collapse margin ratio. Acceptability is measured by comparing the collapse margin ratio, after adjustment for the effects of spectral shape, to acceptable values that depend on the quality of the information used to define the system, total system uncertainty, and established limits on collapse probability.

Performance evaluation follows the process outlined in Figure 7-1, and includes the following steps:

- Obtain calculated values of system overstrength, Ω, period-based ductility, μ_T, and collapse margin ratio, CMR, for each index archetype, from results of nonlinear analyses (Chapter 6).

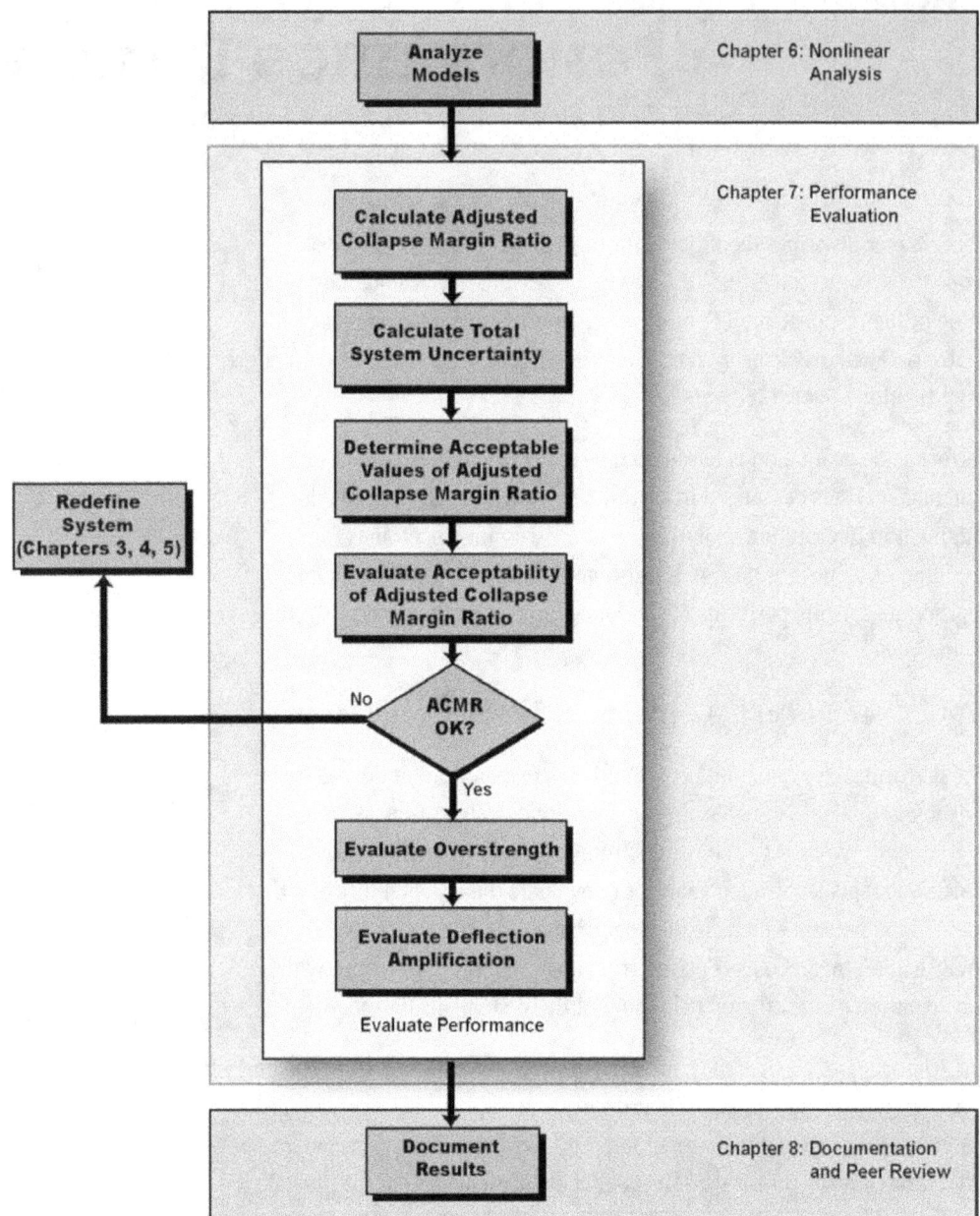

Figure 7-1   Process for performance evaluation

- Calculate the adjusted collapse margin ratio, $ACMR$, for each archetype using the spectral shape factor, SSF, which depends on the fundamental period, T, and period-based ductility, μ_T, as provided in Section 7.2.

- Calculate total system collapse uncertainty, β_{TOT}, based on the quality ratings of design requirements and test data (Chapter 3), and the quality rating of index archetype models (Chapter 5), as provided in Section 7.3.

- Determine acceptable values of adjusted collapse margin ratio, $ACMR_{10\%}$ and $ACMR_{20\%}$, respectively, based on total collapse system uncertainty, β_{TOT}, as provided in Section 7.4.

- Evaluate the adjusted collapse margin ratio, $ACMR$, for each archetype and average values of $ACMR$ for each archetype performance group relative to acceptable values as provided in Section 7.5.

- Evaluate the system overstrength factor, Ω_O, as provided in Section 7.6.

- Evaluate the displacement amplification factor, C_d, as provided in Section 7.7.

If the evaluation of $ACMR$ finds trial values of seismic performance factors to be unacceptable, then the system should be redefined and reanalyzed, as needed, and then re-evaluated by repeating performance evaluation process. Systems could be redefined by adjusting the design requirements (Chapter 3), re-characterizing behavior (Chapter 4), or redesigning with new trial values of seismic performance factors (Chapter 5). In general, it is expected that more than one iteration of the evaluation process will be required to determine optimal (and acceptable) values of the seismic performance factors for the system of interest.

7.1.1 Performance Group Evaluation Criteria

In Chapter 4, index archetype configurations are assembled into performance groups that reflect major differences in configuration, design and seismic load intensity and structural period. The binning of index archetype configurations provides the basis for statistical assessment of minimum and average properties of seismic performance factors.

In general, trial values of seismic performance factors are evaluated for each performance group. Results within each performance group are averaged to determine the value for the group, which is the primary basis for judging acceptability of each trial value. The trial value of the response modification factor, R, must be found acceptable for all performance groups. The system overstrength factor, Ω_O, is based on the largest average value of overstrength, Ω, for all performance groups (subject to certain limits). The deflection amplification factor, C_d, is derived from the acceptable value of R and consideration of the effective damping of the system.

The governing performance group for the response modification factor, R, is the one with the smallest average value of $ACMR$. The governing performance group for the overstrength factor, Ω_O, is the one with the largest average value of Ω. It is likely that the response modification factor, R, and the system overstrength factor, Ω_O, will be governed by different performance groups.

Results are also evaluated to identify potential outliers within each performance group (i.e., individual index archetypes that perform significantly worse than the average performance of the group). Outliers can be accommodated by adopting more conservative values of seismic performance factors, or they can be eliminated from the archetype design space by revising the design requirements (e.g., implementation of height limits or other restrictions on use). Revision of seismic performance factors or design requirements will necessitate re-design and re-analysis of index archetypes, and re-evaluation of system performance.

It is not required that all index archetype configurations be used for evaluation, if it can be shown by selective analysis that certain design combinations (configurations) are not critical and do not control the performance evaluation. Caution should be exercised in removing non-critical index archetype configurations from a governing performance group since their removal could adversely affect the average value of the group used in the evaluation.

7.1.2 Acceptable Probability of Collapse

The fundamental premise of the performance evaluation process is that an acceptably low, yet reasonable, probability of collapse can be established as a criterion for assessing the collapse performance of a proposed system.

In this Methodology, it is suggested that the probability of collapse due to Maximum Considered Earthquake (MCE) ground motions be limited to 10%. Each performance group is required to meet this collapse probability limit, on average, recognizing that some individual archetypes could have collapse probabilities that exceed this value. A limit of twice that value, or 20%, is suggested as a criterion for evaluating the acceptability of potential "outliers" within a performance group.

It should be noted that these limits were selected based on judgment. Within the performance evaluation process, these values could be adjusted to reflect different values of acceptable probabilities of collapse that are deemed appropriate by governing jurisdictions, or other authorities employing this Methodology to establish seismic design requirements and seismic performance factors for a proposed system.

7.2 Adjusted Collapse Margin Ratio

Collapse capacity, and the calculation of collapse margin ratio, can be significantly influenced by the frequency content (spectral shape) of the ground motion record set. To account for the effects of spectral shape, the collapse margin ratio, CMR, is modified to obtain an adjusted collapse margin ratio, $ACMR$, for each index archetype, i:

$$ACMR_i = SSF_i \times CMR_i \qquad (7\text{-}1)$$

This adjustment is in addition to the adjustment of the CMR made to account for three-dimensional nonlinear dynamic analysis effects specified in Chapter 6.

7.2.1 Effect of Spectral Shape on Collapse Margin

Baker and Cornell (2006) have shown that rare ground motions in the Western United States, such as those corresponding to the MCE, have a distinctive spectral shape that differs from the shape of the design spectrum used for structural design in ASCE/SEI 7-05. In essence, the shape of the spectrum of rare ground motions is peaked at the period of interest, and drops off more rapidly (and has less energy) at periods that are longer or shorter than the period of interest. Where ground motion intensities are defined based on the spectral acceleration at the first-mode period of a structure, and where structures have sufficient ductility to inelastically soften into longer periods of vibration, this peaked spectral shape, and more rapid drop at other periods, causes rare records to be less damaging than would otherwise be expected based on the shape of the standard design spectrum.

The most direct approach to account for spectral shape would be to select a unique set of ground motions that have the appropriate shape for each site, hazard level, and structural period of interest. This, however, is not feasible in a generalized procedure for assessing the collapse performance of a class of structures, with a range of possible configurations, located in different geographic regions, with different soil site classifications. To remove this conservative bias, simplified spectral shape factors, SSF, which depend on fundamental period and period-based ductility, are used to adjust collapse margin ratios. Background and development of spectral shape factors are described in Appendix B.

7.2.2 Spectral Shape Factors

Spectral shape factors, SSF, are a function of the fundamental period, T, the period-based ductility, μ_T, and the applicable Seismic Design Category.

Table 7-1a and Table 7-1b provide values of *SSF* for use in adjusting the collapse margin ratio, *CMR*.

Table 7-1a   Spectral Shape Factor (*SSF*) for Archetypes Designed for SDC B, SDC C, or SDC D$_{min}$

T (sec.)	Period-Based Ductility, μ_T							
	1.0	1.1	1.5	2	3	4	6	≥ 8
≤ 0.5	1.00	1.02	1.04	1.06	1.08	1.09	1.12	1.14
0.6	1.00	1.02	1.05	1.07	1.09	1.11	1.13	1.16
0.7	1.00	1.03	1.06	1.08	1.10	1.12	1.15	1.18
0.8	1.00	1.03	1.06	1.08	1.11	1.14	1.17	1.20
0.9	1.00	1.03	1.07	1.09	1.13	1.15	1.19	1.22
1.0	1.00	1.04	1.08	1.10	1.14	1.17	1.21	1.25
1.1	1.00	1.04	1.08	1.11	1.15	1.18	1.23	1.27
1.2	1.00	1.04	1.09	1.12	1.17	1.20	1.25	1.30
1.3	1.00	1.05	1.10	1.13	1.18	1.22	1.27	1.32
1.4	1.00	1.05	1.10	1.14	1.19	1.23	1.30	1.35
≥ 1.5	1.00	1.05	1.11	1.15	1.21	1.25	1.32	1.37

Table 7-1b   Spectral Shape Factor (*SSF*) for Archetypes Designed using SDC D$_{max}$

T (sec.)	Period-Based Ductility, μ_T							
	1.0	1.1	1.5	2	3	4	6	≥ 8
≤ 0.5	1.00	1.05	1.1	1.13	1.18	1.22	1.28	1.33
0.6	1.00	1.05	1.11	1.14	1.2	1.24	1.3	1.36
0.7	1.00	1.06	1.11	1.15	1.21	1.25	1.32	1.38
0.8	1.00	1.06	1.12	1.16	1.22	1.27	1.35	1.41
0.9	1.00	1.06	1.13	1.17	1.24	1.29	1.37	1.44
1.0	1.00	1.07	1.13	1.18	1.25	1.31	1.39	1.46
1.1	1.00	1.07	1.14	1.19	1.27	1.32	1.41	1.49
1.2	1.00	1.07	1.15	1.2	1.28	1.34	1.44	1.52
1.3	1.00	1.08	1.16	1.21	1.29	1.36	1.46	1.55
1.4	1.00	1.08	1.16	1.22	1.31	1.38	1.49	1.58
≥ 1.5	1.00	1.08	1.17	1.23	1.32	1.4	1.51	1.61

Since spectral shape factors are considerably different between SDC D$_{max}$ and other Seismic Design Categories, the governing performance group for the adjusted collapse margin ratio, *ACMR*, may not be the same as the governing performance group for the collapse margin ratio, *CMR*, before adjustment.

7.3   Total System Collapse Uncertainty

Many sources of uncertainty contribute to variability in collapse capacity. Larger variability in the overall collapse prediction will necessitate larger collapse margins in order to limit the collapse probability to an acceptable level at the MCE intensity. It is important to evaluate all significant sources of uncertainty in collapse response, and to incorporate their effects in the collapse assessment process.

7.3.1   Sources of Uncertainty

The following sources of uncertainty are considered in the collapse assessment process:

- **Record-to-Record Uncertainty (RTR).** Record-to-record uncertainty is due to variability in the response of index archetypes to different ground motion records. Record-to-record variability is evident in incremental dynamic response plots (as shown in Figure 6-5). Variability in response is due to the combined effects of: (1) variations in frequency content and dynamic characteristics of the various records; and (2) variability in the hazard characterization as reflected in the Far-Field ground motion record set. Values of record-to-record variability, β_{RTR}, ranging from 0.35 to 0.45 are fairly consistent among various building types (Haselton, 2006; Ibarra and Krawinkler, 2005a and 2005b; Zareian et al., 2006; Zareian, 2006). Based on available research and studies of example archetype evaluations using the Far-Field ground motion record set in Appendix A, a fixed value of $\beta_{RTR} = 0.40$ is assumed in the performance evaluation of systems with significant period elongation (i.e., period-based ductility, $\mu_T \geq 3$). Most systems, even those with limited ductile capacity, have significant period elongation before collapse, and are appropriately evaluated using this value.

  Studies in Appendix A also found that record-to-record variability can be significantly less than $\beta_{RTR} = 0.40$ for systems that have little, or no, period elongation (e.g., systems with very limited ductility and certain base-isolated systems). For these systems, values of record-to-record variability can be reduced as follows:

$$\beta_{RTR} = 0.1 + 0.1\mu_T \leq 0.40 \qquad (7\text{-}2)$$

  where β_{RTR} must be greater than or equal to 0.20.

- **Design Requirements Uncertainty (DR).** Design requirements uncertainty is related to the completeness and robustness of the design requirements, and the extent to which they provide safeguards against unanticipated failure modes. Design requirements-related uncertainty is quantified in terms of the quality of design requirements, rated in accordance with the requirements in Chapter 3.

- **Test Data Uncertainty (*TD*).** Test data uncertainty is related to the completeness and robustness of the test data used to define the system. Uncertainty in test data is closely associated with, but distinct from, modeling-related uncertainty. Test data-related uncertainty is quantified in terms of the quality of test data, rated in accordance with the requirements in Chapter 3.

- **Modeling Uncertainty (*MDL*).** Modeling uncertainty is related to how well index archetype models represent the full range of structural response characteristics and associated design parameters of the archetype design space, and how well the analysis models capture structural collapse behavior through direct simulation or non-simulated component checks. Modeling-related uncertainty is quantified in terms of the quality of index archetype models, rated in accordance with the requirements in Chapter 5.

7.3.2 Combining Uncertainties in Collapse Evaluation

The total uncertainty is obtained by combining *RTR, DR, TD,* and *MDL* uncertainties. Formally, the collapse fragility of each index archetype is defined by the random variable, S_{CT}, assumed to be equal to the product of the median value of the collapse ground motion intensity, $\hat{S}_{CT}$, as calculated by nonlinear dynamic analysis, and the random variable, λ_{TOT}:

$$S_{CT} = \hat{S}_{CT}\lambda_{TOT} \tag{7-3}$$

where λ_{TOT} is assumed to be lognormally distributed with a median value of unity and a lognormal standard deviation of β_{TOT}. The lognormal random variable, λ_{TOT}, is defined as the product of four component random variables as:

$$\lambda_{TOT} = \lambda_{RTR}\lambda_{DR}\lambda_{TD}\lambda_{MDL} \tag{7-4}$$

where $\lambda_{RTR}, \lambda_{MDL}, \lambda_{DR},$ and λ_{TD} are assumed to be independent and lognormally distributed with median values of unity, and lognormal standard deviation parameters, $\beta_{RTR}, \beta_{DR}, \beta_{TD},$ and β_{MDL} respectively. Since the four component random variables are assumed to be statistically independent, the lognormal standard deviation parameter, β_{TOT}, describing total collapse uncertainty, is given by:

$$\beta_{TOT} = \sqrt{\beta_{RTR}^2 + \beta_{DR}^2 + \beta_{TD}^2 + \beta_{MDL}^2} \qquad (7\text{-}5)$$

Where: β_{TOT} = total system collapse uncertainty (0.275 - 0.950)

β_{RTR} = record-to-record collapse uncertainty (0.20 - 0.40)

β_{DR} = design requirements-related collapse uncertainty (0.10 – 0.50)

β_{TD} = test data-related collapse uncertainty (0.10 – 0.50)

β_{MDL} = modeling-related collapse uncertainty (0.10 – 0.50).

The performance evaluation process does not require explicit calculation of the lognormal distribution given by Equation 7-3 and Equation 7-4. Acceptance criteria, however, are based on the composite uncertainty, β_{TOT}, developed on the basis of Equation 7-5.

7.3.3 Effect of Uncertainty on Collapse Margin

Uncertainty influences the shape of a collapse fragility curve plotted from incremental dynamic analysis results. Figure 7-2 shows two collapse fragility curves reflecting two different levels of uncertainty. The dashed curve "a" reflects a $\beta_{RTR} = 0.4$, and the solid curve "b" reflects a $\beta_{TOT} = 0.65$. As indicated in the figure, additional uncertainty has the effect of flattening the curve. While the median collapse intensity, $\hat{S}_{CT}$, is unchanged, additional uncertainty causes a large increase in the probability of collapse at the MCE intensity, S_{MT}.

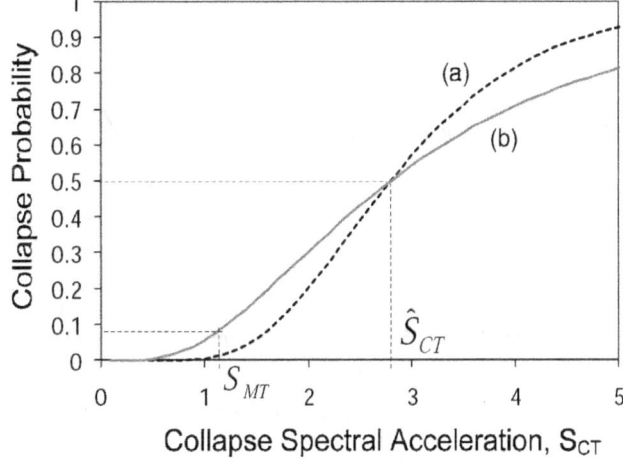

Figure 7-2   Collapse fragility curves considering (a) ground motion record-to-record uncertainty, (b) modifications for total uncertainty.

Changes in the probability of collapse at the MCE intensity will affect the collapse margin ratio, *CMR*. Figure 7-3 shows collapse fragility curves for two hypothetical seismic-force-resisting systems that have different levels of collapse uncertainty. In this example, both systems have been designed for the same seismic response coefficient, C_S, and happen to have the same median collapse intensity, $\hat{S}_{CT}$. System No. 1, however, has a larger uncertainty and a "flatter" collapse fragility curve. To achieve the same 10% probability of collapse for MCE ground motions, a larger collapse margin ratio is required for System No. 1 than is required for System No. 2 (i.e., $CMR_1 > CMR_2$). Thus, System No. 1 would have to be designed using a smaller response modification coefficient, R, than System No. 2.

Figure 7-4 shows collapse fragility curves for another set of hypothetical seismic-force-resisting systems with different levels of collapse uncertainty. As in Figure 7-3, both systems are designed for the same seismic response coefficient, C_S, but in this case, the two systems are also designed for the same response modification coefficient, R. The difference in uncertainty, however, requires different collapse margin ratios to achieve the same median probability of collapse. In order to utilize the same response modification coefficient, R, System No. 1, with larger uncertainty and flatter collapse fragility, is required to have a larger collapse margin ratio than System No. 3 (i.e., $CMR_1 > CMR_3$).

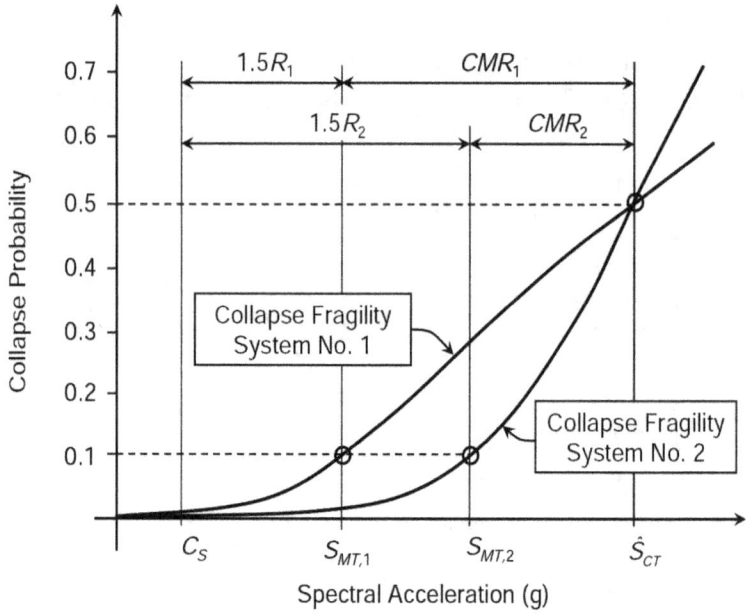

Figure 7-3   Illustration of fragility curves and collapse margin ratios for two hypothetical seismic-force-resisting systems – same median collapse level.

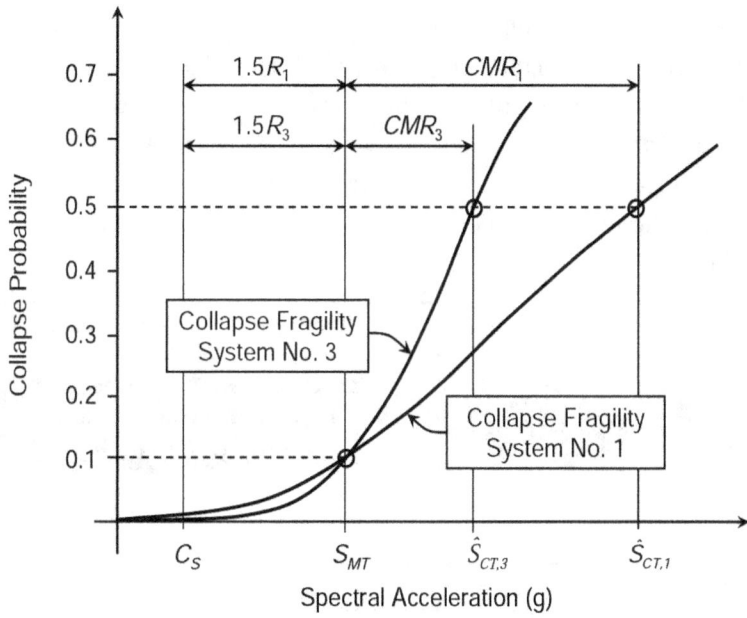

Figure 7-4   Illustration of fragility curves and collapse margin ratios for two hypothetical seismic-force-resisting systems – same R factor.

7.3.4 Total System Collapse Uncertainty

Total system collapse uncertainty is calculated based on Equation 7-5, and is a function of record-to-record (RTR) uncertainty, design requirements-related (DR) uncertainty, test data-related (TD) uncertainty, and modeling (MDL) uncertainty.

Quality ratings for design requirements, test data, and nonlinear models are translated into quantitative values of uncertainty based on the following scale: (A) Superior, $\beta = 0.10$; (B) Good, $\beta = 0.20$; (C) Fair, $\beta = 0.35$; and (D) Poor, $\beta = 0.50$. A broad range of subjective values of uncertainty are associated with the four quality ratings (Superior, Good, Fair and Poor) to reward "higher quality" systems that have more robust design requirements, more comprehensive test data, and more reliable nonlinear analysis models with lower values of total system collapse uncertainty.

Record-to-record uncertainty and, hence, total system collapse uncertainty, depends on period-based ductility. While subjective in nature, uncertainty values associated with quality ratings, when combined with record-to-record uncertainty, yield reasonable values of total system collapse uncertainty. Resulting values range from $\beta_{TOT} = 0.275$ for the most certain of systems (all Superior quality ratings and $\beta_{RTR} = 0.2$), to $\beta_{TOT} = 0.950$ for the least certain of systems (all Poor quality ratings and $\beta_{RTR} = 0.4$).

Values of total system collapse uncertainty, β_{TOT}, for index archetype models with a period-based ductility, $\mu_T \geq 3$ are provided in Table 7-2a through Table 7-2d, based on record-to-record uncertainty, $\beta_{RTR} = 0.4$. Each of these four tables is specific to a different model quality rating of (A) Superior, (B) Good, (C) Fair, or (D) Poor. Values in each table are based on Equation 7-5 and the applicable combination of quality ratings for test data and design requirements.

Table 7-2a   Total System Collapse Uncertainty (β_{TOT}) for Model Quality (A) Superior and Period-Based Ductility, $\mu_T \geq 3$

Quality of Test Data	Quality of Design Requirements			
	(A) Superior	(B) Good	(C) Fair	(D) Poor
(A) Superior	0.425	0.475	0.550	0.650
(B) Good	0.475	0.500	0.575	0.675
(C) Fair	0.550	0.575	0.650	0.725
(D) Poor	0.650	0.675	0.725	0.825

Table 7-2b   Total System Collapse Uncertainty (β_{TOT}) for Model Quality (B) Good and Period-Based Ductility, $\mu_T \geq 3$

Quality of Test Data	Quality of Design Requirements			
	(A) Superior	(B) Good	(C) Fair	(D) Poor
(A) Superior	0.475	0.500	0.575	0.675
(B) Good	0.500	0.525	0.600	0.700
(C) Fair	0.575	0.600	0.675	0.750
(D) Poor	0.675	0.700	0.750	0.825

Table 7-2c   Total System Collapse Uncertainty (β_{TOT}) for Model Quality (C) Fair and Period-Based Ductility, $\mu_T \geq 3$

Quality of Test Data	Quality of Design Requirements			
	(A) Superior	(B) Good	(C) Fair	(D) Poor
(A) Superior	0.550	0.575	0.650	0.725
(B) Good	0.575	0.600	0.675	0.750
(C) Fair	0.650	0.675	0.725	0.800
(D) Poor	0.725	0.750	0.800	0.875

7: Performance Evaluation

Table 7-2d Total System Collapse Uncertainty (β_{TOT}) for Model Quality (D) Poor and Period-Based Ductility, $\mu_T \geq 3$

Quality of Test Data	Quality of Design Requirements			
	(A) Superior	(B) Good	(C) Fair	(D) Poor
(A) Superior	0.650	0.675	0.725	0.825
(B) Good	0.675	0.700	0.750	0.825
(C) Fair	0.725	0.750	0.800	0.875
(D) Poor	0.825	0.825	0.875	0.950

In general, most archetypes are expected to have a period-based ductility of $\mu_T \geq 3$, and Tables 7-2a through 7-2d will be used for collapse performance evaluation of most systems. For index archetype models that have a period-based ductility of $\mu_T < 3$, tables have not been provided. Values of record-to-record uncertainty, β_{RTR}, should be calculated using Equation 7-2, and total collapse system uncertainty, β_{TOT}, should be calculated using Equation 7-5, rounded to the nearest 0.025. Values of design requirements uncertainty, β_{DR}, test data uncertainty, β_{TD}, and model uncertainty, β_{MDL}, should be based on their respective quality ratings.

7.4 Acceptable Values of Adjusted Collapse Margin Ratio

Acceptable values of adjusted collapse margin ratio are based on total system collapse uncertainty, β_{TOT}, and established values of acceptable probabilities of collapse. They are based on the assumption that the distribution of collapse level spectral intensities is lognormal, with a median value, $\acute{S}_{CT}$, and a lognormal standard deviation equal to the total system collapse uncertainty, β_{TOT}.

Table 7-3 provides acceptable values of adjusted collapse margin ratio, $ACMR_{10\%}$ and $ACMR_{20\%}$, based on total system collapse uncertainty and values of acceptable collapse probability, taken as 10% and 20%, respectively. Other values of collapse probability ranging from 5% - 25% are shown for comparison and reference. Lower values of acceptable collapse probability and higher levels of collapse uncertainty result in higher required values of adjusted collapse margin ratio.

Table 7-3   Acceptable Values of Adjusted Collapse Margin Ratio ($ACMR_{10\%}$ and $ACMR_{20\%}$)

Total System Collapse Uncertainty	Collapse Probability				
	5%	10% ($ACMR_{10\%}$)	15%	20% ($ACMR_{20\%}$)	25%
0.275	1.57	1.42	1.33	1.26	1.20
0.300	1.64	1.47	1.36	1.29	1.22
0.325	1.71	1.52	1.40	1.31	1.25
0.350	1.78	1.57	1.44	1.34	1.27
0.375	1.85	1.62	1.48	1.37	1.29
0.400	1.93	1.67	1.51	1.40	1.31
0.425	2.01	1.72	1.55	1.43	1.33
0.450	2.10	1.78	1.59	1.46	1.35
0.475	2.18	1.84	1.64	1.49	1.38
0.500	2.28	1.90	1.68	1.52	1.40
0.525	2.37	1.96	1.72	1.56	1.42
0.550	2.47	2.02	1.77	1.59	1.45
0.575	2.57	2.09	1.81	1.62	1.47
0.600	2.68	2.16	1.86	1.66	1.50
0.625	2.80	2.23	1.91	1.69	1.52
0.650	2.91	2.30	1.96	1.73	1.55
0.675	3.04	2.38	2.01	1.76	1.58
0.700	3.16	2.45	2.07	1.80	1.60
0.725	3.30	2.53	2.12	1.84	1.63
0.750	3.43	2.61	2.18	1.88	1.66
0.775	3.58	2.70	2.23	1.92	1.69
0.800	3.73	2.79	2.29	1.96	1.72
0.825	3.88	2.88	2.35	2.00	1.74
0.850	4.05	2.97	2.41	2.04	1.77
0.875	4.22	3.07	2.48	2.09	1.80
0.900	4.39	3.17	2.54	2.13	1.83
0.925	4.58	3.27	2.61	2.18	1.87
0.950	4.77	3.38	2.68	2.22	1.90

7.5 Evaluation of the Response Modification Coefficient, R

Acceptable performance is defined by the following two basic collapse prevention objectives:

- The probability of collapse for MCE ground motions is approximately 10%, or less, on average across a performance group.

- The probability of collapse for MCE ground motions is approximately 20%, or less, for each index archetype within a performance group.

Acceptable performance is achieved when, for each performance group, adjusted collapse margin ratios, ACMR, for each index archetype meet the following two criteria:

- the average value of adjusted collapse margin ratio for each performance group exceeds $ACMR_{10\%}$:

$$\overline{ACMR_i} \geq ACMR10\% \qquad (7\text{-}6)$$

- individual values of adjusted collapse margin ratio for each index archetype within a performance group exceeds $ACMR_{20\%}$:

$$ACMR_i \geq ACMR20\% \qquad (7\text{-}7)$$

7.6 Evaluation of the Overstrength Factor, Ω_0

The average value of archetype overstrength, Ω, is calculated for each performance group. The value of the system overstrength factor, Ω_0, for use in design should not be taken as less than the largest average value of calculated archetype overstrength, Ω, from any performance group. The system overstrength factor, Ω_0, should be conservatively increased to account for variation in overstrength results of individual index archetypes, and judgmentally rounded to half unit intervals (e.g., 1.5, 2.0, 2.5, and 3.0).

The system overstrength factor, Ω_0, need not exceed 1.5 times the response modification coefficient, R. A practical limit on the value of Ω_0 is about 3.0, consistent with the largest value of this factor specified in Table 12.2-1 of ASCE/SEI 7-05 for all current approved seismic-force-resisting systems.

Example applications (Chapter 9) show that values of archetype overstrength, Ω, can be as large as $\Omega = 6.0$ for certain configurations, and are highly variable. Limiting system overstrength to $\Omega_0 = 3.0$, as specified in ASCE/SEI 7-05, was considered necessary for practical design considerations.

7.7 Evaluation of the Deflection Amplification Factor, C_d

The deflection amplification factor, C_d, is based on the acceptable value of the response modification factor, R, reduced by the damping factor, B_I, corresponding to the inherent damping of the system of interest:

$$C_d = \frac{R}{B_I} \qquad (7\text{-}8)$$

where: C_d = deflection amplification factor

R = system response modification factor

B_I = numerical coefficient as set forth in Table 18.6-1 of ASCE/SEI 7-05 for effective damping, β_I, and period, T

β_I = component of effective damping of the structure due to the inherent dissipation of energy by elements of the structure, at or just below the effective yield displacement of the seismic-force-resisting system, Section 18.6.2.1 of ASCE/SEI 7-05.

In general, inherent damping may be assumed to be 5 percent of critical, and a corresponding value of the damping coefficient, $B_I = 1.0$ (Table 18.6-1, ASCE/SEI 7-05). Thus, for most systems the value of C_d will be equal to the value of R.

Equating the deflection amplification factor, C_d, to the R factor is based on the "Newmark rule," which assumes that inelastic displacement is approximately equal to elastic displacement (at the roof). This is consistent with research findings for systems with nominal (5% of critical) damping and fundamental periods greater than the transition period, T_s. It is recognized that for short-period systems ($T < T_s$) inelastic displacement generally exceeds elastic displacement, but it was not considered appropriate to base the deflection amplification factor on response of short-period systems, unless the systems are displacement sensitive. Short-period, displacement sensitive systems should incorporate the consequences of these larger inelastic displacements.

Chapter 8

Documentation and Peer Review

This chapter describes documentation and peer review requirements for a proposed seismic-force-resisting system. It identifies recommended qualifications, expertise, and responsibilities for personnel involved with the development and review of a proposed system. It lists information that should be included in a report documenting the development of a system, and discusses requirements for review at each step of the developmental process.

8.1 Recommended Qualifications, Expertise and Responsibilities for a System Development Team

In order to collect the necessary data and apply the procedures of this Methodology, a system development team will need to have certain qualifications, experience, and expertise. These include the ability to: (1) adequately test materials, components and assemblies; (2) develop comprehensive design and construction requirements; (3) develop archetype designs; and (4) analyze archetype models.

A development team is responsible for following the procedures of this Methodology in determining seismic performance factors for a proposed system, defining the limits under which a system will be applicable, documenting results, and obtaining approval of the peer review panel.

8.1.1 System Sponsor

The system sponsor is a person or organization that has conceived a new seismic-force-resisting system and will benefit from its use. The system sponsor is responsible for assembling a development team, selecting an independent peer review panel, and submitting a proposed system for approval and use.

8.1.2 Testing Qualifications, Expertise and Responsibilities

The testing laboratories engaged in the development of a proposed system must have the capability to perform material, component, connection, assembly, and system tests necessary for quantifying the material and behavioral properties of the system. Testing laboratories used to conduct the experimental investigation program should comply with national or international accreditation criteria, such as ISO/IEC 17025, *General Requirements for the Competence of Testing and Calibration Laboratories* (ISO/IEC, 2005). Testing laboratories that are not accredited may still be used for an experimental investigation program, subject to the approval of the peer review panel.

Testing facility staff should have the necessary expertise to establish and execute an experimental program, conduct the tests, and mine existing research from other relevant available test data.

8.1.3 Engineering and Construction Qualifications, Expertise and Responsibilities

Member(s) of a development team must have sufficient experience and expertise to develop comprehensive design and construction requirements, and to perform trial system designs. This should include familiarity with seismic design requirements specified in ASCE/SEI 7, and material-specific reference standards for design and detailing requirements for other similar systems. To be viable for use, a proposed system must be feasible to construct. Familiarity with proposed construction techniques, or established construction techniques for other similar systems is needed.

8.1.4 Analytical Qualifications, Expertise and Responsibilities

Member(s) of a development team must have sufficient experience and expertise to interpret test data and develop sophisticated nonlinear models capable of simulating the potential failure modes and collapse behaviors of a proposed system. This should include knowledge and experience in the analytical approaches specified in the Methodology, knowledge of material, component, connection, and overall system performance, and experience with analysis software capable of simulating system response.

8.2 Documentation of System Development and Results

The results of system development efforts must be thoroughly documented at each step of the process for: (1) review and approval by the peer review panel; (2) review and approval by an authority having jurisdiction over its eventual use; and (3) use in design and construction.

Documentation of the development of seismic performance factors for a proposed system should include, but is not necessarily limited to, the following:

- Description of the intended system applications and expected performance

- Limitations on system use

- Typical horizontal and vertical geometric configurations

- Clear and complete design requirements and specifications for the system, providing enough information to quantify strength limit states, proportion and detail components, analyze predicted response, and confirm satisfactory behavior

- Summary of test data and other supporting evidence from an experimental investigation program validating material properties and component behavior, calibrating nonlinear analysis models, and establishing performance acceptance criteria

- Description of index archetype configurations and extent of archetype design space

- Identification of performance groups, applicable Seismic Design Categories, and gravity load intensities

- Idealized model configurations, nonlinear modeling parameters, documentation of analysis software, and information used in model calibration

- Criteria for non-simulated collapse modes

- Summary of nonlinear model results, demand parameters, and response quantities

- Quality ratings for design requirements, test data, and nonlinear models

- Summary of performance evaluation results, derived quantities, and acceptance criteria

- Proposed seismic performance factors (R, Ω_O, and C_d)

8.3   Peer Review Panel

Implementation of this Methodology involves much uncertainty, judgment and potential for variation. Deciding on an appropriate level of detail to adequately characterize performance of a proposed system should be performed at each step in the process in collaboration with an independent peer review panel.

It is recommended that a peer review panel consisting of knowledgeable experts be retained for this purpose. The peer review panel should be familiar with the procedures of this Methodology, should have sufficient knowledge to render an informed opinion on the developmental process, and should include expertise in each of the following areas:

- Material, component, and assembly testing

- Engineering design and construction

- Nonlinear collapse simulation

Members of the peer review panel must be qualified to critically evaluate the development of the proposed system including testing, design, and analysis. If a unique computer code is developed by the development team, the peer review panel should be capable of performing independent analyses of the proposed system using other analysis platforms.

8.3.1 Peer Review Panel Selection

It is envisioned that the cost of the peer review panel will be borne by the system sponsor. As such, it is expected that members of the peer review panel could be selected by the system sponsor. An alternative arrangement could be made in which the system sponsor submits funding to the authority having jurisdiction, which then uses the funding to implement an independent peer review process. Such an arrangement would be similar to the outside plan check process currently used in some building departments.

It is intended that the peer review panel be an independent set of reviewers who will advise and guide the development team at each step in the process. It is recommended that other stakeholders, including authorities with jurisdiction over the eventual use of the system in design and construction, be consulted in the selection of peer review panel members, and in the deliberation on their findings.

8.3.2 Peer Review Roles and Responsibilities

The peer review panel is responsible for reviewing and commenting on the approach taken by the development team including the extent of the experimental program, testing procedures, design requirements, development of structural system archetypes, analytical approaches, extent of the nonlinear analysis investigation, and the final selection of the proposed seismic performance factors.

The peer review panel is responsible for reporting their opinion on the work performed by the developmental team, their findings, recommendations, and conclusions. All documentation from the peer review panel should be made available for review by the authority having jurisdiction over approval of the proposed system.

If there are any areas where concurrence between the peer review panel and the development team was not reached, or where the peer review panel was not satisfied with the approach or extent of the work performed, this information should be made available as part of the peer review documentation, and reflected in the total uncertainty used in calculating the system acceptance criteria, and in determining the final values of proposed seismic performance factors.

8.4 Submittal

It is expected that a system sponsor will wish to submit a proposed system to an authority for approval and use. For national building codes and standards, one such authority is the National Institute of Building Science's Building Seismic Safety Council (BSSC) Provisions Update Committee (PUC), which has jurisdiction over the FEMA's National Earthquake Hazards Reduction Program *(NEHRP) Recommended Provisions for Seismic Regulations for New Buildings and Other Structures* (*NEHRP Recommended Provisions*). BSSC's PUC, along with its technical subcommittees, is a nationally recognized leader in reviewing and endorsing new seismic force-resisting systems for ultimate inclusion in national building codes and standards.

In some cases, a proposed system could be submitted to the ASCE/SEI 7 standard development committee, but this committee would normally only accept systems similar to systems that are already listed in the standard. Systems can also be submitted directly to model building codes through the code change process.

Another approach is to promote a new system through a relevant material standard organization, such as the American Concrete Institute (ACI), American Institute of Steel Construction (AISC), American Iron and Steel Institute (AISI), or the American Forest & Paper Association (AF&PA). Approval through one of these organizations, however, will still require adoption by national building codes or standards before use.

If a proposed system is intended for a single project application, then documentation should be submitted, along with drawings and calculations for the single application, to the authority having jurisdiction over the site where the system is being proposed for use.

Chapter 9

Example Applications

This chapter presents examples illustrating the application of the Methodology to reinforced concrete special moment frame, reinforced concrete ordinary moment frame, and wood light-frame shear wall seismic-force-resisting systems.

9.1   General

In the following sections three seismic-force-resisting systems are evaluated using the methods outlined Chapters 3 through 7. The examples span different system types, design requirements, test data, archetype models, and analysis software. Models include both simulated and non-simulated collapse modes. Each system is currently contained in Table 12.2-1 of ASCE/SEI 7-05, and the examples utilize design requirements and test data currently available for these approved systems. For new (proposed) systems, design requirements will generally not exist, and would need to be developed.

These examples illustrate the application of the Methodology, and one example illustrates the process of iteratively modifying the system design requirements so that the proposed structural system meets the prescribed collapse performance objectives of the Methodology. These examples also demonstrate consistency between the acceptance criteria of the Methodology and the inherent safety against collapse intended by current seismic codes.

These examples were completed in parallel with the development of the Methodology. As such, they are consistent with the procedures contained herein, but are not necessarily in complete compliance with every requirement. Where they occur, deviations are noted, and explanations are provided as to how the example could be completed in accordance with the Methodology.

In contrast to the process for developing structural system archetypes in Chapter 4, these examples begin with the development of a representative index archetype model, which is used as a basis for generating a set of archetypical configurations that do not attempt to rigorously interrogate the limits of what is permitted within the governing design requirements. As such, they do not necessarily include archetypes that bound the full extent of the design space. Additionally, the archetype designs used in these examples do not account for potential overstrength caused by wind load requirements.

While representative examples from current code-approved systems have been selected for developmental studies, the results are not intended to propose specific changes to current building code requirements for any currently approved system.

9.2 Example Application - Reinforced Concrete Special Moment Frame System

9.2.1 Introduction

In this example, a reinforced concrete special moment frame system, as defined by ACI 318-05 *Building Code Requirements for Structural Concrete* (ACI, 2005), is considered as if it were a new system proposed for inclusion in ASCE/SEI 7-05.

9.2.2 Overview and Approach

In this example, detailing requirements of ACI 318-05 are assumed to be given. The system design requirements of ASCE/SEI 7-05 are used as the framework, and seismic performance factors (SPFs) are determined by iteration until the acceptance criteria of the Methodology are met. Seismic performance factors under consideration in this example include the R factor, C_d factor, and Ω_0 factor.

All pertinent design requirements of ASCE/SEI 7-05, including drift limits and minimum base shear requirements are assumed to apply initially. In the Methodology, the user has full flexibility to define and modify any aspect of the proposed system design requirements, as long as modifications are tested within the index archetype configurations. This includes the R factor, stiffness requirements, detailing requirements, capacity design requirements, minimum base shear requirements, height limits, drift limits, and any other requirements that control the design of the structural system.

The iterative assessment process begins with initial assumptions of $R = 8$, $C_d = 5.5$[1], story drift limits of 2%, and minimum design base shear requirements consistent with ASCE/SEI 7-05. Overstrength, Ω_0, is not assumed initially, but is determined from the computed lateral overstrength

---

[1] $C_d = 5.5$ is used in this example, based on the value specified for reinforced concrete special moment frames in ASCE/SEI 7-05. In actual applications of the Methodology, $C_d = R$ should be used unless $C_d < R$ can be substantiated in accordance with the criteria of Section 7.7.

9: Example Applications

of the archetype designs. A set of structural system archetypes are developed for reinforced concrete special moment frame buildings, nonlinear models are developed to simulate structural collapse, models are analyzed to predict the collapse capacities of each design, and the adjusted collapse margin ratios are evaluated and compared to acceptance criteria.

After completing an assessment using the initial set of SPFs, certain archetypes (taller building configurations) did not meet the acceptance criteria, and were found to have inadequate collapse safety. Aspects of the structural design requirements were modified, and the system was reassessed and found to pass the acceptance criteria.

This example has been adapted from collaborative research on the development of structural archetypes for reinforced concrete special moment frames, calibration of nonlinear element models for collapse simulation, simulation of structural response to collapse, spectral shape considerations, and treatment of uncertainties (Haselton and Deierlein, 2007).

9.2.3 Structural System Information

Design Requirements

This example utilizes ACI 318-05 design requirements in place of the requirements that would be developed for a newly proposed system. For the purpose of assessing uncertainty according to Section 3.4, ACI 318-05 design requirements are categorized as (A) Superior since they represent many years of development and include lessons learned from a number of major earthquakes.

In the process of completing the assessment of the class of reinforced concrete special moment frame buildings, it was found that often seemingly subtle design requirements have important effects on the design and resulting structural performance; the various design requirements often interact and affect the design and performance differently than one might expect. Therefore, for newly proposed systems, it is important that the set of design requirements is well developed and clearly specified, and that the requirements are applied in their totality when designing the archetype structures.

Test Data

This example assessment relies on existing published test data in place of test data that would be developed for a newly proposed system. Specifically, column tests reported in Pacific Earthquake Engineering Research Center's Structural Performance Database developed by Berry, Parrish, and Eberhard

(PEER, 2006b; Berry et al., 2004) are utilized. To develop element models, the data are utilized from cyclic and monotonic tests of 255 rectangular columns failing in flexure and flexure-shear.

The quality of the test data is an important consideration when quantifying the uncertainty in the overall collapse assessment process. The test data used in this example cover a wide range of column design configurations and contain both monotonic and cyclic loading protocols. Even so, many of the loading protocols are not continued to deformations large enough to observe loss of strength, and it is difficult to use such data to calibrate models for structural collapse assessment. These test data also do not include beam elements with attached slabs. Additionally, data include no systematic test series that both (1) subject similar specimens to different loading protocols (e.g., monotonic and cyclic) and (2) continue the loading to deformations large enough for the capping behavior to be observed. Lastly, only column element tests were utilized when used to calibrate the element model, while sub-assemblage tests and full-scale tests were not used. Based on the guidelines of Section 3.6 and considering the above observations, this test data set is categorized as (B) Good.

9.2.4 Identification of Reinforced Concrete Special Moment Frame Archetype Configurations

Figure 9-1 shows the two-dimensional three-bay multi-story frame that is considered an appropriate index archetype model for reinforced concrete frame buildings. This archetype model includes joint panel elements, beam and column elements, elastic foundation springs, and a leaning column to account for the P-delta effect from loads on the gravity system. This two-dimensional model, not accounting for torsional effects, is considered acceptable because most reinforced concrete special moment frame buildings that are regular in plan will not be highly sensitive to torsional effects, and the goal is to verify the performance of a full class of buildings, rather than one specific building with a unique torsional issue. This index archetype model was used as a basis for postulating the index archetype configurations covering the archetype design space for reinforced concrete special moment frame buildings. Appendix C provides more background on the development of this archetype configuration and model.

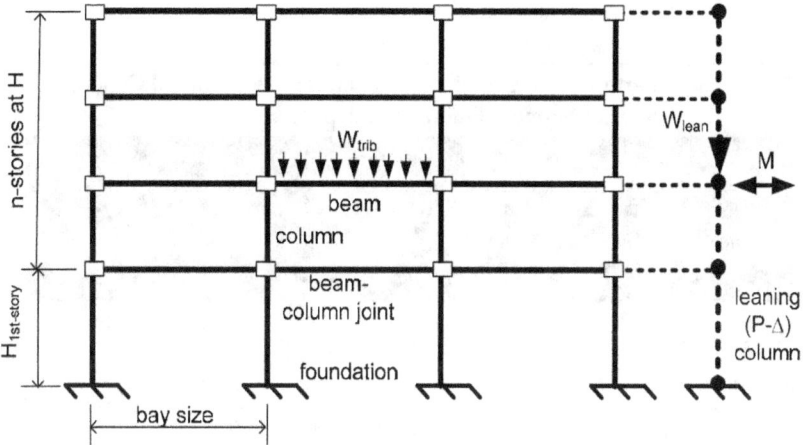

Figure 9-1   Index archetype model for moment frame buildings (after
              Haselton and Deierlein, 2007, Chapter 6).

Using the above index archetype model, a set of structural archetype designs
are developed to represent the archetype design space, following the design
configuration and performance group requirements of Chapter 4. Chapter 4
specifies consideration of up to 16 archetype performance groups for a
structural system whose performance can be adequately evaluated using two
basic configurations, as shown in Table 9-1. Two basic configurations are
considered in the example evaluation of the reinforced concrete special
moment frame system, archetypes with 20-foot and 30-foot bay widths,
respectively. While two configurations are sufficient to illustrate the
Methodology, additional configurations would likely be required to fully
investigate performance of the reinforced concrete special moment frame
system.

High and low gravity load intensities are represented by space frame and
perimeter frame systems, respectively. For the archetypes used in this
example, the ratio of gravity load tributary area to lateral load tributary area
is typically six times larger for space frame systems.

Table 9-1 Performance Groups for Evaluation of Reinforced Concrete Special Moment Frame Archetypes

Performance Group Summary					
Group No.	**Grouping Criteria**				**Number of Archetypes**
	Basic Config.	**Design Load Level**		**Period Domain**	
		Gravity	**Seismic**		
PG-1	20-foot Bay Width	High (Space Frame)	SDC D_{max}	Short	$2+1^1$
PG-2				Long	4
PG-3			SDC D_{min}	Short	0
PG-4				Long	1^2
PG-5		Low (Perimeter Frame)	SDC D_{max}	Short	$2+1^1$
PG-6				Long	4
PG-7			SDC D_{min}	Short	0
PG-8				Long	3^2
PG-9	30-foot Bay Width	High (Space Frame)	SDC D_{max}	Short	0
PG-10				Long	1^3
PG-11			SDC D_{min}	Short	0
PG-12				Long	
PG-13		Low (Perimeter Frame)	SDC D_{max}	Short	0
PG-14				Long	1^3
PG-15			SDC D_{min}	Short	0
PG-16				Long	

1. Example includes only two archetypes for each short-period performance group (PG-1 and PG-5); full implementation of the Methodology requires a total of 3 (2+1) archetypes in each performance group.

2. Example evaluates a selected number of low seismic (SDC D_{min}) archetypes to determine that high seismic (SDC D_{max}) archetypes control the R factor.

3. Example evaluates two 30-foot bay width archetypes to determine that 30-foot bay width archetypes do not control performance.

Archetypes are designed within a range of building heights (six heights between 1-story and 20-stories, as expected for reinforced concrete frame buildings with no walls), and then separated into the short-period and long-period performance groups. At a minimum, three buildings are needed for each performance group, so if 16 complete performance groups were evaluated, then at least 48 archetypes would need to be designed and assessed. Instead of designing and assessing all 48 buildings, initial pilot studies were used to find the more critical design cases: in this case, high-seismic (SDC D) designs with 20-foot bay spacing. By utilizing these pilot studies and then focusing on the critical design cases, it was possible to reduce the number of required archetypes from 48 to 18 (for a complete exercise of the Methodology, a few additional archetypes would be needed, as described below). Appendix C provides a more detailed discussion on the development of structural system archetypes for the reinforced concrete special moment frame system.

Table 9-1 shows archetypes used in assessing the reinforced concrete special moment frame structural system, and provides the rationale for why each archetype was chosen. As indicated in the table, full implementation of the Methodology would require two additional three-story buildings to be added to PG-1 and PG-5 to meet the required minimum three archetypes per group.

The approach utilized here, focusing on the critical design cases, is not required in the Methodology. The benefit of this approach is that it can significantly reduce the number of required archetype designs (this example needed only 20, instead of 48 archetypes) and allow a wider range of design conditions to be considered in the assessment. When this approach is utilized, the peer review panel should closely review choice of critical design cases.

Table 9-2 shows the properties for each of the archetype designs used in this evaluation. Seismic demands are represented by the maximum and minimum seismic criteria of Seismic Design Category (SDC) D, in accordance with Section 5.2.1: $S_{DS} = 1.0$ g and $S_{DI} = 0.60$ g for SDC D_{max} and $S_{DS} = 0.50$ g and $S_{DI} = 0.20$ g for SDC D_{min}[2]. The space frame buildings are denoted by "S" and the perimeter frame buildings are denoted by "P." The computed value of the fundamental period, T, in Table 9-2, is based on Equation 5-5.

Each archetype was fully designed in accordance with the governing design requirements. Additional information on index archetype designs is provided in Appendix C. Figure 9-2 shows example design documentation for the four-story, SDC D_{max} design with 30-foot bay width of Archetype ID 1010. This archetype will be used throughout this illustrative assessment, to clearly show how the Methodology should be applied in assessing each archetype design.

---

[2] In this example, archetypes designed for low seismic (SDC D_{min}) loads, assumed $S_{DI} = 0.167$ g based on interim criteria, which differs slightly from the $S_{DI} = 0.20$ g required by the Methodology.

Table 9-2   Reinforced Concrete Special Moment Frame Archetype Structural Design Properties

Archetype ID	No. of Stories	Framing (Gravity Loads)	Key Archetype Design Parameters					$S_{MT}(T)$ [g]
			Seismic Design Criteria					
			SDC	R	T [sec]	T_1 [sec]	V/W [g]	
Performance Group No. PG-5 (Short Period, 20' Bay Width Configuration)								
2069	1	P	D_{max}	8	0.26	0.71	0.125	1.50
2064	2	P	D_{max}	8	0.45	0.66	0.125	1.50
--	3	P	D_{max}	8	0.63	--	0.119	1.43
Performance Group No. PG-6 (Long Period, 20' Bay Width Configuration)								
1003	4	P	D_{max}	8	0.81	1.12	0.092	1.11
1011	8	P	D_{max}	8	1.49	1.71	0.050	0.60
5013	12	P	D_{max}	8	2.13	2.01	0.035	0.42
5020	20	P	D_{max}	8	3.36	2.63	0.022	0.27
Performance Group No. PG-1 (Short Period, 20' Bay Width Configuration)								
2061	1	S	D_{max}	8	0.26	0.42	0.125	1.50
1001	2	S	D_{max}	8	0.45	0.63	0.125	1.50
--	3	S	D_{max}	8	0.63	--	0.119	1.43
Performance Group No. PG-3 (Long Period, 20' Bay Width Configuration)								
1008	4	S	D_{max}	8	0.81	0.94	0.092	1.11
1012	8	S	D_{max}	8	1.49	1.80	0.050	0.60
5014	12	S	D_{max}	8	2.13	2.14	0.035	0.42
5021	20	S	D_{max}	8	3.36	2.36	0.022	0.27
Selected Archetypes - Performance Group Nos. PG-4 and PG-8 (20' Bay Width)								
6011	8	P	D_{min}	8	1.60	3.00	0.013	0.15
6013	12	P	D_{min}	8	2.28	3.35	0.010	0.10
6020	20	P	D_{min}	8	3.60	4.08	0.010	0.065
6021	20	S	D_{min}	8	3.60	4.03	0.010	0.065
Selected Archetypes - Performance Group Nos. PG-10 and PG-14 (30' Bay Width)								
1009	4	P-30	D_{max}	8	1.03	1.16	0.092	1.03
1010	4	S-30	D_{max}	8	1.03	0.86	0.092	1.03

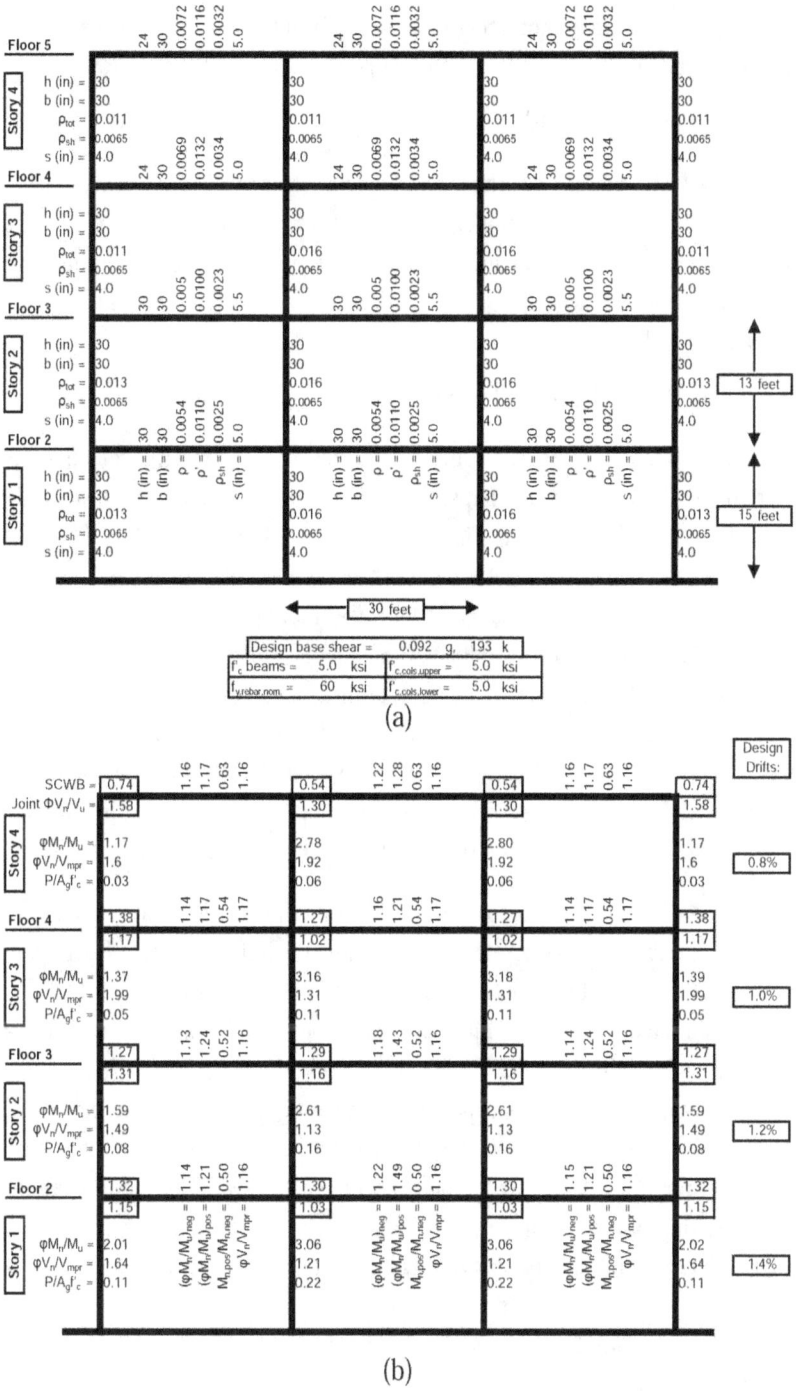

Figure 9-2   Design documentation for a four-story space frame archetype
with 30-foot bay spacing (Archetype ID 1010) (after Haselton
and Deierlein 2007, Chapter 6). The notation used in this
figure is defined in the list of symbols at the end of the
document.

9.2.5 Nonlinear Model Development

This section summarizes explicit modeling of structural collapse. A more complete discussion on nonlinear modeling of reinforced concrete special moment frame systems is provided in Appendices D and E.

Index Archetype Models

System-level modeling uses the three-bay multi-story frame configuration shown in Figure 9-1. This model consists of elastic joint elements, plastic hinge reinforced concrete beam and column elements, a leaning column to account for P-delta effects, and elastic foundation springs to account for foundation and soil flexibility.

Nonlinear Beam-Column Element Models

Even though many reinforced concrete element models exist, most cannot be used to simulate structural collapse. Recent research by Ibarra, Medina, and Krawinkler (2005) has resulted in an element model that is capable of capturing the severe deterioration that precipitates sideway collapse (Appendix E). Figure 9-3 shows the tri-linear monotonic backbone curve and associated hysteretic behavior of this model. This model includes important aspects, such as the "capping point," where monotonic strength loss begins, and the post-capping negative stiffness. These features enable modeling of the strain-softening behavior associated with concrete crushing, rebar buckling and fracture, or bond failure. In general, accurate simulation of sidesway structural collapse requires modeling this post-capping behavior.

Researchers have also used a variety of other methods to simulate cyclic response of reinforced concrete beam-columns, including creating fiber models that can capture cracking behavior and the spread of plasticity throughout the element (e.g., Filippou, 1999). The decision to use a lumped-plasticity approach was based on the observation that currently available fiber element models are not capable of simulating the strain-softening associated with rebar buckling, and thus cannot reliably simulate collapse of flexurally dominated reinforced concrete frames. Research is ongoing, and this modeling limitation may be overcome in the near future. In future applications of this Methodology, the choice of element model should be carefully evaluated for any given structural system, with consideration of available simulation technologies and their capabilities for simulating structural collapse.

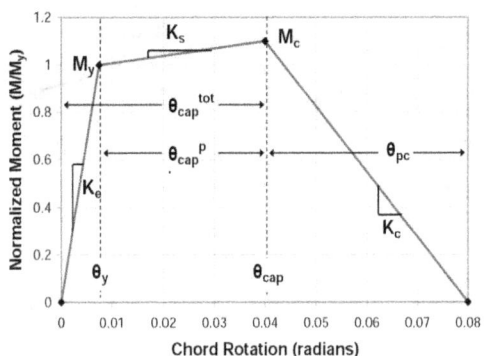

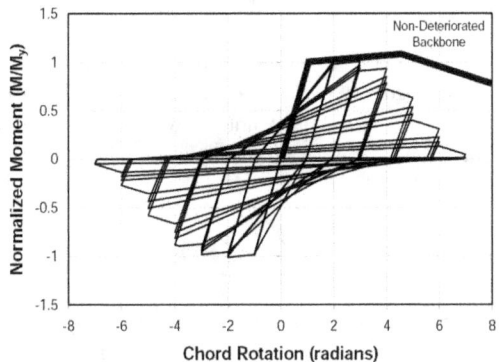

Figure 9-3   Monotonic and cyclic behavior of component model used in this study (after Haselton and Deierlein 2007, Chapter 4).

Figure 9-4 shows an example of the experimental and calibrated response of a reinforced concrete column specimen. Appendix E presents a detailed discussion of how this element model was calibrated using cyclic and monotonic tests of 255 rectangular reinforced concrete elements. Such calibration needs to be completed carefully, in order to avoid errors in the collapse capacity prediction. Often various deterioration modes are improperly mixed together in the calibration process, such as cyclic strength deterioration versus in-cycle strength loss, and this can lead to substantial errors in collapse predictions.

The calibration results for the 255 tests (Appendix E) were subsequently used to create empirical equations to predict the element model parameters (as shown in Figure 9-3), based on the element design information such as axial load ratio and confinement ratio. These equations were used to predict the modeling parameters for the archetype designs used in this example. For illustration, Figure 9-5 shows the predicted modeling parameters for each element of Archetype ID 1010 which is a four-story SDC D_{max} design with 30-foot bay width.

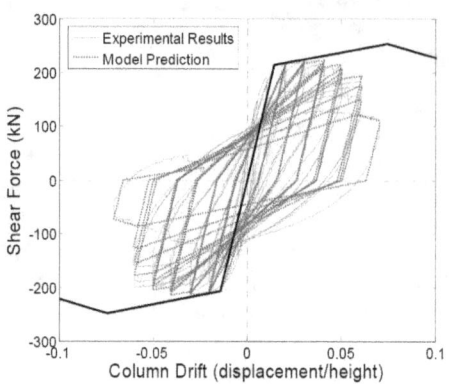

Figure 9-4   Illustration of experimental and calibrated element response (Saatciolgu and Grira, specimen BG-6) (after Haselton et al., 2007). The solid black line shows the calibrated monotonic backbone. The notation is defined in the list of symbols at the end of this document.

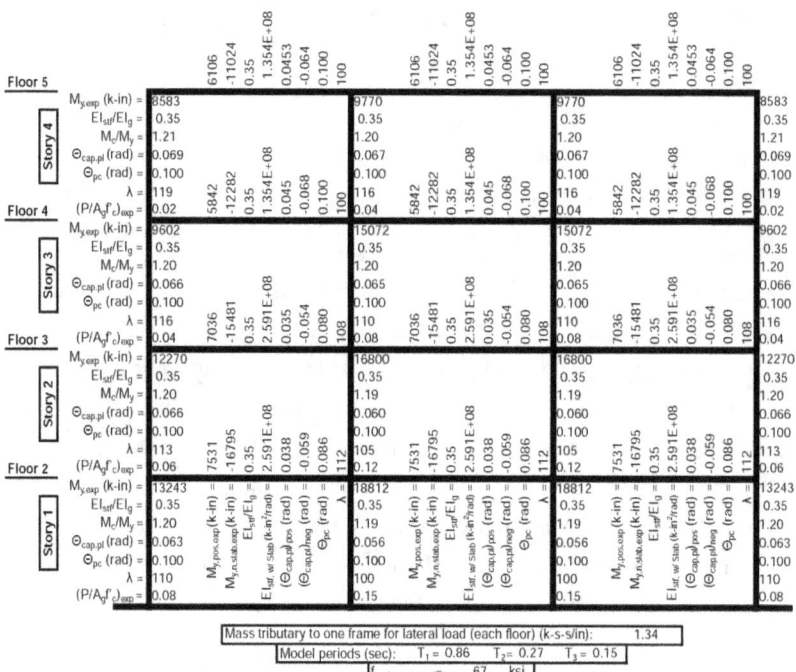

Figure 9-5   Structural modeling documentation for a four-story space frame archetype with 30-foot bay spacing (Archetype ID 1010) (after Haselton and Deierlein, 2007, Chapter 6).

Uncertainty due to Model Quality

For the purpose of assessing uncertainty, this modeling is rated as (B) Good, according to the guidelines of Section 5.7. Reinforced concrete special moment frame buildings are controlled by many detailing and capacity design requirements, which limit possible failure modes. The primary expected failure mode is flexural hinging leading to sidesway collapse, which the modeling approach can simulate reasonably well by capturing post-peak degrading response under both monotonic and cyclic loading. The modeling approach is able to directly simulate structural response up to collapse by simulating all expected modes of damage that could lead to collapse, and is well calibrated to large amounts of data. Even so, this model is not given the (A) Superior rating because there is still room for improvement in the model: the model was calibrated using column data and is not well-calibrated to beam-slabs, and the model does not capture axial-flexural interaction in columns. Additionally, for a complete assessment, the archetype design space would likely need to be expanded to include a wider range of basic configurations beyond 20-foot and 30-foot bay spacings.

9.2.6   Nonlinear Structural Analysis

The structural analysis software selected for completing this example is the Open Systems for Earthquake Engineering Simulation (OpenSees, 2006), which was developed by the Pacific Earthquake Engineering Research (PEER) Center. This software includes the modeling aspects required for collapse simulation of reinforced concrete special moment frame buildings, such as the Ibarra et al. element model, joint models, a large deformation geometric transformation, and several numerical algorithms for solving the systems of equations associated with nonlinear dynamic and static analyses.

Nonlinear static (pushover) analysis is performed in accordance with Section 6.3[3], in order to compute the system overstrength factor, Ω_0, and period-based ductility, μ_T, and to help verify the structural model. Figure 9-6 shows an example of the pushover curve and story drift distributions for Archetype ID 1010.

For the reinforced concrete special moment frame archetype, yielding occurs at story drift ratio 0.005 and the effective yield roof displacement, $\delta_{y,eff}$, is computed according to Equation 6-7 as $0.0042h_r$. A maximum strength, V_{max}, of 635 kips is reached and followed by the onset of negative stiffness which occurs at a story drift ratio of 0.040 and a roof displacement of $0.035h_r$. This is then followed with 20% strength loss at $\delta_u = 0.056h_r$. Using these values, the overstrength factor of Archetype ID 1010 can be computed as $\Omega = 635k/193k = 3.3$, and the period-based ductility, which will be used

---

[3] In this example, pushover analysis is based on the lateral load pattern prescribed by Equation 12.8-13 of ASCE/SEI 7-05.

later to adjust the *CMR* according to Section 7.2, can be computed as $\mu_T = \delta_u / \delta_{y,eff} = 0.056h_r / 0.0042h_r = 13.2$.

To compute the collapse capacity for each archetype design, the incremental dynamic analysis (IDA) approach is used (Section 6.4), with the Far-Field ground motion set and ground motion scaling method presented in Section 6.2.3 (and Appendix A). Note that Section 6.4 does not require a full IDA (as is shown in Figure 9-7) but only a simplified version, which has the goal of quantifying the median collapse capacity of each archetype model.

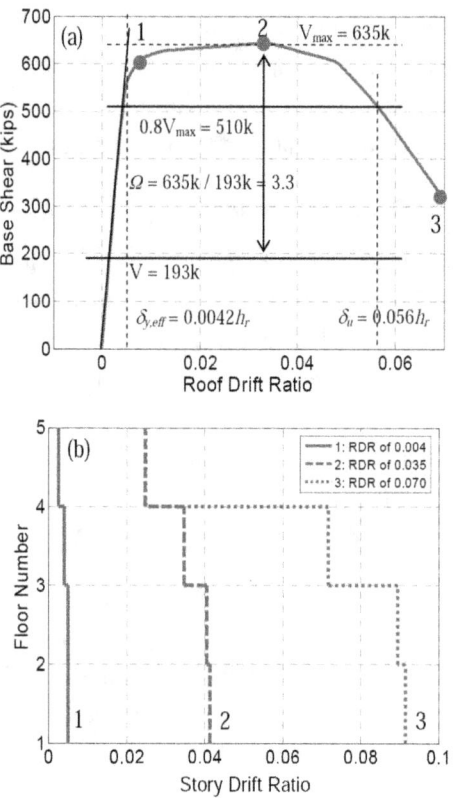

Figure 9-6　　(a) Monotonic static pushover, and (b) peak story drift ratios at three deformation levels during pushover for Archetype ID 1010. The pushover is based on the building code specified lateral load distribution (ASCE, 2005) (after Haselton and Deierlein, 2007, Chapter 6).

Figure 9-7 illustrates how the IDA method is used to compute the collapse capacity of Archetype ID 1010. For each of the 44 ground motions of the Far-Field Set the spectral acceleration at collapse (S_{CT}) is computed. Next the median collapse level ($\hat{S}_{CT}$) is computed to be 2.58 g. The collapse margin ratio (*CMR*), defined as the ratio of $\hat{S}_{CT}$ to the Maximum Considered Earthquake (MCE) ground motion demand (S_{MT}), is 2.50 for Archetype ID

1010. Although a full IDA was utilized in this example, it is not required to quantify *CMR*, as discussed in Section 6.4.2.

In this example, it is assumed that reinforced concrete special moment frame buildings collapse in a sideway mechanism, which can be directly simulated using the structural analysis model. This assertion is made due to the many detailing, continuity, and capacity design provisions preventing other collapse modes (Appendix D). For structural systems where some collapse modes are not simulated by the structural model, these additional modes must be accounted for using component limit state checks for non-simulated collapse modes (Section 5.5).

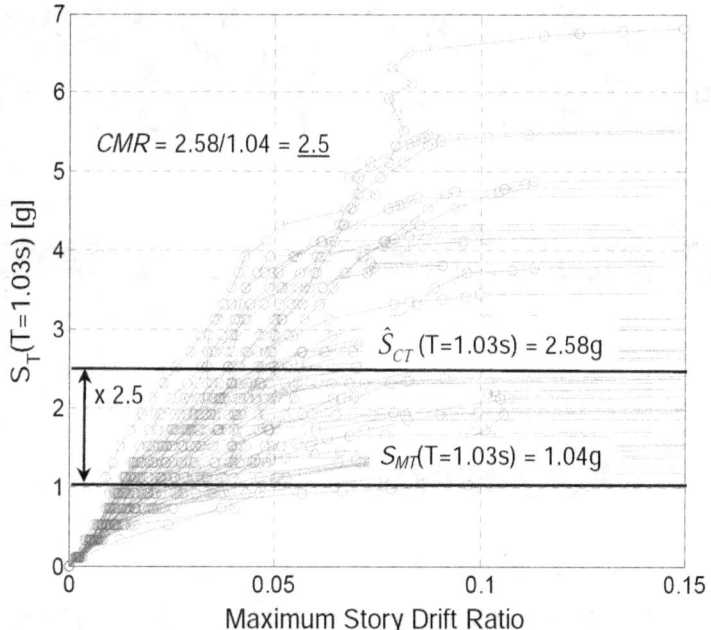

Figure 9-7   Incremental dynamic analysis to collapse, showing the Maximum Considered Earthquake ground motion intensity (S_{MT}), median collapse capacity intensity ($\hat{S}_{CT}$), and collapse margin ratio (*CMR*) for Archetype ID 1010.

Table 9-3 summarizes the results of pushover and IDA analyses. These IDA results show that the average *CMR* is 1.34 and 1.26 for perimeter frame performance groups and is 2.01 and 1.56 for the space frame performance groups. These values have not yet been adjusted for the beneficial effects of spectral shape (Section 7.2). Allowable *CMR* values and acceptance criteria are discussed later.

In addition, these results verify that the buildings designed in low-seismic regions have higher *CMR*s (i.e., lower collapse risk) than those designed in high-seismic regions. Compared to a building with 20-foot bay width, buildings with 30-foot bay width tend to have slightly higher *CMR*, due to gravity load effects causing some increases in overstrength, and due to joint shear provisions making columns slightly larger and more ductile. Therefore, the remaining assessment focuses on 12 more critical archetype buildings which are designed for high-seismic sites (SDC D_{max}) and with 20-foot bay width. This approach substantially reduces the number of required structural designs and analyses.

Table 9-3   Summary of Collapse Results for Reinforced Concrete Special Moment Frame Archetype Designs

Archetype ID	Design Configuration			Pushover and IDA Results			
	No. of Stories	Framing (Gravity Loads)	Seismic SDC	Static Ω	$S_{MT}[T]$ (g)	$S_{CT}[T]$ (g)	CMR
Performance Group No. PG-5 (Short Period, 20' Bay Width Configuration)							
2069	1	P	D_{max}	1.6	1.5	1.77	1.18
2064	2	P	D_{max}	1.8	1.5	2.25	1.50
--	3	P	D_{max}	--	--	--	--
Performance Group No. PG-6 (Long Period, 20' Bay Width Configuration)							
1003	4	P	D_{max}	1.6	1.11	1.79	1.61
1011	8	P	D_{max}	1.6	0.6	0.76	1.25
5013	12	P	D_{max}	1.7	0.42	0.51	1.22
5020	20	P	D_{max}	2.6	0.27	0.22	0.82
Mean of Performance Group:				1.9	NA	NA	1.23
Performance Group No. PG-1 (Short Period, 20' Bay Width Configuration)							
2061	1	S	D_{max}	4.0	1.5	2.94	1.96
1001	2	S	D_{max}	3.5	1.5	3.09	2.06
--	3	S	D_{max}	--	--	--	--
Performance Group No. PG-3 (Long Period, 20' Bay Width Configuration)							
1008	4	S	D_{max}	2.7	1.11	1.97	1.78
1012	8	S	D_{max}	2.3	0.60	0.98	1.63
5014	12	S	D_{max}	2.8	0.42	0.67	1.59
5021	20	S	D_{max}	3.5	0.27	0.34	1.25
Mean of Performance Group:				2.8	NA	NA	1.56
Selected Archetypes - Performance Group Nos. PG-4 and PG-8 (20' Bay Width)							
6011	8	P	D_{min}	1.8	0.15	0.32	2.12
6013	12	P	D_{min}	1.8	0.1	0.21	2.00
6020	20	P	D_{min}	1.8	0.07	0.11	1.73
6021	20	S	D_{min}	3.4	0.07	0.24	3.70
Selected Archetypes - Performance Group Nos. PG-10 and PG-14 (30' Bay Width)							
1009	4	P-30	D_{max}	1.6	1.04	2.05	1.98
1010	4	S-30	D_{max}	3.3	1.04	2.58	2.50

9.2.7 Performance Evaluation

The previous section discussed how to simulate structural collapse, compute median collapse level, $\hat{S}_{CT}$, and compute the collapse margin ratio, CMR. However, CMR does not account for the unique spectral shape of rare ground motions. Chapter 7 discusses spectral shape and how it affects the predicted collapse capacity, and provides simplified spectral shape factors, $SSFs$, that are used to adjust the median collapse level, $\hat{S}_{CT}$, to account for spectral shape effects. Table 9-4 and Table 9-5 (taken from Chapter 7) show these factors for Seismic Design Categories B, C, and D_{min} and SDC D_{max}, respectively. The values in these tables depend on the fundamental period, T, period-based ductility, μ_T, and the seismic design category. The tables show that the SSF values range from 1.0 to 1.37 for Seismic Design Categories B, C, and D_{min} and from 1.00 to 1.61 for SDC D_{max}.

The adjusted collapse margin ratio, $ACMR$, is computed by multiplying SSF (from Table 9-4 or Table 9-5) and CMR (from Table 9-3). Later, Table 9-8 will show this margin adjustment for the reinforced concrete special moment frame archetypes, and the resulting $ACMR$ values.

Table 9-4   Spectral Shape Factor *(SSF)* for Archetypes Designed for Seismic Design Categories B, C, or D_{min} Seismic Criteria (from Table 7-1a)

T (sec.)	Period-Based Ductility, μ_T							
	1.0	1.1	1.5	2	3	4	6	≥ 8
≤ 0.5	1.00	1.02	1.04	1.06	1.08	1.09	1.12	1.14
0.6	1.00	1.02	1.05	1.07	1.09	1.11	1.13	1.16
0.7	1.00	1.03	1.06	1.08	1.10	1.12	1.15	1.18
0.8	1.00	1.03	1.06	1.08	1.11	1.14	1.17	1.20
0.9	1.00	1.03	1.07	1.09	1.13	1.15	1.19	1.22
1.0	1.00	1.04	1.08	1.10	1.14	1.17	1.21	1.25
1.1	1.00	1.04	1.08	1.11	1.15	1.18	1.23	1.27
1.2	1.00	1.04	1.09	1.12	1.17	1.20	1.25	1.30
1.3	1.00	1.05	1.10	1.13	1.18	1.22	1.27	1.32
1.4	1.00	1.05	1.10	1.14	1.19	1.23	1.30	1.35
≥ 1.5	1.00	1.05	1.11	1.15	1.21	1.25	1.32	1.37

Table 9-5 Spectral Shape Factor (*SSF*) for Archetypes Designed using Seismic Design Category D$_{max}$ Seismic Criteria (from Table 7-1b)

T (sec.)	Period-Based Ductility, μ_T							
	1.0	1.1	1.5	2	3	4	6	≥ 8
≤ 0.5	1.00	1.05	1.1	1.13	1.18	1.22	1.28	1.33
0.6	1.00	1.05	1.11	1.14	1.2	1.24	1.3	1.36
0.7	1.00	1.06	1.11	1.15	1.21	1.25	1.32	1.38
0.8	1.00	1.06	1.12	1.16	1.22	1.27	1.35	1.41
0.9	1.00	1.06	1.13	1.17	1.24	1.29	1.37	1.44
1.0	1.00	1.07	1.13	1.18	1.25	1.31	1.39	1.46
1.1	1.00	1.07	1.14	1.19	1.27	1.32	1.41	1.49
1.2	1.00	1.07	1.15	1.2	1.28	1.34	1.44	1.52
1.3	1.00	1.08	1.16	1.21	1.29	1.36	1.46	1.55
1.4	1.00	1.08	1.16	1.22	1.31	1.38	1.49	1.58
≥ 1.5	1.00	1.08	1.17	1.23	1.32	1.4	1.51	1.61

In addition to quantifying the *ACMR* by Equation 7-1, the composite uncertainty, β_{TOT}, in collapse capacity is also needed. Table 9-6 (from Table 7-2b) shows composite uncertainties, which account for the variability between ground motion records of a given intensity (defined as a constant β_{RTR} = 0.40 for systems with $\mu_T \geq 3$), the uncertainty in the nonlinear structural modeling ((B) Good), the quality of data used to calibrate the element models ((B) Good), and the quality of the structural system design requirements ((A) Superior). For this example assessment, the composite uncertainty is β_{TOT} = 0.500 and is shown in bold.

Table 9-6 Total System Collapse Uncertainty (β_{TOT}) for Model Quality (B) Good and Period-Based Ductility, $\mu_T \geq 3$ (from Table 7-2b)

Quality of Test Data	Quality of Design Requirements			
	(A) Superior	(B) Good	(C) Fair	(D) Poor
(A) Superior	0.475	0.500	0.575	0.675
(B) Good	**0.500**	0.525	0.600	0.700
(C) Fair	0.575	0.600	0.675	0.750
(D) Poor	0.675	0.700	0.750	0.825

The acceptable collapse margin ratio is determined from the composite uncertainty and the acceptable conditional probability of collapse under the MCE ground motions. Chapter 7 defines the collapse performance objectives as: (1) a conditional collapse probability of 20% for each archetype building, and (2) a conditional collapse probability of 10% for each performance group. Table 9-7 (from Table 7-3) shows values of acceptable *ACMR* computed assuming a lognormal distribution of collapse capacity. Based on β_{TOT} = 0.500, the acceptable *ACMR*$_{20\%}$ value is 1.52 for each individual archetype and the acceptable $ACMR_{10\%}$ value is 1.90 for the average of each performance group.

Table 9-7   Acceptable Values of Adjusted Collapse Margin Ratio ($ACMR_{10\%}$ and $ACMR_{20\%}$) (from Table 7-3)

Total System Collapse Uncertainty	Collapse Probability				
	5%	10% ($ACMR_{10\%}$)	15%	20% ($ACMR_{20\%}$)	25%
0.400	1.93	**1.67**	1.51	**1.40**	1.31
0.425	2.01	**1.72**	1.55	**1.43**	1.33
0.450	2.10	**1.78**	1.59	**1.46**	1.35
0.475	2.18	**1.84**	1.64	**1.49**	1.38
0.500	2.28	**1.90**	1.68	**1.52**	1.40
0.525	2.37	**1.96**	1.72	**1.56**	1.42
0.550	2.47	**2.02**	1.77	**1.59**	1.45
0.575	2.57	**2.09**	1.81	**1.62**	1.47
0.600	2.68	**2.16**	1.86	**1.66**	1.50

Table 9-8 presents the final results and acceptance criteria for each of the 18 archetype designs. This shows the collapse margin ratio (CMR) computed directly from IDA, the SSF, and the final adjusted collapse margin ratio ($ACMR$). The acceptable margins are then shown, and each archetype is shown to either pass or fail the acceptance criteria.

The results in Table 9-8 show that high-seismic archetypes (SDC D_{max} designs) have lower $ACMR$ values (as compared to SDC D_{min} designs) and control the collapse performance. Additionally, the 20-foot bay width designs have slightly lower $ACMR$ values. Therefore, this example focuses on the performance groups containing archetypes designed for high seismic (SDC D_{max}) with 20-foot bay widths (PG-1, PG-3, PG-5, and PG-6). Based on the results shown in Table 9-8, the perimeter frame performance groups are more critical than the space frame performance groups, so PG-5 or PG-6 will govern the response modification factor, R.

Table 9-8   Summary of Final Collapse Margins and Comparison to Acceptance Criteria for Reinforced Concrete Special Moment Frame Archetypes

Arch. ID	Design Configuration			Computed Overstrength and Collapse Margin Parameters					Acceptance Check	
	No. of Stories	Framing (Gravity Loads)	SDC	Static Ω	CMR	μ_T	SSF	ACMR	Accept. ACMR	Pass/ Fail
Performance Group No. PG-5 (Short Period, 20' Bay Width Configuration)										
2069	1	P	D_{max}	1.6	1.18	14.0	1.33	1.57	1.52	Pass
2064	2	P	D_{max}	1.8	1.50	19.6	1.33	2.00	1.52	Pass
--	3	P	D_{max}	1.7*	--	--	--	2.13*	--	--
Mean of Performance Group:				1.7*	1.34	16.8	1.33	1.90*	1.90	Pass
Performance Group No. PG-6 (Long Period, 20' Bay Width Configuration)										
1003	4	P	D_{max}	1.6	1.61	10.9	1.41	2.27	1.52	Pass
1011	8	P	D_{max}	1.6	1.25	9.8	1.61	2.01	1.52	Pass
5013	12	P	D_{max}	1.7	1.22	7.4	1.58	1.93	1.52	Pass
5020	20	P	D_{max}	2.6	0.82	4.1	1.40	1.15	1.52	Fail
Mean of Performance Group:				1.9	1.23	8.1	1.50	1.84	1.90	Fail
Performance Group No. PG-1 (Short Period, 20' Bay Width Configuration)										
2061	1	S	D_{max}	4.0	1.96	16.1	1.33	2.61	1.52	Pass
1001	2	S	D_{max}	3.5	2.06	14.0	1.33	2.74	1.52	Pass
--	3	S	D_{max}	3.1[1]	--	--	--	2.63*	--	--
Mean of Performance Group:				3.5[1]	2.01	15.0	1.33	2.66*	1.90	Pass
Performance Group No. PG-3 (Long Period, 20' Bay Width Configuration)										
1008	4	S	D_{max}	2.7	1.78	11.3	1.41	2.51	1.52	Pass
1012	8	S	D_{max}	2.3	1.63	7.5	1.58	2.58	1.52	Pass
5014	12	S	D_{max}	2.8	1.59	8.6	1.61	2.56	1.52	Pass
5021	20	S	D_{max}	3.5	1.25	4.4	1.42	1.78	1.52	Pass
Mean of Performance Group:				2.8	1.56	8.0	1.51	2.36	1.90	Pass
Selected Archetypes - Performance Group Nos. PG-4 and PG-8 (20' Bay Width)										
6011	8	P	D_{min}	1.8	2.12	3.0	1.21	2.56	1.52	Pass
6013	12	P	D_{min}	1.8	2.00	3.7	1.24	2.47	1.52	Pass
6020	20	P	D_{min}	1.8	1.73	2.8	1.20	2.08	1.52	Pass
6021	20	S	D_{min}	3.4	3.70	3.3	1.22	4.51	1.52	Pass
Selected Archetypes - Performance Group Nos. PG-10 and PG-14 (30' Bay Width)										
1009	4	P-30	D_{max}	1.6	1.98	13.4	1.41	2.79	1.52	Pass
1010	4	S-30	D_{max}	3.3	2.50	13.2	1.41	3.53	1.52	Pass

1. For completeness, the reinforced concrete special moment frame example assumes values of static overstrength and *ACMR* of missing 3-story archetypes (based on the average of respective 2-and 4-story values).

The overstrength factor, Ω_O, will likely not be governed by the same performance group that governs the response modification factor, R. Table 9-8 shows that the space frame performance groups have higher overstrength,

Ω, values than perimeter frame groups, short-period space frames have higher overstrength values than long-period space frames, and the four-story 30-foot bay width design has a higher overstrength than 20-foot bay width designs (e.g., Archetype ID 1010 versus 1008). This suggests that either PG-9 or PG-11 (defined in Table 9-1) would govern the overstrength factor, Ω_O, but the archetypes in these performance groups were not designed and assessed in this example application.

The results for the two performance groups that may govern the response modification factor (PG-5 or PG-6), show that the majority of archetype buildings have acceptable $ACMR$, but a trend is evident: for space and perimeter frame buildings taller than four-stories, the $ACMR$ decreases substantially with increased building height. This causes the 20-story perimeter frame archetype to have an unacceptable $ACMR$ and causes the average $ACMR$ of the long period perimeter frame performance group (PG-6) to also be unacceptable.

As currently defined' the "newly proposed" reinforced concrete special moment frame system does not meet the collapse performance objectives of this Methodology, and needs adjusted design requirements in order to meet the acceptance criteria. One simple alternative would be to limit the proposed system to a maximum height of 12-stories (or 160 feet), or to require a space frame system for buildings taller than 160 feet. This solution is not ideal, so the next section looks more closely at other possible solutions.

9.2.8 Iteration: Adjustment of Design Requirements to Meet Performance Goals

The reinforced concrete special moment frame system did not meet the performance criteria with the initial set of design requirements. For the initially assessed designs, the *ACI 318-05* design requirements were used along with $R = 8$, $C_d = 5.5$, a story drift limit of 2%, and minimum design base shear requirements of ASCE/SEI 7-05. These requirements must now be modified in some way that will improve the reinforced concrete special moment frame collapse performance and cause the system to meet the performance criteria of Section 7.4.

According to Table 9-8, the adjusted collapse margin ratio ($ACMR$) decreases with increasing height. Haselton and Deierlein (2007) showed that this type of poor performance is caused by the damage localizing more for taller moment frames since damage localization is driven primarily by higher P-delta effects as the building height increases. In order to ensure better collapse performance in taller reinforced concrete frame buildings, this issue could be addressed in various ways. More conservative strong-column-weak-beam ratios, i.e., larger than 1.2, could be developed for taller buildings, to spread more uniformly over the height of the building. Instead, strength requirements could also be increased for taller buildings, by using a period-dependent R factor. Krawinkler and Zareian (2007) illustrate how the R factor would need to change, as a function of building period, in order to create uniform collapse probabilities for moment frame buildings of varying height (more strength required for longer period frame buildings).

In this example, another simple approach is attempted and the minimum design base shear is increased to solve this problem. This solution is not the most direct way to solve the specifically identified problems of damage localization and P-delta for taller moment-resisting frame buildings, but it is a simple solution that works. Specifically, the ASCE/SEI 7-05 minimum base shear requirement, $C_s = 0.01$ (ASCE, 2006a, Equation 12.8-5) is replaced by Equation 9.5.5.2.1-3 of ASCE 7-02 (ASCE, 2002). The ASCE 7-02 equation is shown here as Equation 9-1.

$$C_s \geq 0.044 S_{DS} I \qquad\qquad (9\text{-}1)$$

This change to the design requirements impacts only the design of the 12- and 20-story archetypes in SDC D_{max} and the design of the 8-, 12-, and 20-story archetypes in SDC D_{min}.

Table 9-9 shows the design information for the redesigned buildings. A comparison to Table 9-2 shows that the design base shear coefficient (V/W) increased from 0.022 to 0.044 for the 20-story archetype in SDC D_{max} and increased from 0.010 to 0.017 for the 20-story archetype in SDC D_{min}. The base shear coefficient also increased, to a lesser extent, for the other buildings shown in Table 9-9. The minimum base shear requirement is governing the design of the taller frames.

The building designs were revised and the collapse assessments were completed for the revised designs. Table 9-10 shows the updated collapse performance results, with the italic lines showing the designs that were affected by the change to the minimum base shear requirement. This shows that each archetype building meets the performance requirement of $ACMR \geq$ 1.52 (i.e., 20% conditional collapse probability) and the average $ACMR \geq$ 1.90 for each performance group (i.e., 10% conditional collapse probability). This shows that after modifying the minimum design base shear requirement, the "newly proposed" reinforced concrete special moment frame system attains the required collapse performance and could be added as a "new system" in the building code provisions.

Table 9-9 Structural Design Properties for Reinforced Concrete Special Moment Frame Archetypes Redesigned Considering a Minimum Base Shear Requirement

Archetype ID	No. of Stories	Framing (Gravity Loads)	Key Archetype Design Parameters						
			Seismic Design Criteria						$S_{MT}(T)$ [g]
			SDC	R_{eff}[1]	T [sec]	T_1 [sec]	V/W [g]	
Re-Designed Archetypes - Performance Group No. PG-6 (Long Period, 20' Bay Width)									
1013	12	P	D_{max}	6.4	2.13	2.01	0.044	0.42
1020	20	P	D_{max}	4.1	3.36	2.63	0.044	0.27
Re-Designed Archetypes - Performance Group No. PG-3 (Long Period, 20' Bay Width)									
1014	12	S	D_{max}	6.4	2.13	2.14	0.044	0.42
1021	20	S	D_{max}	4.1	3.36	2.36	0.044	0.27
Re-Designed Archetypes - Performance Group No. PG-4, PG-8 (Long Period, 20' Bays)									
4011	8	P	D_{min}	5.8	1.6	3.00	0.017	0.15
4013	12	P	D_{min}	6.6	2.28	3.35	0.017	0.10
4020	20	P	D_{min}	4.2	3.6	4.08	0.017	0.065
4021	20	S	D_{min}	8.0	3.6	4.03	0.017	0.065

1. Effective value of R due to limits on the seismic coefficient, C_s.

Notice that the short-period perimeter frame high-seismic performance group (PG-5) is now the governing group that controls the value of the response modification factor, R. It is common that short-period structures have higher strength demands than moderate period structures, and therefore can control the collapse performance assessment; this has been documented in many research publications, starting with Newmark and Hall (1973).

Table 9-10 shows that building archetypes were only assessed up to a height of 20 stories. Taller buildings were not assessed because of the limitations of the ground motion record set, which is applicable only to buildings with elastic fundamental periods below 4.0 seconds (Chapter 6). Even so, Table 9-10 shows a trend that the collapse safety increases with building height. As long as this trend is observed for buildings with fundamental periods below 4.0 seconds, and the peer review committee believes that the trend is stable and defensible, then a height limit would not need to be imposed for this reinforced concrete special moment frame system.

Table 9-10 Summary of Final Collapse Margins and Comparison to Acceptance Criteria for Archetypes Redesigned with an Updated Minimum Base Shear Requirement

Arch. ID	Design Configuration			Computed Overstrength and Collapse Margin Parameters					Acceptance Check	
	No. of Stories	Framing (Gravity Loads)	SDC	Static Ω	CMR	μ_T	SSF	ACMR	Accept. ACMR	Pass/ Fail
Performance Group No. PG-5 (Short Period, 20' Bay Width Configuration)										
2069	1	P	D_{max}	1.6	1.18	14.0	1.33	1.57	1.52	Pass
2064	2	P	D_{max}	1.8	1.50	19.6	1.33	2.00	1.52	Pass
--	3	P	D_{max}	1.7*	--	--	--	2.13*	--	--
Mean of Performance Group:				1.7*	1.34	16.8	1.33	1.90*	1.90	**Pass**
Performance Group No. PG-6 (Long Period, 20' Bay Width Configuration)										
1003	4	P	D_{max}	1.6	1.61	10.9	1.41	2.27	1.52	Pass
1011	8	P	D_{max}	1.6	1.25	9.8	1.61	2.01	1.52	Pass
1013	12	P	D_{max}	1.7	1.45	11.4	1.61	2.33	1.52	Pass
1020	20	P	D_{max}	1.6	1.66	5.6	1.49	2.47	1.52	Pass
Mean of Performance Group:				1.6	1.49	9.4	1.53	2.27	1.90	**Pass**
Performance Group No. PG-1 (Short Period, 20' Bay Width Configuration)										
2061	1	S	D_{max}	4.0	1.96	16.1	1.33	2.61	1.52	Pass
1001	2	S	D_{max}	3.5	2.06	14.0	1.33	2.74	1.52	Pass
--	3	S	D_{max}	3.1[1]	--	--	--	2.63*	--	--
Mean of Performance Group:				3.5[1]	2.01	15.0	1.33	2.66*	1.90	**Pass**
Performance Group No. PG-3 (Long Period, 20' Bay Width Configuration)										
1008	4	S	D_{max}	2.7	1.78	11.3	1.41	2.51	1.52	Pass
1012	8	S	D_{max}	2.3	1.63	7.5	1.58	2.58	1.52	Pass
1014	12	S	D_{max}	2.1	1.59	7.7	1.60	2.54	1.52	Pass
1021	20	S	D_{max}	2.0	1.98	5.7	1.50	2.96	1.52	Pass
Mean of Performance Group:				2.3	1.75	8.1	1.52	2.65	1.90	**Pass**
Selected Archetypes - Performance Group Nos. PG-4 and PG-8 (20' Bay Width)										
4011	8	P	D_{min}	1.8	1.93	3.6	1.23	2.38	1.52	Pass
4013	12	P	D_{min}	1.8	2.29	4.3	1.26	2.89	1.52	Pass
4020	20	P	D_{min}	1.8	2.36	3.9	1.24	2.94	1.52	Pass
4021	20	S	D_{min}	2.8	3.87	3.8	1.24	4.81	1.52	Pass
Selected Archetypes - Performance Group Nos. PG-10 and PG-14 (30' Bay Width)										
1009	4	P-30	D_{max}	1.6	1.98	13.4	1.41	2.79	1.52	Pass
1010	4	S-30	D_{max}	3.3	2.5	13.2	1.41	3.53	1.52	Pass

1. For completeness, the reinforced concrete special moment frame example assumes values of static overstrength and *ACMR* of missing 3-story archetypes (based on the average of respective 2- and 4-story values).

9.2.9 Evaluation of Ω_0 Using Final Set of Archetype Designs

At this point, the overstrength, Ω_0, value can be established for use in the proposed design provisions. Chapter 7 specifies that the Ω_0 value should not be taken as less than the largest average value of overstrength, Ω, from any performance group, and should be conservatively increased to account for variations in individual Ω values. The final Ω_0 value should be rounded to the nearest 0.5, and limited to a maximum value of 3.0.

Section 9.2.7 explained that the governing performance group for the overstrength factor, Ω_0, will either be PG-9 or PG-11 (as defined in Table 9-1). This portion of this example is not entirely complete, because the archetypes in these performance groups were not designed and assessed. Even so, the results from other performance groups are used to determine the appropriate overstrength factor, Ω_0.

The average Ω values for the high-seismic 20-foot bay width performance groups are 1.7 and 1.6 for the perimeter frame performance groups (PG-5 and PG-6) and 3.8 and 2.3 for the space frame performance groups (PG-1 and PG-3). Even though all performance groups were not completed, these results are enough to show that the upper-bound value of $\Omega_0 = 3.0$ is warranted, due to the average Ω value of 3.8 observed for PG-1.

9.2.10 Summary Observations

This example shows that current seismic provisions for reinforced concrete special moment frame systems in ACI 318-05 and ASCE/SEI 7-05 provide an acceptable level of collapse safety in SDC D, with an important modification of imposing the minimum base shear requirement from the ASCE 7-02 provisions. In addition, it demonstrates that the Methodology is reasonably well calibrated because recent building code design provisions lead to acceptable collapse safety, as defined by the Methodology (Section 7.4). This example also illustrates how the Methodology could be used as a tool for testing possible changes to building code requirements and evaluating code change proposals.

9.3 Example Application - Reinforced Concrete Ordinary Moment Frame System

9.3.1 Introduction

In this example, a reinforced concrete ordinary moment frame system, as defined by ACI 318-05 and ASCE/SEI 7-05, is evaluated as if it were a new system proposed for inclusion in ASCE/SEI 7-05.

This example illustrates the Methodology for limited ductility systems, which are only permitted in lower seismic design categories, are typical of construction in the Central and Eastern United States, and are designed for a much lower ratio of lateral to gravity loads. Since these systems lack the capacity design and ductile detailing provisions of special moment frames, reinforced concrete ordinary moment frame systems are susceptible to additional modes of damage, such as shear failure in columns, leading to rapid strength and stiffness deterioration. Because there are many possible failure modes in reinforced concrete ordinary moment frame systems, this example incorporates limit state checks for collapse modes that are not simulated directly in nonlinear analysis.

9.3.2 Overview and Approach

The structural system for this example is defined by the design and detailing provisions of ASCE/SEI 7-05 and ACI 318-05, and is evaluated using the methods of Chapters 3 through 7. A set of structural system archetypes are developed for reinforced concrete ordinary moment frame buildings, nonlinear models are developed to simulate structural collapse, models are analyzed to predict the collapse capacities of each archetype, and adjusted collapse margin ratios are evaluated and compared to acceptance criteria.

This example has been adapted from collaborative research on the development of structural archetypes for reinforced concrete ordinary moment frames, calibration of nonlinear element models for collapse simulation, simulation of structural response to collapse, and treatment of uncertainties (Liel and Deierlein, 2008).

9.3.3 Structural System Information

Design Requirements

This example utilizes ACI 318-05 design requirements, which are extremely detailed and represent years of accumulated research and building code development. For the purpose of assessing composite uncertainty, the design requirements are categorized as (A) Superior to reflect the high degree of confidence in the design equations for reinforced concrete ordinary moment frames. This rating is in agreement with the reinforced concrete special moment frame example.

Test Data

The element models used in this study are the same as those in the reinforced concrete special moment frame study and based on published test data from the Pacific Earthquake Engineering Research Center's Structural

Performance Database (PEER, 2006b, Berry et al., 2004). Although extensive data is available for reinforced concrete elements, there are still limitations as discussed in Section 9.2.3. Accordingly, for the purpose of assessing the total uncertainty, the test data is categorized as (B) Good, in agreement with the reinforced concrete special moment frame example.

Seismic Design Criteria

Reinforced concrete ordinary moment frames are permitted only in Seismic Design Categories B and below. Following the requirements of Section 4.2.1, the highest allowable SDC is the focus of the performance evaluation. Accordingly, reinforced concrete ordinary moment frames are evaluated at the lower limit of SDC B, where $S_{D1} = 0.067$ g and $S_{DS} = 0.167$ g (B_{min}), and the upper limit of SDC B, where $S_{D1} = 0.133$ g and $S_{DS} = 0.33$g (B_{max}). A subset of archetype models is also evaluated at the limits of SDC C, where reinforced concrete ordinary moment frames are not permitted, to illustrate how the Methodology can be used to evaluate current code limits.

9.3.4 Identification of Reinforced Concrete Ordinary Moment Frame Archetype Configurations

Figure 9-8 shows the two-dimensional three-bay multi-story frame that is considered an appropriate index archetype model for reinforced concrete ordinary moment frame buildings. This is the same general archetype model that was used to evaluate the reinforced concrete special moment frame system, and includes joint panel elements, beam and column elements, elastic foundation springs, and a leaning column to account for P-delta effects due to the seismic mass on the gravity system.

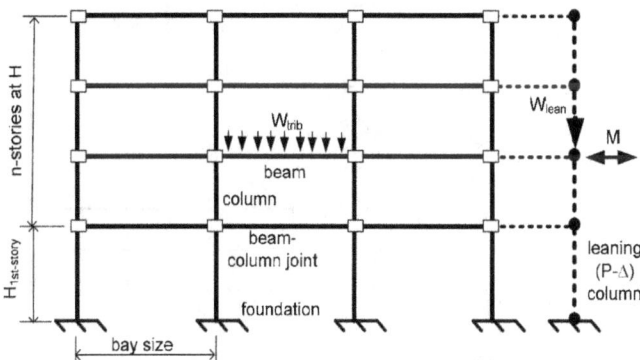

Figure 9-8   Index archetype model for reinforced concrete ordinary moment frames.

Using the above index archetype model, a set of structural archetype designs are developed to represent the archetype design space, following the design configuration and performance group requirements of Chapter 4. Since this example is intended primarily to demonstrate that the Methodology is applicable to reinforced concrete ordinary moment frame systems, a partial group of archetypes is considered and it is recognized that this partial group does not interrogate all basic design configurations that would be permitted by the code. Therefore, a complete FEMA P695 assessment of the reinforced concrete ordinary moment frame system would include a wider range of design configurations, in addition to the subset of archetypes evaluated here.

Table 9-11 shows how archetypes are organized into performance groups, according to the requirements of Section 4.3. These groups represent the two basic configurations considered in this example, the range of allowable gravity loads and design ground motion intensities, and the building period range (short- and long-period systems).

Table 9-11   Performance Groups for Evaluation of Reinforced Concrete Ordinary Moment Frames

Performance Group Summary					
Group No.	Grouping Criteria				Number of Archetypes
	Basic Config.	Design Load Level		Period Domain	
		Gravity	Seismic		
PG-1	20-foot Bay Width	High (Space Frame)	SDC B_{max}	Short	0^1
PG-2				Long	
PG-3			SDC B_{min}	Short	
PG-4				Long	
PG-5		Low (Perimeter Frame)	SDC B_{max}	Short	
PG-6				Long	
PG-7			SDC B_{min}	Short	
PG-8				Long	
PG-9	30-foot Bay Width	High (Space Frame)	SDC B_{max}	Short	0^2
PG-10				Long	4
PG-11			SDC B_{min}	Short	0^2
PG-12				Long	4
PG-13		Low (Perimeter Frame)	SDC B_{max}	Short	0^2
PG-14				Long	4
PG-15			SDC B_{min}	Short	0^2
PG-16				Long	4

1. Performance of reinforced concrete ordinary moment frame archetypes with 20-foot beam span not evaluated, because 30-foot spans judged to be representative for reinforced concrete ordinary moment frames.

2. No short-period reinforced concrete ordinary moment frame archetypes are considered in this example; for a complete reinforced concrete ordinary moment frame assessment, a 1-story would need to be investigated.

Table 9-11 shows that only four performance groups are utilized in this example, accounting for the variations in seismic design level and gravity load level. For reinforced concrete ordinary moment frame buildings, the bay width is not varied, and the rationale for this is explained below. Since reinforced concrete ordinary moment frames tend to be flexible systems, all designs fall into the long-period performance groups. For a full application of the Methodology, at least one one-story building would need to be evaluated in each of the short-period performance groups (PG-9, PG-11, PG-13, and PG-15).

Space frames, which are typical for OMF designs are "high gravity" systems. "Low gravity" systems could represent either perimeter frames of a flat plate system, or the perimeter framing of a one-way joist system. Figure 9-9 and Figure 9-10 show the layout of the archetype space and perimeter frame systems. Buildings have a bay spacing of 30-feet and cover a range of building heights (2-, 4-, 8-, or 12-stories). The bay spacing is different from the default of 20-feet used in the special moment frame example, to better reflect the typical configurations of gravity-dominated reinforced concrete ordinary moment frame designs. For space-frames, a transverse span of 35 feet was used to maximize the gravity load contribution, whereas a smaller transverse span of 30 feet was used for the lightly loaded perimeter-frames.

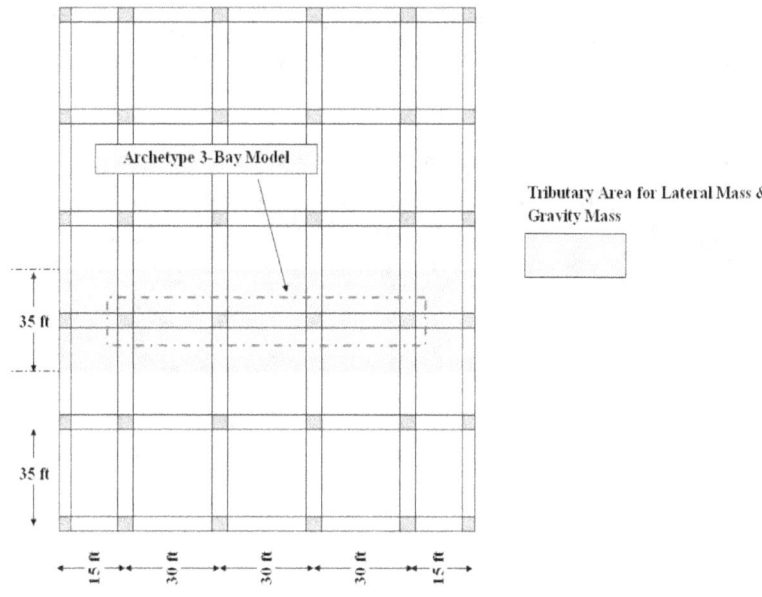

Figure 9-9   High gravity (space frame) layout.

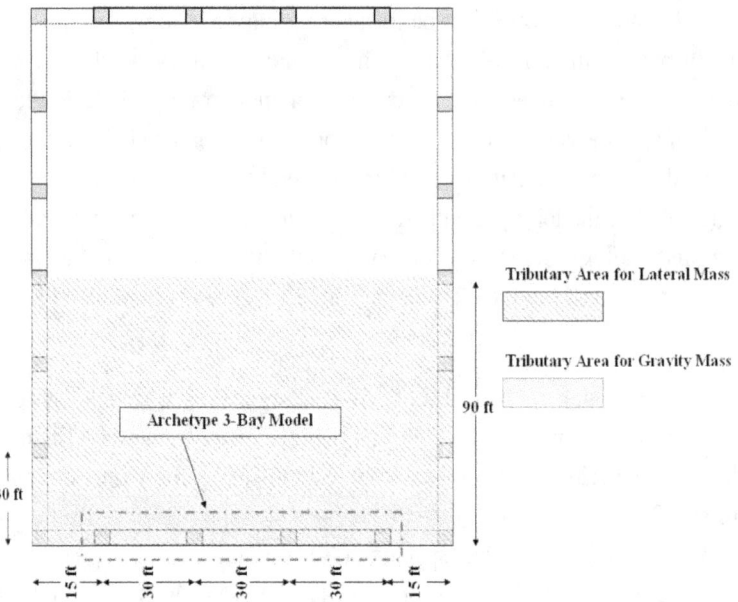

Figure 9-10   Low gravity (perimeter frame) layout. Interior (gravity) columns
are not shown for clarity.

Table 9-12 summarizes the design properties of the reinforced concrete
ordinary moment frame archetype designs needed to evaluate SDC B,
including the design base shear and code-calculated structural period. Each
of the archetypes was fully designed in accordance with the governing design
requirements (ASCE/SEI 7-05 and ACI 318-05). Additional information on
archetype designs is provided in Appendix C.

The design documentation provided is consistent with that provided for the
special moment frame models and an example for Archetype ID 9203 is
shown in Figure 9-11. Similar documentation has been maintained for all
other archetype designs.

Table 9-12   Reinforced Concrete Ordinary Moment Frame Archetype Design Properties, SDC B

Archetype ID	No. of Stories	Framing (Gravity Loads)	Key Archetype Design Parameters					$S_{MT}(T)$ [g]
			Seismic Design Criteria					
			SDC	R	T [sec]	T_1 [sec]	V/W [g]	
Performance Group No. PG-16 (Long Period)								
9101	2	P	B_{min}	3	0.55	1.56	0.041	0.18
9103	4	P	B_{min}	3	0.99	2.81	0.023	0.10
9105	8	P	B_{min}	3	1.81	4.58	0.012	0.06
9107	12	P	B_{min}	3	2.59	5.80	0.010	0.04
Performance Group No. PG-12 (Long Period)								
9102	2	S	B_{min}	3	0.55	0.85	0.041	0.18
9104	4	S	B_{min}	3	0.99	1.49	0.023	0.10
9106	8	S	B_{min}	3	1.81	2.53	0.012	0.06
9108	12	S	B_{min}	3	2.59	2.85	0.010	0.04
Performance Group No. PG-14 (Long Period)								
9201	2	P	B_{max}	3	0.51	1.23	0.087	0.39
9203	4	P	B_{max}	3	0.93	1.93	0.048	0.22
9205	8	P	B_{max}	3	1.70	3.39	0.026	0.12
9207	12	P	B_{max}	3	2.44	4.43	0.018	0.08
Performance Group No. PG-10 (Long Period)								
9202	2	S	B_{max}	3	0.51	0.81	0.087	0.39
9204	4	S	B_{max}	3	0.93	1.36	0.048	0.22
9206	8	S	B_{max}	3	1.70	2.35	0.026	0.12
9208	12	S	B_{max}	3	2.44	2.85	0.018	0.08

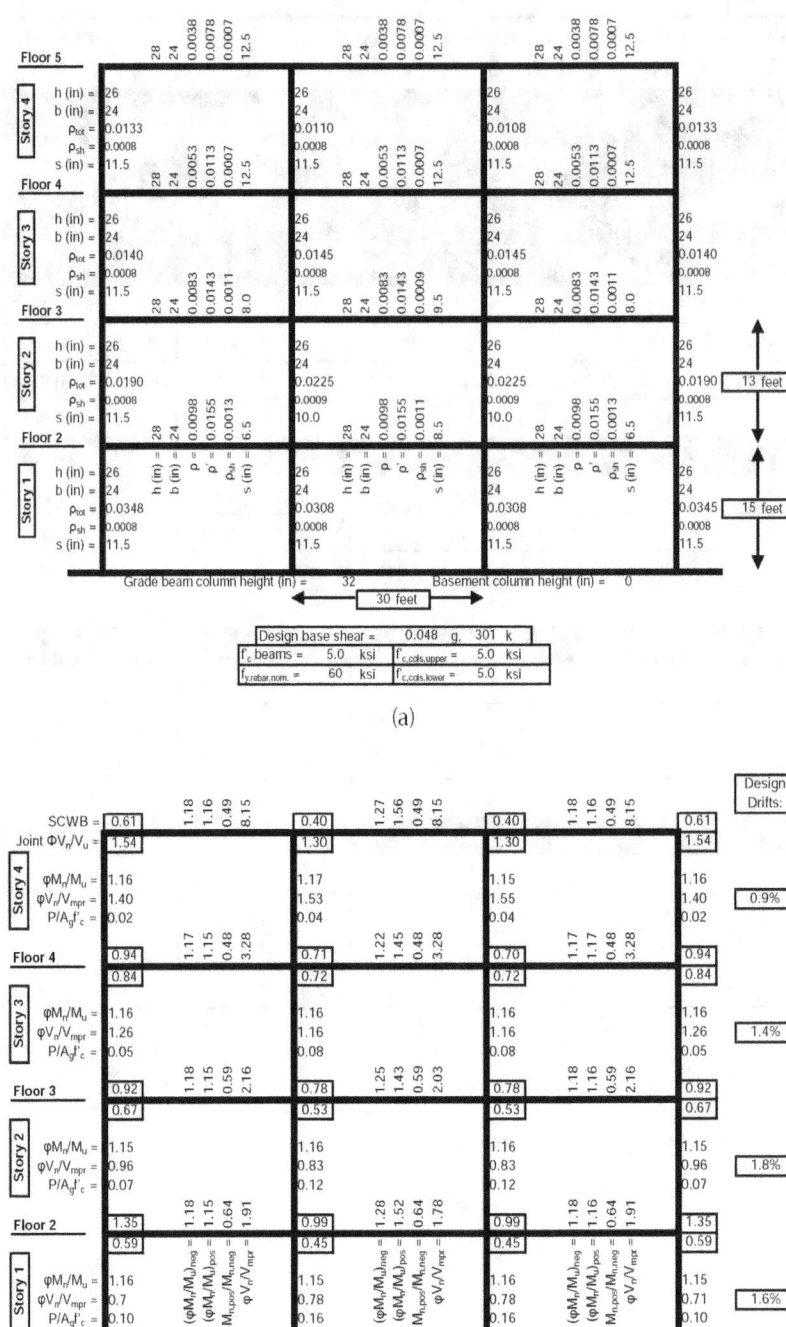

Figure 9-11   Design documentation for a 4-story reinforced concrete ordinary moment frame archetype with perimeter frame.

9.3.5 Nonlinear Model Development

The nonlinear models of reinforced concrete ordinary moment frames are similar to those developed in the reinforced concrete special moment frame system example; model development was discussed in Section 9.2.5 and is also discussed in more exhaustive detail in Appendix E. For illustration, Figure 9-12 shows the predicted modeling parameters for each element of Archetype ID 9203. As expected, reinforced concrete beams and columns in these structures have substantially less ductility than their special moment frame counterparts, as reflected in the deformation capacity ($\theta_{cap,pl}$ and θ_{pc}) and cyclic deterioration parameters (λ). As with the special moment frame example, these models were implemented in the OpenSees software platform, developed by the Pacific Earthquake Engineering Research Center (OpenSees, 2006).

Shear failure and subsequent loss of gravity-load bearing capacity in columns is not explicitly included in the analysis models, but according to requirements of Section 5.5, it is incorporated through post-processing as a non-simulated failure mode. Shear-induced axial failure of columns is difficult to simulate using available technologies and test data, and is accounted for in post-processing for this reason; however, if possible, it would be better to incorporate this failure mode directly into the nonlinear structural simulation.

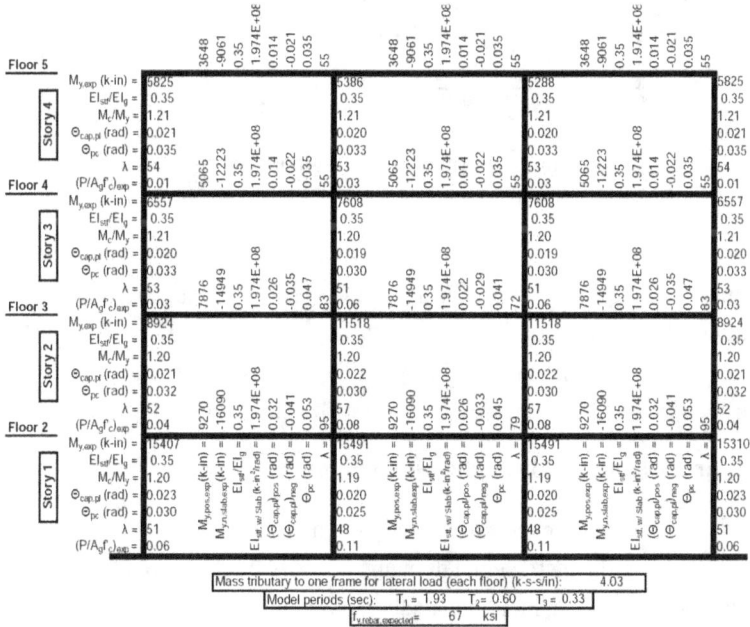

Figure 9-12   Structural modeling documentation for a 4-story reinforced concrete ordinary moment frame archetype (Archetype ID 9203).

9.3.6 Nonlinear Structural Analysis

Static Pushover Analysis

To compute the system overstrength, Ω, of each archetype design, a monotonic static pushover analysis is utilized with the lateral load pattern prescribed in ASCE/SEI 7-05. Figure 9-13 illustrates results of static pushover analysis for Archetype ID 9203. For this reinforced concrete ordinary moment frame archetype, the effective yield roof displacement, $\delta_{y,eff}$, is computed according to Equation 6-7 as $0.0047h_r$. The capping displacement (the onset of negative stiffness) occurs at roof displacement of about $0.0125h_r$ and the displacement at 20% strength loss is $\delta_u = 0.018h_r$. The period-based ductility (which will later be used to adjust the *CMR* according to Section 7.2), can be computed as $\mu_T = \delta_u / \delta_{y,eff} = 3.8$. These structures have substantially less ductility than their reinforced concrete special moment frame counterparts (e.g., Figure 9-13 compared to Figure 9-6).

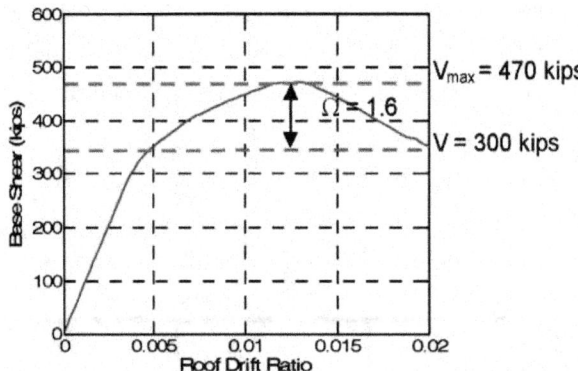

Figure 9-13   Monotonic static pushover for reinforced concrete ordinary moment frame archetype (Archetype ID 9203).

Results from pushover analysis are tabulated below in Table 9-13 for all performance groups in SDC B. Pertinent results include computed static overstrength (Ω) and period-based ductility ($\mu_T = \delta_u / \delta_{y,eff}$). When non-simulated collapse modes are considered in the collapse assessment, δ_u should account for the occurrence of the non-simulated collapse mode.

Nonlinear Dynamic Analysis and Simulation

The collapse capacity for each archetype design is computed according to Section 6.4, using the incremental dynamic analysis (IDA) approach. Ground motions (the Far-Field record set) and scaling procedures are based on requirements of Sections 6.2.2 and 6.2.3, respectively. Figure 9-14 illustrates how the IDA method is used to compute the sidesway collapse margin ratio of Archetype ID 9203. It should be noted that a full IDA is not required to quantify *CMR*, as discussed in Section 6.4.2.

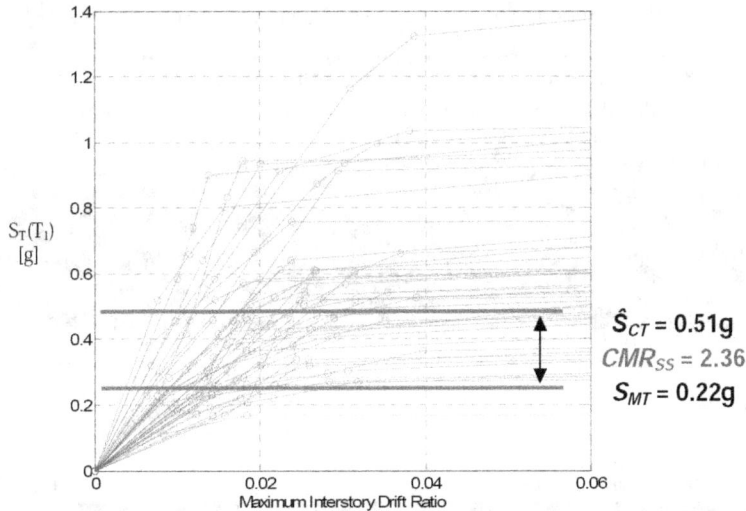

Figure 9-14   Incremental dynamic analysis to collapse showing the Maximum Considered Earthquake ground motion intensity (S_{MT}), median collapse capacity intensity ($\hat{S}_{CT}$), and sidesway collapse margin ratio (CMR_{SS}) for Archetype ID 9203.

Table 9-13 summarizes the IDA results for the selected 16 archetype designs. These results reflect only the sidesway collapse mechanism, which is directly simulated in the structural analysis model. The subscript "SS" denotes consideration of sidesway collapse only and not the non-simulated collapse modes. As such, CMR_{SS} is *not used for system evaluation*, but is presented here for illustrative purposes only.

Non-Simulated Collapse Modes

In this example, only sidesway collapse mechanisms based on strength and stiffness degradation due to flexure and flexure-shear are simulated directly in the analysis. Nonlinear models do not predict the occurrence of column shear failure and subsequent rapid deterioration and loss of gravity-load bearing capacity, which may occur because reinforced concrete ordinary moment frame columns have light transverse reinforcement and are not subject to capacity design requirements. A detailed discussion of possible failure modes in reinforced concrete moment frames is included in Appendix D.

To account for failure modes not directly included in analysis modes, according to Section 5.5, further post-processing of the nonlinear dynamic results is required. Results of static pushover analysis also require post-processing so that the computed period-based ductility, μ_T, (which is used to compute the *SSF*) accounts for non-simulated failure modes. The ultimate roof displacement, δ_u, should be based on the roof displacement corresponding to a 20% loss of base shear or the occurrence of the non-simulated collapse mode, whichever occurs at a smaller displacement.

Table 9-13   Summary of Pushover Analysis and IDA Sidesway Collapse Results for Reinforced Concrete Ordinary Moment Frame Archetype Designs, SDC B

Archetype ID	Design Configuration			Pushover and IDA Results			
	No. of Stories	Framing (Gravity Loads)	Seismic SDC	Static Ω	$\mu_T{}'$	$S_{CT}[T]$ (g)	CMR_{SS}
Performance Group No. PG-16 (Long Period)							
9101	2	P	B_{min}	2.0	3.7	0.91	4.99
9103	4	P	B_{min}	1.8	3.0	0.33	3.30
9105	8	P	B_{min}	2.6	3.1	0.18	3.31
9107	12	P	B_{min}	2.3	2.5	0.14	3.68
Mean of Performance Group:				2.2	3.1	NA	3.82
Performance Group No. PG-12 (Long Period)							
9102	2	S	B_{min}	6.6	3.0	1.40	7.69
9104	4	S	B_{min}	5.3	2.1	0.68	6.70
9106	8	S	B_{min}	6.0	3.0	0.39	7.05
9108	12	S	B_{min}	6.0	4.2	0.35	9.07
Mean of Performance Group:				6.0	3.1	NA	7.63
Performance Group No. PG-14 (Long Period)							
9201	2	P	B_{max}	1.6	3.5	1.07	2.72
9203	4	P	B_{max}	1.6	3.8	0.51	2.36
9205	8	P	B_{max}	1.5	2.8	0.25	2.11
9207	12	P	B_{max}	1.7	3.0	0.18	2.22
Mean of Performance Group:				1.6	3.3	NA	2.36
Performance Group No. PG-10 (Long Period)							
9202	2	S	B_{max}	2.9	3.3	1.36	3.47
9204	4	S	B_{max}	3.0	2.3	0.83	3.87
9206	8	S	B_{max}	3.1	3.0	0.41	3.49
9208	12	S	B_{max}	3.8	5.8	0.38	4.65
Mean of Performance Group:				3.2	3.6	NA	3.87

1. Due to time constraints, μ_T values do not account for non-simulated collapse modes, Even so, these must be included when computing period-based ductility.

Fragility functions (adopted from Aslani, 2005) are introduced to determine the drift at which the column loses its ability to carry gravity loads following shear failure. To determine if the shear-induced axial failure mode has occurred, the column drift level (demand) is extracted from the analysis data and compared to the median drift level associated with column axial failure from the fragility functions (capacity). If the median drift level has been exceeded in any column, the non-simulated collapse mode is assumed to have occurred. This approach is likely conservative because is assumes that when the non-simulated collapse limit is exceeded in *one* element, it triggers collapse of the entire building. In many cases, gravity loads can be redistributed to nearby elements, and the axial failure of a single column will not cause complete collapse of an entire structure. However, because the structural simulation model cannot represent system behavior and redistribution after the vertical collapse of a column, this is taken as the non-simulated collapse state.

Using the approach to non-simulated collapse modes in Section 5.5, the collapse fragility is adjusted, increasing the probability of collapse to include both the simulated and non-simulated failure modes. This change reduces the collapse margin ratio. This reduction is more significant in space frame structures that have higher column axial loads. The effect of non-simulated failure modes on the computed collapse margin ratios for the 16 archetype structures is shown in Table 9-14.

Table 9-14   Effect of Non-Simulated Collapse Modes on Computed Collapse Margin Ratios for Reinforced Concrete Ordinary Moment Frame Archetypes, SDC B

Archetype ID	Design Configuration			Collapse Margin Ratios			
	No. of Stories	Framing (Gravity Loads)	Seismic SDC	S_{CT} [T] (g)	CMR_{SS}	$CMR_{non\text{-}simulated}$	Percent Decrease
Performance Group No. PG-16 (Long Period, 30' Bay Width)							
9101	2	P	B_{min}	0.91	4.99	4.96	0.6%
9103	4	P	B_{min}	0.33	3.30	3.08	6.7%
9105	8	P	B_{min}	0.18	3.31	2.57	22.3%
9107	12	P	B_{min}	0.14	3.68	2.96	19.5%
Mean of Performance Group:				NA	3.82	3.39	11.2%
Performance Group No. PG-12 (Long Period, 30' Bay Width)							
9102	2	S	B_{min}	1.40	7.69	3.98	48.2%
9104	4	S	B_{min}	0.68	6.70	2.79	58.4%
9106	8	S	B_{min}	0.39	7.05	4.36	38.2%
9108	12	S	B_{min}	0.35	9.07	4.19	53.8%
Mean of Performance Group:				NA	7.63	3.83	49.8%
Performance Group No. PG-14 (Long Period, 30' Bay Width)							
9201	2	P	B_{max}	1.07	2.72	2.04	25.0%
9203	4	P	B_{max}	0.51	2.36	1.99	15.6%
9205	8	P	B_{max}	0.25	2.11	1.68	20.5%
9207	12	P	B_{max}	0.18	2.22	1.93	13.4%
Mean of Performance Group:				NA	2.36	1.91	18.9%
Performance Group No. PG-10 (Long Period, 30' Bay Width)							
9202	2	S	B_{max}	1.36	3.47	1.79	48.5%
9204	4	S	B_{max}	0.83	3.87	2.08	46.3%
9206	8	S	B_{max}	0.41	3.49	2.48	28.9%
9208	12	S	B_{max}	0.38	4.65	1.95	58.0%
Mean of Performance Group:				NA	3.87	2.08	46.4%

9.3.7 Performance Evaluation for SDC B

The collapse margin ratio is adjusted according to Section 7.2 to account for the proper spectral shape of rare ground motions through the spectral shape factor, SSF. According to Table 9-4 (from Section 7.2.2), the spectral shape factor is computed based on the SDC, the period-based ductility, μ_T, and the structural periods, T and T_1; these parameter values are documented in Table 9-12 and later in Table 9-15. These structures have an average period-based ductility, μ_T, of 3.2, resulting in an average SSF of 1.17, as shown below in Table 9-15. Since the ordinary moment frame archetypes are designed for SDC B, where the benefit of spectral shape is more limited, and they have limited deformation capacity, the SSF values are smaller than for the special moment frame example.

To assess the ordinary moment frame system, the composite uncertainty, β_{TOT}, in collapse capacity is needed. As described above, the quality of test data is rated (B) Good and the quality of structural system design requirements is rated (A) Superior. According to Section 5.7, the uncertainty in the archetype model is based on (1) the completeness of the set of index archetypes and (2) how well the structural collapse behavior is captured either by direct simulation or use of non-simulated component checks. Regarding the completeness of the set of archetypes, it has already been stated that this example is a partial example, and the archetype set is not complete; however, for the purpose of quantifying model uncertainty in this example, it is assumed that the set of archetypes is complete. Regarding the collapse modeling, although the component model is calibrated to columns that fail in flexure-shear, the structural simulation model may lose some fidelity after the occurrence of shear failure, because shear failure is not directly predicted. Based on this rationale, the archetype model quality is rated as (C) Fair. Based on these individual values for the ordinary moment frame example, the composite uncertainty determined to be $\beta_{TOT} = 0.575$, according to Section 7.3.4.

The acceptable collapse margin ratio is determined from the composite uncertainty and the acceptable conditional probability of collapse under the MCE ground motions. Chapter 7 defines the collapse performance objectives as: (1) a conditional collapse probability of 20% for each archetype building, and (2) a conditional collapse probability of 10% for each performance group. For the reinforced concrete ordinary moment frame systems with a composite uncertainty of 0.575, this corresponds to a required $ACMR_{20\%}$ of 1.62 for each archetype building, with a required average $ACMR_{10\%}$ of 2.09 for each performance group. These values are taken from Table 7-3 (also shown in Table 9-7).

Table 9-15 presents the final results and acceptance criteria for the 16 reinforced concrete ordinary moment frame archetypes in SDC B. As shown in Table 9-15, the reinforced concrete ordinary moment frame structural system passes for each of the two performance groups in both B_{max} and B_{min}. Although the average $ACMR$ is close to the limit for PG-14, in some cases the $ACMR$s are considerably above the required values. The large $ACMR$s tend to occur for structures with substantial overstrength, which resulted from the dominance of gravity loading in the design, especially where seismic design forces are low compared to gravity loading (e.g., spaceframes in SDC B_{min}).

Table 9-15 Summary of Collapse Margins and Comparison to Acceptance Criteria for Reinforced Concrete Ordinary Moment Frame Archetypes, SDC B

Arch. ID	Design Configuration			Computed Overstrength and Collapse Margin Parameters					Acceptance Check	
	No. of Stories	Framing (Gravity Loads)	SDC	Static Ω	CMR	μ_T^1	SSF	ACMR	Accept. ACMR	Pass/ Fail
Performance Group No. PG-16 (Long Period)										
9101	2	P	B_{min}	2.0	4.96	3.7	1.09	5.40	1.62	Pass
9103	4	P	B_{min}	1.8	3.08	3.0	1.12	3.44	1.62	Pass
9105	8	P	B_{min}	2.6	2.57	3.1	1.21	3.10	1.62	Pass
9107	12	P	B_{min}	2.3	2.96	2.5	1.18	3.50	1.62	Pass
Mean of Performance Group:				2.2	3.39	3.1	1.15	3.86	2.09	Pass
Performance Group No. PG-12 (Long Period)										
9102	2	S	B_{min}	6.6	3.98	3.0	1.08	4.29	1.62	Pass
9104	4	S	B_{min}	5.3	2.79	2.1	1.09	3.03	1.62	Pass
9106	8	S	B_{min}	6.0	4.36	3.0	1.21	5.26	1.62	Pass
9108	12	S	B_{min}	6.0	4.19	4.2	1.37	5.75	1.62	Pass
Mean of Performance Group:				6.0	3.83	3.1	1.19	4.58	2.09	Pass
Performance Group No. PG-14 (Long Period)										
9201	2	P	B_{max}	1.6	2.04	3.5	1.09	2.21	1.62	Pass
9203	4	P	B_{max}	1.6	1.99	3.8	1.13	2.24	1.62	Pass
9205	8	P	B_{max}	1.5	1.68	2.8	1.20	2.01	1.62	Pass
9207	12	P	B_{max}	1.7	1.93	3.0	1.21	2.33	1.62	Pass
Mean of Performance Group:				1.6	1.91	3.3	1.15	2.20	2.09	Pass
Performance Group No. PG-10 (Long Period)										
9202	2	S	B_{max}	2.9	1.79	3.3	1.08	1.94	1.62	Pass
9204	4	S	B_{max}	3.0	2.08	2.3	1.09	2.28	1.62	Pass
9206	8	S	B_{max}	3.1	2.48	3.0	1.21	2.99	1.62	Pass
9208	12	S	B_{max}	3.8	1.95	5.8	1.31	2.56	1.62	Pass
Mean of Performance Group:				3.2	2.08	3.6	1.17	2.44	2.09	Pass

1. Due to time constraints, μ_T values do not account for non-simulated collapse modes. Even so, these must be included when computing period-based ductility.

9.3.8 Performance Evaluation for SDC C

A smaller subset of archetypes was utilized to assess whether reinforced concrete ordinary moment frame buildings designed in SDC C would meet the collapse safety criteria of Section 7.4. The archetype designs and results for B_{max} are used in this assessment, since B_{max} is identical to C_{min}. In addition, we consider four additional archetype reinforced concrete ordinary moment frames, designed for C_{max}, as shown in Table 9-16. These four designs compose two partial performance groups for SDC C_{max}; complete performance groups would include at least three designs in each group.

Seismic performance is assessed using the same procedure as described for the evaluation of reinforced concrete ordinary moment frames in SDC B. Table 9-17 shows the computed overstrength, Ω, and period-based ductility, μ_T, factors from static pushover analysis. The computed collapse margin ratios, showing only the results that include non-simulated failure modes, are also reported in Table 9-17. This table also compares the adjusted collapse margin ratios to the acceptance criteria.

Table 9-16 Reinforced Concrete Ordinary Moment Frame Archetype Design Properties for SDC C Seismic Criteria

Archetype ID	No. of Stories	Framing (Gravity Loads)	Seismic Design Criteria					$S_{MT}(T)$ [g]
			SDC	R	T [sec]	T_1 [sec]	V/W [g]	
SDC C_{min} Performance Group (Long Period) - PG-14								
9201	2	P	C_{min}	3	0.51	1.23	0.087	0.39
9203	4	P	C_{min}	3	0.93	1.93	0.048	0.22
9205	8	P	C_{min}	3	1.70	3.39	0.026	0.12
9207	12	P	C_{min}	3	2.44	4.43	0.018	0.08
SDC C_{min} Performance Group (Long Period) - PG-10								
9202	2	S	C_{min}	3	0.51	0.81	0.087	0.39
9204	4	S	C_{min}	3	0.93	1.36	0.048	0.22
9206	8	S	C_{min}	3	1.70	2.35	0.026	0.12
9208	12	S	C_{min}	3	2.44	2.85	0.018	0.08
SDC C_{max} Performance Group (Long Period)								
9303	4	P	C_{max}	3	0.87	1.51	0.077	0.34
9307	12	P	C_{max}	3	2.29	3.72	0.029	0.13
SDC C_{max} Performance Group (Long Period)								
9304	4	S	C_{max}	3	0.87	1.30	0.077	0.34
9308	12	S	C_{max}	3	2.29	2.57	0.029	0.13

Table 9-17 Summary of Pushover Results, Collapse Margins, and Comparison to Acceptance Criteria for Reinforced Concrete Ordinary Moment Frame Archetypes, SDC C

Arch. ID	Design Configuration			Computed Overstrength and Collapse Margin Parameters					Acceptance Check	
	No. of Stories	Framing (Gravity Loads)	SDC	Static Ω	CMR	$\mu_T{}^1$	SSF	ACMR	Accept. ACMR	Pass/ Fail
SDC C$_{min}$ Performance Group (Long Period) - PG-14										
9201	2	P	C$_{min}$	1.6	2.04	3.5	1.09	2.21	1.62	Pass
9203	4	P	C$_{min}$	1.6	1.99	3.8	1.13	2.24	1.62	Pass
9205	8	P	C$_{min}$	1.5	1.68	2.8	1.20	2.01	1.62	Pass
9207	12	P	C$_{min}$	1.7	1.93	3.0	1.21	2.33	1.62	Pass
Mean of Performance Group:				1.6	1.91	3.3	1.15	2.20	2.09	Pass
SDC C$_{min}$ Performance Group (Long Period) - PG-10										
9202	2	S	C$_{min}$	2.9	1.79	3.3	1.08	1.94	1.62	Pass
9204	4	S	C$_{min}$	3.0	2.08	2.3	1.09	2.28	1.62	Pass
9206	8	S	C$_{min}$	3.1	2.48	3.0	1.21	2.99	1.62	Pass
9208	12	S	C$_{min}$	3.8	1.95	5.8	1.31	2.56	1.62	Pass
Mean of Performance Group:				3.2	2.08	3.6	1.17	2.44	2.09	Pass
SDC C$_{max}$ Performance Group (Long Period)										
9303	4	P	C$_{max}$	1.5	1.55	3.6	1.14	1.76	1.62	Pass
9307	12	P	C$_{max}$	1.4	1.03	2.1	1.16	1.19	1.62	Fail
Mean of Performance Group:				1.5	1.29	2.9	1.15	1.48	2.09	Fail
SDC C$_{max}$ Performance Group (Long Period)										
9304	4	S	C$_{max}$	2.1	1.97	3.3	1.13	2.23	1.62	Pass
9308	12	S	C$_{max}$	2.7	1.58	4.7	1.27	2.01	1.62	Pass
Mean of Performance Group:				2.4	1.78	4.0	1.20	2.12	2.09	Fail

1. Due to time constraints, μ_T values do not account for non-simulated collapse modes. Even so, these must be included when computing period-based ductility.

As shown in Table 9-17, archetype structures with a perimeter configuration fail the acceptance criteria for C$_{max}$ seismic criteria for both the average *ACMR* of the long-period performance group, as well as individual Archetype ID 9307 (i.e., taller archetype). These results indicate that the Methodology would not allow the use of reinforced concrete ordinary moment frames from SDC C, consistent with current Code restrictions. In order to be permitted in SDC C a lower *R* factor or other changes in design requirements would be necessary.

9.3.9 Evaluation of Ω_o Using Set of Archetype Designs

Development of the overstrength factor, Ω_o, is based on SDC B archetype designs, since this is the highest SDC that is currently allowed for this system. If the system were being approved for use in SDC C, then the SDC C archetypes would instead be used for establishing Ω_o.

The first step is to compute the overstrength values (Ω) for each individual archetype building. There is a relatively wide range of overstrength observed for the set of archetype designs (Ω ranges from 1.5 to 6.6, as reported in Table 9-17). The PG-14 archetypes (SDC B_{max}, perimeter frame) that govern the R factor have computed Ω values ranging from 1.5 to 1.7, with an average of 1.6. The PG-10 archetypes (SDC B_{max}, space frame) have higher computed overstrength values, between 2.9 and 3.8, with an average of 3.2. The PG-12 archetypes (SDC B_{min}, space frame) have the largest overstrength values, ranging up to 6.6, with an average of 6.0. Due to the average values being greater than 3.0 for one or more of the performance groups, the upper-bound value of $\Omega_o = 3.0$ is recommended for reinforced concrete ordinary moment frames, based on the requirements of Section 7.6.

9.3.10 Summary Observations

This example shows that current seismic provisions for reinforced concrete ordinary moment frame systems provide an acceptable level of collapse safety for SDC B, but not for SDC C. These results are consistent with the provisions for use of reinforced concrete ordinary moment frame in ASCE/SEI 7-05. Levels of collapse safety observed for reinforced concrete ordinary moment frames in SDC B_{max} are comparable to those for reinforced concrete special moment frames in SDC D_{max}, and in B_{min} these systems have a large margin against collapse. Note that in some cases the collapse safety of actual reinforced concrete ordinary moment frame buildings may be higher than calculated in this example, particularly when there are a large number of gravity-designed columns which increase structural strength and stiffness. Even so, for the purpose of establishing seismic design requirements according to this Methodology, Chapter 4 requires that only elements that are designed as part of the seismic-force-resisting system, and are accordingly governed by seismic design requirements, be included in this assessment.

To account for non-simulated failure modes as described in Section 5.5, component limit state checks are incorporated through post-processing of dynamic analysis results. In some cases, incorporation of the non-simulated failure modes significantly affects the collapse margin ratio, demonstrating the importance of carefully considering and including all critical failure modes either explicitly in the simulation models or in non-simulated limit state checks. Use of non-simulated failure modes to account for collapse due to column loss of gravity-load bearing capacity may be conservative, because it does not allow for load redistribution.

9.4 Example Application - Wood Light-Frame System

9.4.1 Introduction

In this example, a wood light-frame system with structural panel sheathing is considered as if it were a new system proposed for inclusion in ASCE/SEI 7-05.

9.4.2 Overview and Approach

Wood light-frame system design requirements of ASCE/SEI 7-05 are used as the framework. A set of structural archetypes are developed for wood light-frame buildings, nonlinear models are developed to simulate structural collapse, models are analyzed to predict the collapse capacities of each design, and the adjusted collapse margin ratio, *ACMR*, is evaluated and compared to acceptance criteria.

Seismic performance factors (SPFs) are determined by iteration until the acceptance criteria of the Methodology are met. This example begins with an initial value of $R = 6$ and checks if such designs pass the acceptance criteria of Section 7.4. This value is different from the current value of $R = 6.5$ for wood light-frame shear wall systems with wood structural panel sheathing in ASCE/SEI 7-05. It has been rounded to the nearest whole number for simplicity, and because developmental studies have shown that there is no discernable difference in collapse performance of structures design for fractional R factors (e.g., $R = 6$ versus $R = 6.5$). The Ω_0 factor is not assumed initially, but is determined from the actual overstrength factors, Ω, calculated for the archetype designs.

9.4.3 Structural System Information

Design Requirements

This example utilizes design requirements for engineered wood light-frame buildings included in ASCE/SEI 7-05, in place of the requirements that would need to be developed for a newly proposed system. For the purpose of assessing uncertainty, the ASCE/SEI 7-05 design requirements are categorized as (A) Superior since they represent many years of development, include lessons learned from a number of major earthquakes, and consider recent results obtained from large research programs on wood light-frame systems, such as the FEMA-funded CUREE-Caltech Woodframe Project and the NSF/NEES-funded NEESWood Project.

Test Data

This example relies on existing published sheathing-to-framing connection test data and wood shear wall assembly test data. Specifically, this example relies on information developed during the CUREE-Caltech Woodframe Project (Fonseca et al., 2002; Folz and Filiatrault, 2001), the NEESWood Project (Ekiert and Hong, 2006), the CoLA wood shear wall test program (CoLA, 2001), and data provided directly by the wood industry (Line et al., 2008).

The quality of the test data is an important consideration when quantifying the uncertainty in the overall collapse assessment process. Cyclic test data were provided by the wood industry for each of the archetypes used later in this example. In addition, more data were used by the authors to calibrate and validate the numerical model; these include monotonic and cyclic tests which cover a wide range of wood sheathing types and thicknesses (e.g., oriented strand board and plywood), framing grades, species, and connector types (e.g., common vs. box nails). All loading protocols were continued to deformations large enough for the capping strength to be observed, which allows better calibration of models for structural collapse assessment. Nevertheless, some uncertainties still exist with these test data sets including: (1) premature failures in some of the CUREE data set caused by specimens with smaller connector edge distances than specified; (2) the use of the Sequential Phased Displacement, SPD, loading protocol in the CoLA tests that tends to cause premature specimen failure by connectors fatigue, which is seldom observed after real earthquakes; (3) the inherent large variability associated with the material properties of wood; and (4) a lack of duplicate tests of the same specimen. Therefore, for the purpose of assessing uncertainty, this test data set is categorized as (B) Good.

9.4.4 Identification of Wood Light-Frame Archetype Configurations

The archetypes are established according to the requirements of Chapter 4, and separated into performance groups according to Section 7.4. The first step in archetype development is to establish the possible building design configurations. Figure 9-15 shows the two different building configurations that are assumed to be representative for the purpose of defining the two-dimensional archetypes for wood light-frame shear wall systems with wood structural panel sheathing. The first configuration is representative of residential buildings, while the second configuration is associated with office, retail, educational, and warehouse/light-manufacturing buildings.

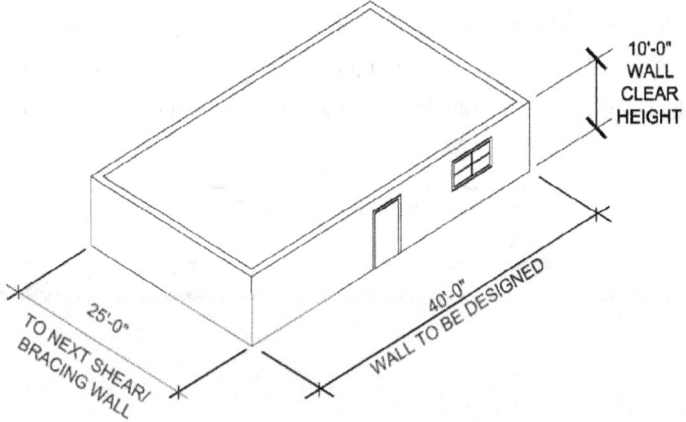

Residential building dimensions

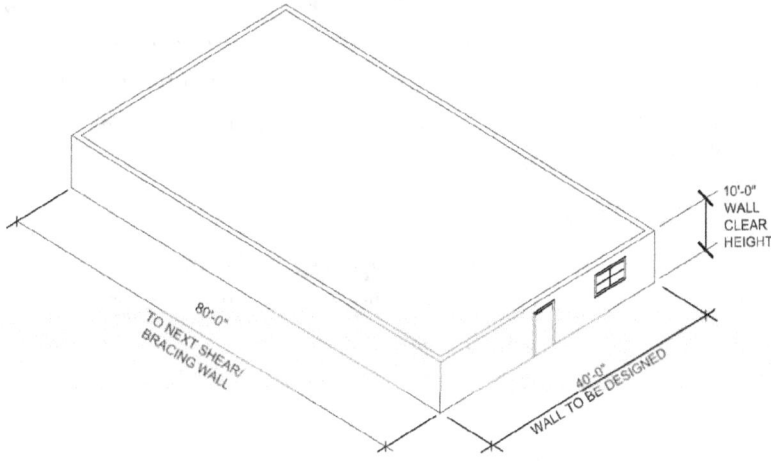

Commercial/educational building dimensions

Figure 9-15   Building configurations considered for the definitions of wood
light-frame archetype buildings.

Table 9-18 lists the range of design parameters considered for the
development of the two-dimensional archetype wall models. According to
Section 5.3, two-dimensional archetype wall models, not accounting for
torsional effects, are considered acceptable because the intended use of the
Methodology is to verify the performance of a full class of buildings, rather
than one specific building with a unique torsional issue. According to the
requirements of Section 4.2.3, nonstructural wall finishes, such as stucco and
gypsum wallboard, were not considered in the modeling of the archetypes.
These finishes are excluded because they are not defined as part of the lateral
structural system, and therefore are not governed by the seismic design

provisions. Depending on their type, wall finishes may greatly influence the seismic response of wood buildings. The Methodology would allow such elements to be included in the structural model, if one defines them as part of the lateral structural system, and design provisions are included to govern their design.

Table 9-18   Range of Variables Considered for the Definition of Wood Light-Frame Archetype Buildings

Variable	Range
Number of stories	1 to 5
Seismic Design Categories (SDC)	D_{max} and D_{min}
Story height	10 ft
Interior and exterior nonstructural wall finishes	Not considered
Wood shear wall pier aspect ratios	High/Low

Following the guidelines of Section 4.3, low aspect ratio (1:1 to 1.43:1) and high aspect ratio (2.70:1 to 3.33:1) walls were used as the two basic configurations in the archetype designs. This was done to evaluate the influence of the aspect ratio strength adjustment factor contained in ASCE/SEI 7-05, which effectively increases the strength of high aspect ratio wood shear walls.

Table 9-19 shows the performance groups (PG) used to evaluate the wood light-frame buildings, consistent with the requirements of Section 4.3.1. To represent these ranges of design parameters, 48 archetypes could have been used to evaluate the system (three designs for each of the 16 performance groups shown in Table 9-19). However, Table 9-19 shows that 16 archetypes were found to be sufficient. The notes in the table explain why these specific archetypes were selected, including the rationale for why these 16 can be used in place of the full set of 48. These 16 wood archetypes were divided among five performance groups: (1) three low aspect ratio wall short-period archetypes designed for SDC D_{max} (PG-1); (2) five SDC D_{max} - high aspect ratio wall short-period archetypes in SDC D_{max} (PG-9); (3) one low aspect ratio shear wall long-period archetype designed for SDC D_{min} (PG-4); and (4) seven SDC D_{min} - high aspect ratio shear wall systems, which are divided into four short-period buildings (PG-11) and three long-period buildings (PG-12). It is believed that this ensemble of 16 archetypes covers the current design space for wood light-frame buildings fairly well, but additional configurations would be required for a complete application of the Methodology. Appendix C provides detailed descriptions of the 16 archetype models developed for wood light-frame buildings.

Table 9-19 Performance Groups Used in the Evaluation of Wood Light-Frame Buildings

| Group No. | Basic Config. | Design Load Level | | Period Domain | Number of Archetypes |
		Gravity	Seismic		
PG-1	Low Wall Aspect Ratio	High (Nominal)	SDC D$_{max}$	Short	3
PG-2				Long	0[1]
PG-3			SDC D$_{min}$	Short	0
PG-4				Long	1[2]
PG-5		Low (NA)	SDC D$_{max}$	Short	0[3]
PG-6				Long	
PG-7			SDC D$_{min}$	Short	
PG-8				Long	
PG-9	High Wall Aspect Ratio	High (Nominal)	SDC D$_{max}$	Short	5
PG-10				Long	0[1]
PG-11			SDC D$_{min}$	Short	4
PG-12				Long	3
PG-13		Low (NA)	SDC D$_{max}$	Short	0[3]
PG-14				Long	
PG-15			SDC D$_{min}$	Short	
PG-16				Long	

Performance Group Summary / Grouping Criteria

1. No long-period SDC D$_{max}$ wood-frame archetypes, because representative designs never exceed $T = 0.6$ s.

2. Only one archetype in low-aspect/SDC D$_{min}$/long-period performance group, because only one representative design exceeds $T = 0.4$ s.

3. No archetypes because light wood-frame archetype design and performance not influenced significantly by gravity loads (i.e., nominal gravity loads used for all designs).

Table 9-20 reports the properties of each of these 16 archetypes. Seismic demands are based on the ground motion intensities of Seismic Design Category D. The archetypes are designed for maximum and minimum seismic criteria of Section 5.2.1: $S_{DS} = 1.0$ g and $S_{DI} = 0.6$ g for SDC D$_{max}$, and $S_{DS} = 0.50$ g and $S_{DI} = 0.20$ g for SDC D$_{min}$[4]. The MCE ground motion spectral response accelerations, S_{MT}, shown in Table 9-20 are based on Table 6-1. In accordance with Section 5.2.4, the periods reported in Table 9-20 are the fundamental period of the archetypes based on Section 12.8.2 of ASCE/SEI 7-05 ($T = C_u T_a$) with a lower bound limit of 0.25 sec.

---

[4] In this example, archetypes designed for low seismic (SDC D$_{min}$) loads, are based on $S_{DS} = 0.38$ g and $S_{DI} = 0.167$ g, based on interim criteria which differ slightly from the final values required by the Methodology.

Table 9-20   Wood Light-Frame Archetype Structural Design Properties

Arch. ID	No. of Stories	Key Archetype Design Parameters						
		Building Configuration	Wall Aspect Ratio	Seismic Design Criteria				$S_{MT}(T)$ [g]
				SDC	T [sec]	T_1 [sec]	V/W [g]	
Performance Group No. PG-1 (Short Period, Low Aspect Ratio)								
1	1	Commercial	Low	D_{max}	0.25	0.40	0.167	1.50
5	2	Commercial	Low	D_{max}	0.26	0.46	0.167	1.50
9	3	Commercial	Low	D_{max}	0.36	0.58	0.167	1.50
Performance Group No. PG-9 (Short Period, High Aspect Ratio)								
2	1	1&2 Family	High	D_{max}	0.25	0.29	0.167	1.50
6	2	1&2 Family	High	D_{max}	0.26	0.37	0.167	1.50
10	3	Multi-Family	High	D_{max}	0.36	0.44	0.167	1.50
13	4	Multi-Family	High	D_{max}	0.45	0.53	0.167	1.50
15	5	Multi-Family	High	D_{max}	0.53	0.62	0.167	1.50
Partial Performance Group No. PG-4 (Long Period, Low Aspect Ratio)								
11	3	Commercial	Low	D_{min}	0.41	0.93	0.063	0.75
Performance Group No. PG-11 (Short Period, High Aspect Ratio)								
3	1	Commercial	High	D_{min}	0.25	0.50	0.063	0.75
4	1	1&2 Family	High	D_{min}	0.25	0.41	0.063	0.75
7	2	Commercial	High	D_{min}	0.30	0.61	0.063	0.75
8	2	1&2 Family	High	D_{min}	0.30	0.62	0.063	0.75
Performance Group No. PG-12 (Long Period, High Aspect Ratio)								
12	3	Multi-Family	High	D_{min}	0.41	0.69	0.063	0.75
14	4	Multi-Family	High	D_{min}	0.51	0.81	0.063	0.75
16	5	Multi-Family	High	D_{min}	0.60	0.91	0.063	0.75

9.4.5   Nonlinear Model Development

Structural modeling of the wood light-frame archetypes is based on a "pancake" approach (Isoda et al., 2001). This system-level modeling approach is capable of simulating the three-dimensional seismic response of a wood light-frame building through a degenerated two-dimensional planar analysis. The computer program SAWS: Seismic Analysis of Woodframe Structures, developed within the CUREE-Caltech Woodframe Project (Folz and Filiatrault, 2004a, b), was used to analyze the wood light-frame archetype models. Because this example does not involve any buildings with torsional irregularities, only a two-dimensional model is utilized by fixing the rotational degree-of-freedom in the SAWS model.

In the SAWS model, the building structure is composed of rigid horizontal diaphragms and nonlinear lateral load resisting shear wall elements. The pinched, strength and stiffness degrading hysteretic behavior of each wood shear wall in the building is characterized using an associated numerical model (Folz and Filiatrault, 2001) that predicts the load-displacement response of the full wall assemblies under general quasi-static cyclic loading, based on sheathing-to-framing connection cyclic test data. Alternatively, cyclic test results from full-scale walls can also be used directly to characterize their hysteretic response. In the SAWS model, the hysteretic behavior of each wall panel is represented by an equivalent nonlinear shear spring element. As shown in Figure 9-16, the hysteretic behavior of this shear spring includes pinching, as well as stiffness and strength degradation, and is governed by 10 different physically identifiable parameters (Folz and Filiatrault 2004a, b). The predictive capabilities of the SAWS program have been demonstrated by comparing its predictions with the results of shake table tests performed on full-scale wood light-frame buildings (Folz and Filiatrault, 2004b; White and Ventura, 2007).

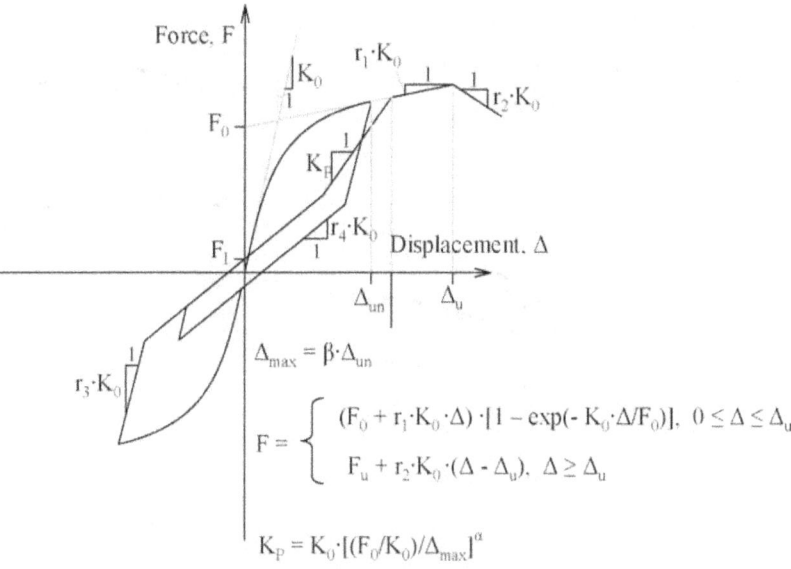

Figure 9-16   Hysteretic model of shear spring element included in SAWS program (after Folz and Filiatrault, 2004a, b).

Table 9-21 shows the sheathing-to-framing connector hysteretic parameters used to construct the equivalent nonlinear shear spring elements of each of the walls contained in the archetype models. The hysteretic model used for these sheathing-to-framing connectors is the same model used for the entire wall panel assemblies, which is shown in Figure 9-16. Figure 9-17 shows the monotonic backbone curves of 8' x 8' shear wall specimens generated using the hysteretic parameters for 8d common nail on 7/16" OSB sheathing shown in Table 9-21, compared with average test data provided by the wood industry (Line et al., 2008). This wood industry data is based on the cyclic envelope averaged from two identical test specimens. Very good agreement is observed between the numerical predictions and the test data shown here, which typically continues up to displacements near the onset of strength loss. The capping displacement and the post-capping behavior of the analytical model are based on additional cyclic test data that is not shown here, which was continued to larger displacements to exhibit strength deterioration.

Table 9-21   Sheathing-to-Framing Connector Hysteretic Parameters Used to Construct Shear Elements for Wood Light-Frame Archetype Models

Connector Type	K_0 (lbs/in)	r_1	r_2	r_3	r_4	F_0 (lbs)	F_I (lbs)	Δ_u (in)	α	β
7/16" OSB - 8d common nails	6,643	0.026	-0.039	1.0	0.008	228	32	0.51	0.7	1.2
19/32" Plywood - 10d common nails	7,777	0.031	-0.056	1.1	0.007	235	39	0.49	0.6	1.2

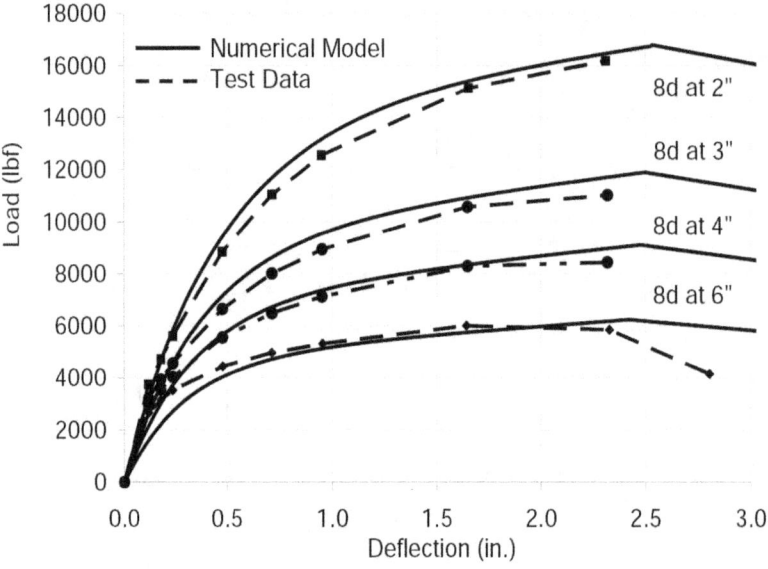

Figure 9-17   Comparison of monotonic backbone curves of 8' x 8' shear wall specimens generated using the hysteretic parameters for 8d common nail on 7/16" OSB sheathing shown in Table 9-21 with average cyclic envelope test data provided by the wood industry (Line et al.,2008).

Uncertainty due to Model Quality

For the purpose of assessing model uncertainty, according to Section 5.7, the archetype designs are assumed to be well representative of the archetype design space, even though a complete assessment may include more basic structural configurations. The structural modeling approach for the wood light-frame archetypes captures the primary shear deterioration modes of the shear walls that precipitate sidesway collapse. However, not all behavioral aspects are captured by this system-level modeling, such as axial-flexural

9: Example Applications

interaction effects of the wall elements, the uplift of narrow wall ends, and the slippage of sill and top plates. These effects are secondary for walls with low aspect ratios, which deform mainly in a shear mode, but are important for archetypes incorporating walls with high aspect ratios. Therefore, the structural model for the archetypes incorporating low-aspect ratio walls is rated as (B) Good, while the same structural model for the archetypes incorporating high-aspect ratio walls is rated as (D) Poor.

9.4.6 Nonlinear Structural Analyses

To compute the system overstrength, Ω, and to help verify the structural model, monotonic static pushover analysis is used with an inverted-triangular lateral load pattern; this approach differs slightly from the final requirements of Section 6.3. Figure 9-18 shows an example of the pushover curve for the two-story Archetype ID 5. For the wood light-frame archetype, the design LRFD seismic coefficient is $V/W = 0.167$. Capping (the onset of negative stiffness) occurs for a seismic coefficient of 0.417 and at a roof drift ratio of 0.0229. Therefore, Ω is calculated to be 2.49 for this archetype model

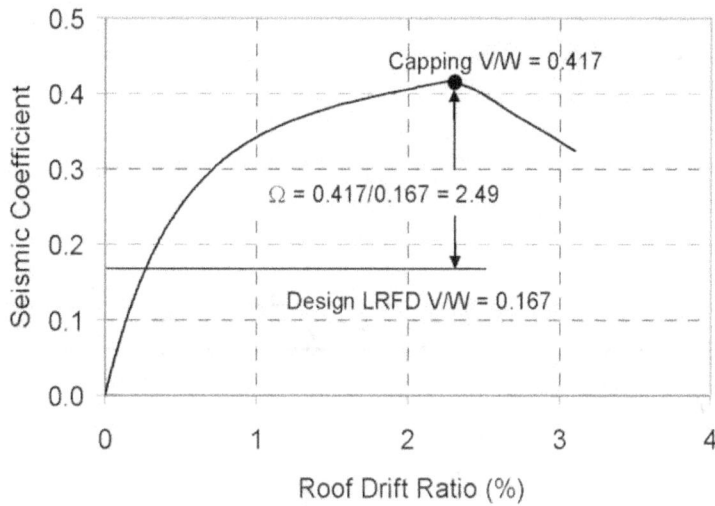

Figure 9-18   Monotonic static pushover curve and computation of Ω for two-story wood light-frame archetype (Archetype ID 5).

Following Section 6.4, to compute the collapse capacity of each wood light-frame archetype design, the incremental dynamic analysis (IDA) approach is used with the Far-Field record set and ground motion scaling method specified in Section 6.2. The intensity of the ground motion causing collapse of the wood light-frame archetype models is defined as the point on the intensity-drift IDA plot having a nearly horizontal slope but without exceeding a peak story drift of 7% in any wall of a model. This collapse story drift limit was selected based on recent collapse shake table testing conducted on full-scale two-story wood buildings in Japan (Isoda et al., 2007). The resulting collapse capacities should not be highly sensitive to this choice of 7% drift, since the IDA curves are relatively flat at such large drifts (see Figure 9-19 below).

Figure 9-19 and Figure 9-20 illustrate how the IDA method is used to compute the collapse margin ratio, *CMR*, for the two-story Archetype ID 5. The spectral acceleration at collapse is computed for each of the 44 ground motions of the Far-Field Set, as shown in Figure 9-19. The collapse fragility curve can then be constructed from the IDA plots, as shown in Figure 9-20. The collapse level earthquake spectral acceleration (spectral acceleration causing collapse in 50% of the analyses) is $S_{CT}(T = 0.26$ sec$) = 2.23$ g for this example. The collapse margin ratio, *CMR*, of 1.49 is then computed as the ratio of S_{CT} and the MCE spectral acceleration value at $T = 0.26$ sec, which is $S_{MT} = 1.50$ g for this building and SDC.

It should be noted that a full IDA is not required to quantify *CMR*, as discussed in Section 6.4.2.

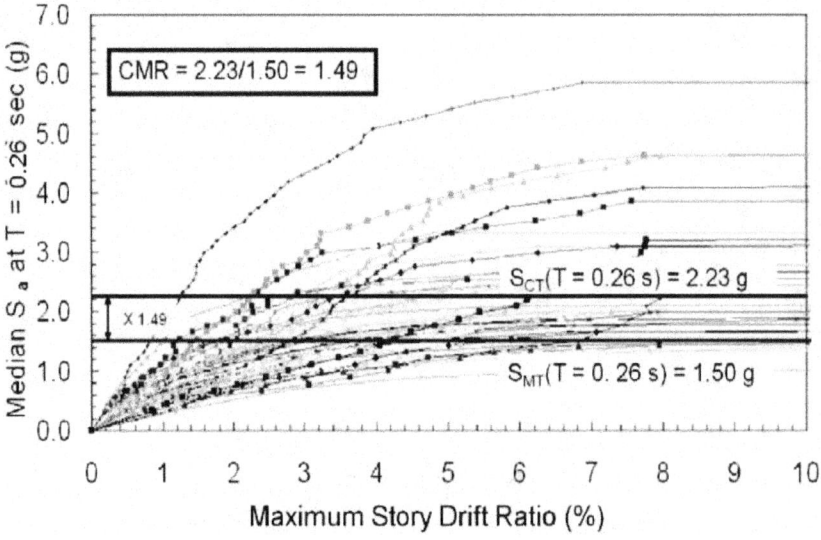

Figure 9-19   Results of incremental dynamic analysis to collapse for the two-story wood light-frame archetype (Archetype ID 5).

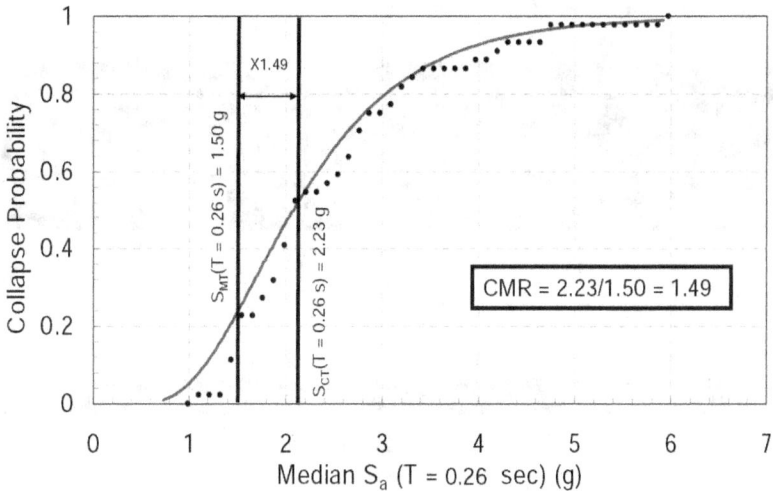

Figure 9-20   Collapse fragility curve for the two-story wood light-frame archetype (Archetype ID 5).

Static pushover analyses were conducted and the IDA method was applied to each of the 16 wood light-frame archetype designs, and Table 9-22 summarizes the results of these analyses. These IDA results indicate that the average collapse margin ratio is 1.43 for the SDC D_{max}, short period – low aspect ratio archetypes (PG-1), 1.90 for the SDC D_{max}, short period – high aspect ratio archetypes (PG-9), 2.64 for the SDC D_{min}, long period – low aspect ratio archetypes (partial PG-4), 2.57 for the SDC D_{min}, short period – high aspect ratio archetypes (PG-11) and 2.82 for the SDC D_{min}, long period – high aspect ratio archetypes (PG-12). These margin values, however, have not yet been adjusted for the beneficial effects of spectral shape (according to Section 7.2). Allowable collapse margins and acceptance criteria are discussed later.

The results shown in Table 9-22 show that the wood light-frame archetypes designed for minimum seismic criteria (SDC D_{min}) have higher collapse margin ratios (lower collapse risk) compared with the archetypes designed for maximum seismic criteria (SDC D_{max}). It is believed that this result originates from the longer vibration periods of archetypes designed for lower levels of seismic load, since the longer periods reduce seismic demands. Also, archetypes incorporating walls with high aspect ratios have higher collapse margin ratios than archetypes with low aspect ratio walls. This is the result of the ASCE/SEI 7-05 strength reduction factor applied to walls with high aspect ratios, which cause an increase in required number of nails to reach a given design strength. This increased nailing density causes an increase in the shear capacity of the walls with high aspect ratios, but the model does not account for the associated increase in flexural deformations.

Table 9-22 Summary of Collapse Results for Wood Light-Frame Archetype Designs

Archetype ID	Design Configuration			Pushover and IDA Results			
	No. of Stories	Building Configuration	Wall Aspect Ratio	Static Ω	$S_{MT}[T]$ (g)	$S_{CT}[T]$ (g)	CMR
Performance Group No. PG-1 (Short Period, Low Aspect Ratio)							
1	1	Commercial	Low	2.0	1.50	2.01	1.34
5	2	Commercial	Low	2.5	1.50	2.23	1.49
9	3	Commercial	Low	2.0	1.50	2.18	1.45
Mean of Performance Group:				2.2	NA	NA	1.43
Performance Group No. PG-9 (Short Period, High Aspect Ratio)							
2	1	1&2 Family	High	4.1	1.50	2.90	1.94
6	2	1&2 Family	High	3.8	1.50	3.20	2.14
10	3	Multi-Family	High	3.7	1.50	2.87	1.91
13	4	Multi-Family	High	2.9	1.50	2.60	1.73
15	5	Multi-Family	High	2.6	1.50	2.67	1.78
Mean of Performance Group:				3.4	NA	NA	1.90
Partial Performance Group No. PG-4 (Long Period, Low Aspect Ratio)							
11	3	Commercial	Low	2.1	0.75	1.98	2.64
Performance Group No. PG-11 (Short Period, High Aspect Ratio)							
3	1	Commercial	High	3.6	0.75	1.71	2.28
4	1	1&2 Family	High	5.4	0.75	2.09	2.78
7	2	Commercial	High	4.0	0.75	1.95	2.60
8	2	1&2 Family	High	3.5	0.75	1.95	2.60
Mean of Performance Group:				4.1	NA	NA	2.57
Performance Group No. PG-12 (Long Period, High Aspect Ratio)							
12	3	Multi-Family	High	4.0	0.75	2.34	3.12
14	4	Multi-Family	High	3.4	0.75	2.09	2.78
16	5	Multi-Family	High	3.3	0.75	1.92	2.56
Mean of Performance Group:				3.6	NA	NA	2.82

9.4.7 Performance Evaluation

Collapse margin ratios computed above do not account for the unique spectral shape of rare ground motions. According to Section 7.2, spectral shape adjustment factors, *SSF*, must be applied to the *CMR* results to account for spectral shape effects. In accordance with Section 7.2.2, the *SSF* can be computed for each archetype based on the SDC and the archetypes' period-based ductility, μ_T, obtained from the pushover curve. Figure 9-21 shows an example of calculating μ_T from the pushover curve for the two-story Archetype ID 5. The period-based ductility, μ_T, of 7.1 is then computed as the ratio of the ultimate roof displacement (defined as the displacement at

80% of the capping strength in the descending branch of the pushover curve) of $\delta_u = 0.0303 h_r$, to the equivalent yield roof displacement of $\delta_{y,eff} = 0.0043 h_r$.

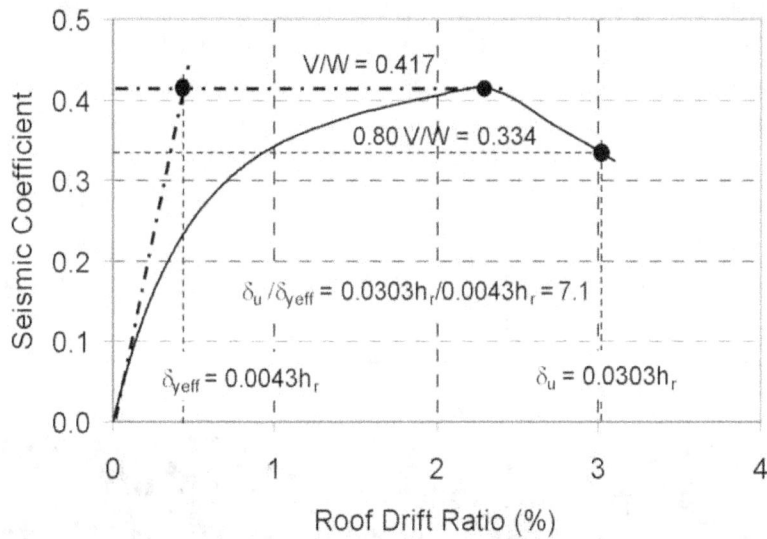

Figure 9-21   Monotonic static pushover curve and computation of $\delta_u / \delta_{y,eff}$ for the two-story wood light-frame Archetype ID 5.

The adjusted collapse margin ratio, *ACMR*, is then computed for each wood light-frame archetype as the multiple of the *SSF* (from Table 7-1b for SDC D) and *CMR* (from Table 9-22). Table 9-23 shows the resulting adjusted collapse margin ratios for the wood light-frame archetypes.

To calculate acceptable values of the adjusted collapse margin ratio, the total system uncertainty is needed. Section 7.3.4 provides guidance for this calculation. Table 7-2 shows these composite uncertainties, which account for the variability between ground motion records of a given intensity (defined as a constant $\beta_{RTR} = 0.40$), the uncertainty in the nonlinear structural modeling, the quality of the test data used to calibrate the element models, and the quality of the structural system design requirements. For this example assessment, the composite uncertainty was based on a (B) Good model quality for archetypes with low aspect ratio walls and a (D) Poor for archetypes with high aspect ratio walls, (A) Superior quality of design requirements and (B) Good quality of test data. Thus, $\beta_{TOT} = 0.500$ for archetype buildings incorporating low aspect ratio walls (Table 7-2b) and $\beta_{TOT} = 0.675$ for archetype buildings incorporating high aspect ratio walls (Table 7-2d).

An acceptable collapse margin ratio must now be selected based on a composite uncertainty, β_{TOT}, and a target collapse probability. Section 7.1.2 defines the collapse performance objectives as: (1) a conditional collapse probability of 20% for all individual wood light-frame archetypes, and (2) a conditional collapse probability of 10% for the average of each of the performance groups of wood light-frame archetypes. . Table 7-3 presents acceptable values of adjusted collapse margin ratio computed assuming a lognormal distribution of collapse capacity. For archetypes incorporating low aspect ratio walls, this corresponds to an acceptable collapse margin ratio $ACMR_{20\%}$ of 1.52 for every wood light-frame archetype and an $ACMR_{10\%}$ of 1.90 for each performance group. For archetype buildings incorporating high aspect ratio walls, this corresponds to an acceptable collapse margin ratio $ACMR_{20\%}$ of 1.76 for every wood light-frame archetype and an $ACMR_{10\%}$ of 2.38 for each performance group.

Table 9-23   Adjusted Collapse Margin Ratios and Acceptable Collapse Margin Ratios for Wood Light-Frame Archetype Designs

Arch. ID	Design Configuration			Computed Overstrength and Collapse Margin Parameters					Acceptance Check	
	No. of Stories	Building Config.	Wall Asp. Ratio	Static Ω	CMR	μ_T	SSF	ACMR	Accept. ACMR	Pass/ Fail
Performance Group No. PG-1 (Short Period, Low Aspect Ratio)										
1	1	Comm.	Low	2.0	1.34	9.9	1.33	1.78	1.52	Pass
5	2	Comm.	Low	2.5	1.49	7.1	1.31	1.95	1.52	Pass
9	3	Comm.	Low	2.0	1.45	12.4	1.33	1.93	1.52	Pass
Mean of Performance Group:				2.2	1.43	9.8	1.32	1.89	1.90	Pass
Performance Group No. PG-9 (Short Period, High Aspect Ratio)										
2	1	1&2-F.	High	4.1	1.94	9.9	1.33	2.57	1.76	Pass
6	2	1&2-F.	High	3.8	2.14	9.6	1.33	2.84	1.76	Pass
10	3	Multi-F.	High	3.7	1.91	7.9	1.33	2.54	1.76	Pass
13	4	Multi-F.	High	2.9	1.73	5.8	1.28	2.21	1.76	Pass
15	5	Multi-F.	High	2.6	1.78	5.4	1.27	2.26	1.76	Pass
Mean of Performance Group:				3.4	1.90	7.7	1.31	2.48	2.38	Pass
Partial Performance Group No. PG-4 (Long Period, Low Aspect Ratio)										
11	3	Comm.	Low	2.1	2.64	7.0	1.13	2.98	1.52	Pass
Performance Group No. PG-11 (Short Period, High Aspect Ratio)										
3	1	Comm.	High	3.6	2.28	9.9	1.14	2.58	1.76	Pass
4	1	1&2-F.	High	5.4	2.78	9.9	1.14	3.16	1.76	Pass
7	2	Comm.	High	4.0	2.60	7.7	1.13	2.95	1.76	Pass
8	2	1&2-F.	High	3.5	2.60	7.7	1.13	2.94	1.76	Pass
Mean of Performance Group:				4.1	2.57	8.8	1.13	2.91	2.38	Pass
Performance Group No. PG-12 (Long Period, High Aspect Ratio)										
12	3	Multi-F.	High	4.0	3.12	7.1	1.13	3.51	1.76	Pass
14	4	Multi-F.	High	3.4	2.78	6.2	1.12	3.12	1.76	Pass
16	5	Multi-F.	High	3.3	2.56	5.7	1.13	2.90	1.76	Pass
Mean of Performance Group:				3.6	2.82	6.3	1.13	3.18	2.38	Pass

Table 9-23 summarizes the final results and acceptance criteria for each of the 16 wood light-frame archetypes. The table presents the collapse margin ratios computed directly from the collapse fragility curves, CMR, the period-based ductility, μ_T, the Spectral Shape Factor, SSF, and the adjusted collapse margin ratio, $ACMR$. The acceptable $ACMR$s are shown and each archetype is shown to either pass or fail the acceptance criteria. Average $ACMR$s are also shown for the four complete performance groups of archetypes.

The results shown in Table 9-23 show that all individual archetypes pass the $ACMR_{20\%}$ criteria and the averages of each performance group pass the $ACMR_{10\%}$ criteria. Therefore, if wood light-frame buildings were a "newly proposed" seismic-force-resisting system with $R = 6$, it would meet the collapse performance objectives of the Methodology, and would be approved as a new system.

The results in Table 9-23 also show that performance groups of archetypes designed for maximum seismic loads (PG-1 and PG-9) have lower adjusted collapse margins ratios than other groups and govern determination of the R factor. Another observation is that archetypes incorporating high-aspect ratio walls have higher collapse margin ratios than those with low-aspect ratio walls. Even so, acceptable $ACMR$ values are also higher for the high-aspect ratio wall archetypes, due to higher composite uncertainty.

9.4.8 Calculation of Ω_0 using Set of Archetype Designs

This section determines the value of the overstrength factor, Ω_0, which would be used in the design provisions for the "newly-proposed" wood light-frame system. Table 9-23 shows the calculated Ω values for each of the archetypes, with a range of values from 2.0 to 5.4. The average values for each performance group are 2.2, 3.4, 4.1, and 3.6, with the largest value of 4.1 being for the high aspect ratio walls in short-period buildings designed for low-seismic demands (PG-11).

According to Section 7.6, the largest possible $\Omega_0 = 3.0$ is warranted, due the average values being greater than 3.0 for three of the performance groups.

9.4.9 Summary Observations

This example shows that current seismic provisions for engineered wood light-frame construction included in ASCE/SEI 7-05 (with use of $R = 6$ rather than $R = 6.5$) to provide an acceptable level of collapse safety. Note that the collapse safety of actual engineering wood light-frame construction is most likely higher than calculated in this example because of the beneficial effects of interior and exterior wall finishes. In accordance with Section

4.2.3, wall finishes were not included in this example because they are not currently defined as part of the lateral structural system, and therefore are not governed by the seismic design provisions.

9.5 Example Applications - Summary Observations and Conclusions

9.5.1 Short-Period Structures

For both reinforced concrete special moment frame and wood light-frame systems, the short-period archetypes (e.g., $T < 0.6$ s for SDC D_{max} designs) were those that had the lowest level of collapse performance. For both of these systems, the short-period performance groups just meet acceptance criteria, and would fail if the Methodology acceptance criteria were made stricter. Thus, it is observed that short-period systems need additional strength (or some other modification that improves the performance) to achieve a level of collapse performance equivalent to systems with longer fundamental periods. This finding is not new, but rather has been reported in research, beginning with Newmark and Hall in 1973. Similar findings have since been reported by a large number of researchers based on analysis of single degree of freedom systems (e.g., Lai and Biggs, 1980; Elghadamsi and Mohraz, 1987; Riddell et al., 1989; Nassar and Krawinkler, 1991; Vidic et al., 1992; Miranda and Bertero, 1994), and for simple multiple degree of freedom systems (Takeda et al., 1998; Krawinkler and Zareian, 2007).

The example applications of this Chapter have verified that strength requirements should be higher for short-period systems, if consistent collapse performance is desired for all systems regardless of fundamental period. These strength requirements suggest the use of a period-dependent R factor as is proposed in many of the referenced papers and reports on this topic. Currently, the ASCE/SEI 7-05 document utilizes period-independent R factors. Future work should look more closely at the question of period-dependent R factors and whether or not they should be considered for use in future versions of ASCE/SEI 7-05.

9.5.2 Tall Moment Frame Structures

The reinforced concrete special moment frame system example in Section 9.2 found that perimeter frame buildings taller than 12-stories high designed on the basis of ASCE/SEI 7-05 do not meet the collapse performance objectives of this Methodology, with the collapse safety worsening with increasing building height. Tall buildings have more damage localization and higher P-delta effects, causing this observed trend in performance.

The issue of worsening collapse safety with increasing building height could be addressed in various ways. Larger column to beam strength ratios could be developed for taller buildings, more restrictive drift limits could be imposed, a period-dependent R factor could be used, or other approaches could be taken. In the example, minimum base shear requirement of ASCE 7-02 was reintroduced into the design requirements, successfully reversing the trends and creating increasing collapse safety with increasing building height.

This information was made available to the ASCE 7 Seismic Committee and a special code change proposal was passed in 2007 (Supplement No. 2), amending the minimum base shear requirements of ASCE/SEI 7-05 to correct this potential deficiency.

9.5.3 Collapse Performance for Different Seismic Design Categories

Example applications generally found lower collapse safety for buildings designed in seismic design categories with stronger ground motion intensity. For example, the $ACMR$ is typically lower for a building designed for SDC D_{max}, as compared to a building design for SDC D_{min}. This trend is primarily caused by the increasing effects of gravity loads for lower levels of seismic demand, which increases the overstrength of the structural system, and in turn increases the collapse capacity of the system.

This finding suggests that the R factor will be governed by the SDC with the strongest ground motion for which the system is proposed. Based on this observation, the Methodology requires that this SDC with the strongest ground motion be used when verifying the R factor. It is expected that such R factors will be conservative for other Seismic Design Categories, but this trend should be confirmed in the archetype investigation.

Chapter 10

Supporting Studies

This chapter describes additional studies performed in support of the development of the Methodology. These studies supplement the illustrative examples presented in Chapter 9, and serve to examine selected aspects of the Methodology as applied to different seismic-force-resisting systems.

10.1 General

Two supporting studies are presented. One study evaluates a 4-story steel special moment frame system. This study illustrates the use of component limit state checks to evaluate failure modes that are not explicitly simulated in the nonlinear dynamic analysis. It also demonstrates the application of the Methodology to steel moment frame systems.

A second study assesses the collapse performance of seismically-isolated systems. This study illustrates application of the Methodology to isolated structural systems, which have fundamentally different dynamic response characteristics, design requirements and collapse failure modes than those of conventional, fixed-base structures. This study also demonstrates the potential use of the Methodology as a tool for validating and improving current design requirements, in this case requirements for isolated structures.

10.2 Assessment of Non-Simulated Failure Modes in a Steel Special Moment Frame System

10.2.1 Overview and Approach

The purpose of this study is to illustrate how component limit state checks can be used to evaluate failure modes that are not explicitly simulated in the nonlinear dynamic analysis. This study follows the approach for non-simulated collapse modes described in Chapter 5.

The procedure for evaluating non-simulated collapse modes is illustrated through the evaluation of a steel special moment frame structure, designed using pre-qualified Reduced Beam Section (RBS) connection details in accordance with current design standards, ASCE/SEI 7-05 and ANSI/AISC 341-05 *Seismic Provisions for Structural Steel Buildings* (AISC, 2005). This study focuses on assessment of a single steel special moment frame building, which in concept could be one of many index archetype configurations serving to describe the archetype design space. In order to evaluate the entire class of steel special moment frames, the procedures applied to this individual building would be extended to the full set of index archetype models.

The primary collapse mechanism of this steel special moment frame occurs through hinging in the RBS regions of the beams and the columns, which can lead to sidesway collapse under large deformations. While gradual deterioration of the inelastic hinges associated with yielding and local buckling is simulated in the analyses, sudden strength and stiffness degradation associated with ductile fractures are not explicitly modeled. In this study, ductile fracture is not simulated because of software limitations.

The use of separate non-simulated limit state checks is supported by a number of related factors. First, through the use of pre-qualified RBS connections, the initiation of ductile fracture is unlikely to occur until large inelastic rotations have been reached and sidesway collapse has occurred or nearly occurred. Hence, the simplified limit state check for fracture is not expected to dominate the results. Second, available test data suggests that the location where ductile fracture may occur and the deformations at which ductile fracture may occur are highly variable, and simulation models would need to define correlations relating fracture probabilities at multiple connections. In this particular structure, the large columns tend to enforce equal rotations across a story. Therefore, even if ductile fracture were modeled, fractures would tend to form simultaneously across a given story and the collapse results obtained would be similar to those obtained using non-simulated limit state checks, unless correlations were explicitly incorporated in the analysis. There is limited data to support estimation of correlations.

In order to ensure that the collapse assessment process represents the behavior of the structural system of interest, the choice to incorporate a particular failure mode using a limit state check, in lieu of direct simulation, should be based on careful consideration of factors like those described above. Where non-simulated failure modes dominate the results, or where their exclusion jeopardizes simulation accuracy before the non-simulated limit state is reached, the appropriateness of the nonlinear model should be re-examined.

10.2.2 Structural System Information

The steel special moment frame archetype analyzed in this study is one of four perimeter moment frames that comprise the seismic-force-resisting system of a four-story building illustrated in Figure 10-1. The four-bay four-story frame provides lateral support to a floor area of 10,800 sq. ft. per floor and gravity support to a tributary area of 1,800 sq. ft. The seismic weight (mass) is equal to 940 kips on the second, third and fourth floors and 1,045 kips on the roof, for a total of 3,865 kips per frame.

The building is designed for a high seismic site located in Seismic Design Category (SDC) D, based on $T = 0.94$ seconds and a Maximum Considered Earthquake (MCE) spectral demand, S_{MT}, of 0.96 g (corresponding to SDC D_{max}). The structure has a design base shear, $V = 0.08\,W$. Designed in accordance with ASCE/SEI 7-05 and ANSI/AISC 341-05, beam sizes range from W24 to W30, and are governed by minimum stiffness requirements (drift limits). The RBS sections have 45% flange reduction. W24 columns are sized to satisfy the connection panel zone strength requirements without the use of web doubler plates. As such, they automatically satisfy other requirements, including the strong-column weak-beam (SCWB) requirement. As a result, the actual SCWB ratio is about 2.5 times larger than the required minimum. This large column overstrength reflects a possible design decision that is representative of current practice in California; however, it implies that this study will not necessarily demonstrate the lower-bound performance of code-conforming steel special moment frames.

Design requirements for this system are well-established, based on experience in past earthquakes, reflecting a high degree of confidence and completeness. For the purpose of assessing system uncertainty, the design requirements are rated (A) Superior.

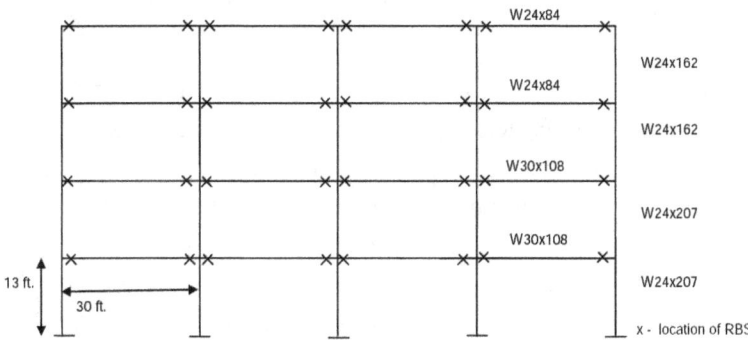

Figure 10-1   Index archetype model of 4-story steel special moment frame seismic-force-resisting system

10.2.3 Nonlinear Analysis Model

This structure is judged to have primarily two collapse modes: (1) sidesway collapse associated with beam and column hinging; and (2) collapse triggered by ductile fracture in one or more RBS connections. Nonlinear dynamic analyses were conducted using the OpenSees (OpenSees, 2006) software, employing elements with concentrated inelastic springs to capture flexural hinging in beams and columns and an inelastic (finite size) joint model for the beam-column panel zone.

Inelastic springs in beams and columns were modeled using the peak-oriented Ibarra element model (Ibarra et al., 2005), which can capture cyclic deterioration and in-cycle negative stiffness in elements as the structure collapses. The monotonic backbone used to model the W24x162 columns is illustrated in Figure 10-2.

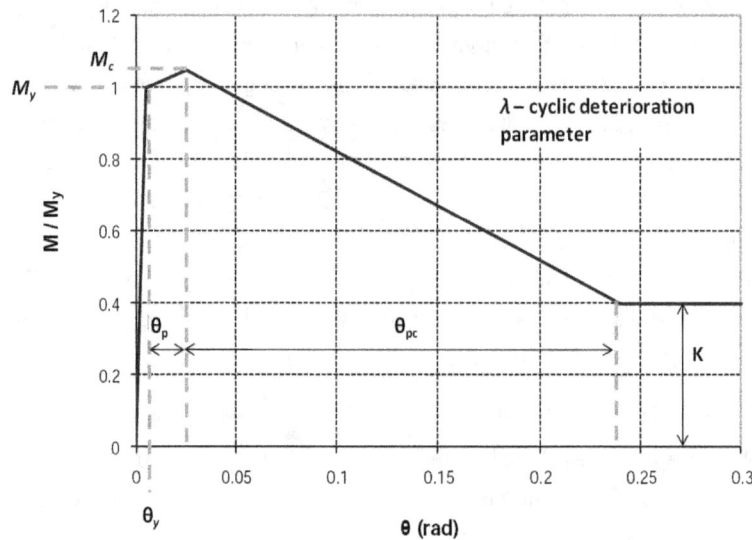

Figure 10-2 Monotonic backbone showing calibrated concentrated plasticity model for a typical column (W24x162).[1]

Model parameters for beams and columns (e.g., plastic rotation capacity, cyclic deterioration parameters) were calibrated to experimental test data (Lignos and Krawinkler, 2007) and reported in Table 10-1. For each beam and column, the model yield point is defined by the plastic moment capacity of the section, M_p, calculated with expected values for the steel yield strength, i.e., 1.1 x F_y. Column initial rotational stiffness is based on Young's modulus for steel and the cross-sectional stiffness. Calculation of beam stiffness includes the contribution of the composite floor slab, and the

---

[1]For more information, model parameters are defined in Appendix E. The discussion in Appendix E deals with the same Ibarra model, but as applied to modeling of reinforced concrete beams and columns.

additional flexibility in beams due to the RBS section is neglected. RBS sections are modeled as adjacent to the panel zone, neglecting offsets from the column face. Beam properties are modeled as asymmetric, depending on the loading direction.

Table 10-1   Model Parameters for Column and Beam Plastic Hinges in 4-Story Steel Special Moment Frame

Section	My (kip-in)	Mc/My +(-)	θp +(-)	θpc +(-)	λ	K
W24x162	2.5 x104	1.05	0.025	0.35	330	0.4
W24x207	3.3 x104	1.05	0.03	0.30	440	0.4
W24x84	8.2 x103	1.1(1.05)	0.025(0.020)	0.17	380	0.4
W30x108	1.3 x104	1.1(1.05)	0.022(0.016)	0.15	260	0.5

As described previously, these backbones do not predict ductile fracture of RBS sections, so that failure mode is evaluated through a non-simulated limit state check. The joint panel zone yield point and hardening parameters are based on Equation 9-1 in ANSI/AISC 341-05 and Krawinkler (1971, 1978). The panel zone spring is modeled as non-deteriorating with a bilinear kinematic hardening model.

Other modeling assumptions are consistent with the requirements of Chapter 5 and Chapter 6. Expected dead and live loads $(1.05D + 0.25L)$ are applied to the structure and used in the computation of the seismic mass. The contribution of the gravity frame is not included in the analysis model, though the leaning P-delta column accounts for gravity loads not tributary to the seismic-force-resisting system. Foundation flexibility is neglected and the foundation is modeled as a fixed-base.

The model is rated (B) Good in accordance with Table 5-3. The model is rated as to how well it captures the behavior of the system up to the point at which the non-simulated collapse mode (ductile fracture) occurs. This rating is the same as that given to the reinforced concrete special moment frame models of Chapter 9, using much of the same rationale. The model is judged to have a high degree of accuracy and robustness, but does not account for effects of overturning on column behavior.

The available test data is also rated (B) Good in accordance with Table 3-2. There is significant test data for steel columns, which has been conducted by a number of different researchers. However, as with the reinforced concrete component test data certain critical configurations are missing, such as tests of steel beams with reinforced concrete slabs.

10.2.4 Procedure for Collapse Performance Assessment, Incorporating Non-Simulated Failure Modes

Accounting for non-simulated failure modes in assessment of collapse margin ratio first requires the identification and calibration of appropriate limit-state models. These limit state models are used to evaluate analysis results to see if the non-simulated limit state was exceeded. The *CMR* is then computed to account for both simulated and non-simulated failure modes, and adjusted for spectral shape effects with the *SSF*. The resulting *ACMR* is compared to the acceptance criteria in Chapter 7.

Identify Non-Simulated Collapse Modes

Properties of beam-column plastic hinges in the analysis model for steel special moment frames are calibrated to predict hinging and gradual deterioration associated with yielding and local buckling. However, experimental data (e.g., Engelhardt et al., 1998; Ricles et al., 2004; Lignos and Krawinkler, 2007) suggest that the steel frame may also experience ductile fracture in RBS sections, or possibly at the joint between the beam and column. For example, in testing done as part of the SAC Steel Project, Engelhardt et al. (1998) reported fractures in qualifying connections at inelastic rotations between 0.05 and 0.07 radians. It should be emphasized that the fracture being considered here is triggered by ductile crack initiation, and occurs after significant inelastic yielding has occurred, in contrast to the connection fractures observed in steel frame structures during the 1994 Northridge earthquake.

Ductile-fracture-induced collapse is conservatively assumed to take place if ductile fracture occurs in any RBS. This collapse limit state is chosen because the steel special moment frame model is able to capture critical aspects of system behavior until the point at which ductile fracture occurs in the first RBS. If the analysis was continued after a ductile fracture is experienced, the fidelity of the results would be in doubt. Since the strong column sidesway mechanism in this frame imposes similar peak rotations in all the RBS hinges, the assumption of equating the first instance of a fractured connection with fracture-induced collapse may not be too unreasonable (overly conservative) for this particular example.

Generally, it would be desirable to incorporate fracture deterioration directly in the nonlinear model. Since it is not directly incorporated in this analysis, a conservative judgment is made about what constitutes collapse for the fracture limit state. A typical representation of the deformations at fracture is shown in Figure 10-3.

Develop Component Fragility

Calculation of the non-simulated collapse mode requires the definition of a fracture fragility function. In this case, the fragility function relates the probability of ductile fracture in the RBS hinge to a response parameter such as the maximum plastic hinge rotation that has occurred in the plastic hinge. Figure 10-4 shows the resulting fragility function, $P[\text{Fracture}|\theta_p]$, which is based on available data (Lignos and Krawinkler, 2007) and engineering judgment.

Figure 10-3   Illustration of fracture behavior (Engelhardt et al., 1998).

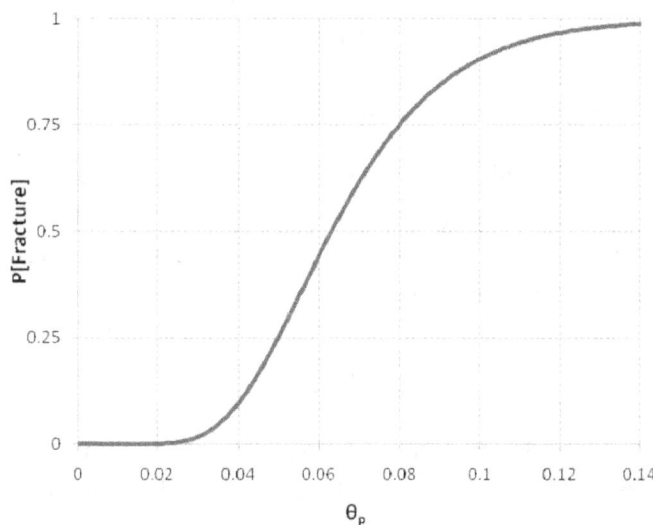

Figure 10-4   Component fragility function, describing probability of ductile fracture occurring as a function of the plastic rotation, θ_p.

The fragility function for fracture is assumed to follow a lognormal distribution, and has a median capacity of $\hat{\theta}_p = 0.063$ radians and a logarithmic standard deviation of $\beta_F = 0.35$. The dispersion, β_F, reflects both test data statistics (from 10 tests) and judgment as to the additional variability that may be encountered in actual buildings. Assuming that the parameters associated with fracture are the same throughout the building, the fragility function is applicable to every RBS in the building.

Identify Collapse Limit based on Component Fragility

The collapse limit point for the non-simulated collapse mode is defined by the median value of the fragility function associated with component failure. Therefore, if the plastic rotation in a RBS exceeds the median of 0.063 radians, that component is assumed to be fractured and the non-simulated limit state is triggered.

Note that the collapse limit state in this formulation ignores the effect of dispersion, β_F, in the collapse fragility. The magnitude of the impact of β_F on the total collapse uncertainty, β_{TOT}, depends on the relative dominance of simulated (sidesway) and non-simulated failure modes. However, the other sources of uncertainty considered in the Methodology (e.g., β_{RTR}, β_{DR}, β_{TD}, β_{MDL}) dominate the uncertainty in the collapse fragility. For explanation of a rigorous approach accounting for the effects of β_F, see Aslani (2005).

Nonlinear Static (Pushover) Analysis

The static pushover analysis results for the steel special moment frame are illustrated in Figure 10-5. The maximum base shear, V_{max}, is 1170 kips, compared to a design base shear of approximately 345 kips, for an overstrength, $\Omega = 3.4$. This large overstrength relative to the design lateral forces is due to the very strong columns sized to avoid doubler plates.

The period-based ductility, given by $\mu_T = \delta_u/\delta_{y.eff}$, is obtained from the pushover analysis as described in Section 6.3. The ultimate roof displacement, δ_u, is taken as either the roof displacement corresponding to a 20% loss in base shear, or the roof displacement at which the non-simulated (ductile fracture) failure mode occurs. Since the strong columns in this structure impose approximately uniform distribution of story drift, the roof displacement at which the non-simulated (fracture) failure mode occurs is approximately $0.063 h_r$, corresponding to a roof displacement of 39.3 inches. For comparison, δ_u associated with a 20% loss in base shear from pushover analysis, is approximately $0.075 h_r$, corresponding to a roof displacement of approximately 46.8 inches. The effective yield displacement, $\delta_{y.eff}$, is computed from Equation 6-7 as $0.009 h_r$ or 5.6 inches, for this structure. Therefore, μ_T is 7.0.

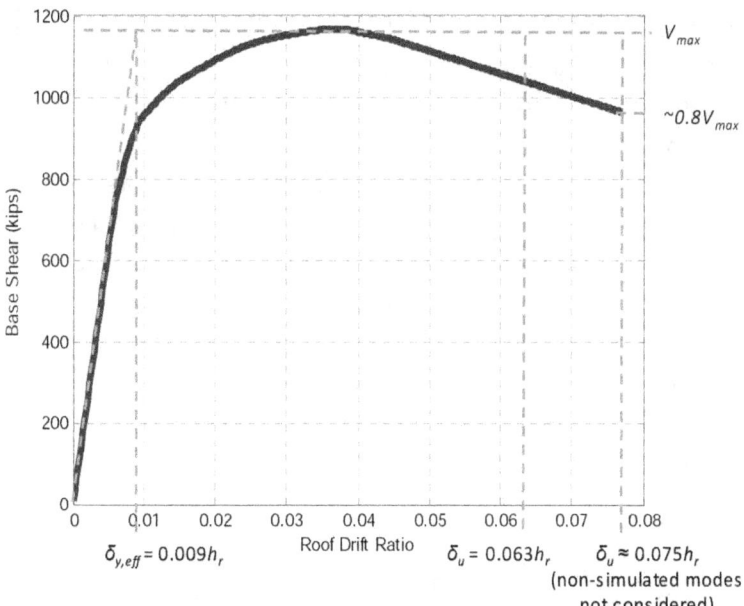

Figure 10-5   Results of nonlinear static (pushover) analysis of a steel special
                moment frame, illustrating computation of period-based
                ductility from non-simulated collapse modes.

Assess Collapse Performance

To assess collapse performance, response of the 4-story steel special moment
frame is calculated using the Far-Field record set and nonlinear dynamic
analysis, as described in Chapters 5 through 7. If only simulated collapse
modes are considered, a median collapse capacity, $\hat{S}_{CT}$, of 2.36 g is obtained.
Due to the strength of the columns relative to the beams, the sidesway
collapse mode is a full four-story mechanism with hinges at all RBS regions
and at the fixed column bases for all ground motions.

The collapse fragility is computed based on both simulated and non-
simulated collapse modes. This procedure is illustrated graphically in Figure
10-6, using representative curves from incremental dynamic analysis for this
structure (each curve contains the dynamic analysis results for one ground
motion, scaled until collapse). The non-simulated fracture mode occurs if the
plastic rotations at any RBS in the building exceed the median capacity
defined by the component fragility function. In the figure, sidesway collapse
points based on results from simulation, $S_{CT(SC)}$, are shown with short bold
pointers. Non-simulated (fracture-induced) collapse points, $S_{CT(NSC)}$, are
shown with pointers from hatched circles corresponding to maximum story
drift at fracture ($\hat{\theta}_p$ = 0.063 radians).

For the purposes of this figure, the component collapse limit of $\hat{\theta}_p$ = 0.063 is
shown as approximately equal to the story drift ratio. This assumption is

made for illustration purposes only, and is not actually used in computing the occurrence of non-simulated failure modes. Of the three curves shown in Figure 10-6, the lowest reaches simulated sidesway collapse and the non-simulated collapse limit state at approximately the same level of ground motion intensity. The other curves reach the non-simulated collapse point under less intense ground motions than suggested by sidesway collapse. Investigation of the occurrence of simulated and non-simulated failure modes is repeated for all ground motions. In computing the collapse fragility, the more critical of these two limit states is taken as the governing collapse point for each ground motion record.

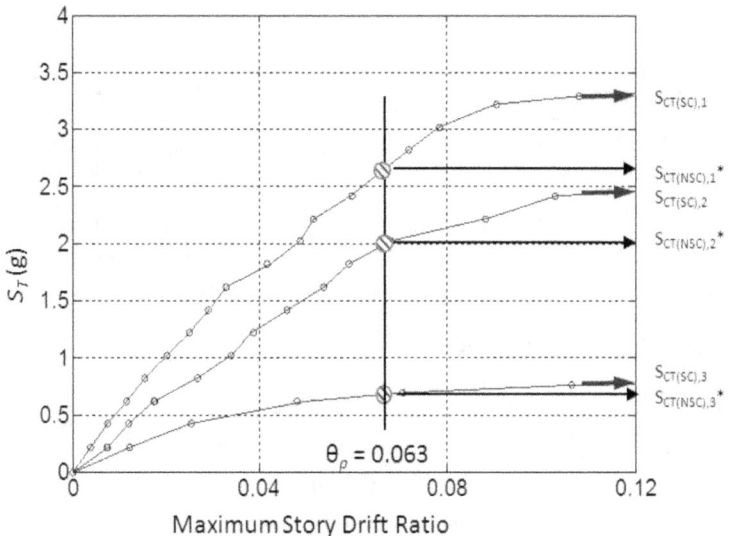

Figure 10-6   Selected simulation results for the steel special moment frame illustrating the identification of non-simulated collapse modes. The governing collapse point for each ground motion record is identified with an asterisk (*).

Given ratings of (B) Good for modeling, (A) Superior for design requirements, and (B) Good for test data, the total system collapse uncertainty is $\beta_{TOT} = 0.500$. Since $\mu_T \geq 3$ for this building, this value includes record-to-record uncertainty, $\beta_{RTR} = 0.40$, in accordance with Section 7.3.4. It is noted that for systems driven by very brittle non-simulated collapse modes, β_{RTR} can potentially be reduced in accordance with Equation 7-2. This reduction should be exercised with caution. The approach to non-simulated failure modes is based only on the median component limit state, neglecting the underlying uncertainty associated with the occurrence of the component failure mode, β_F. If record-to-record variability is significantly reduced with Equation 7-2, the total uncertainty in the collapse fragility could be too low, and non-conservative.

The combined collapse fragility is illustrated in Figure 10-7. In this figure, the horizontal axis fragility parameter, S_{CT}, is normalized by MCE demand, S_{MT} to permit direct comparison of the collapse margin ratio, CMR, for the structure with and without consideration of non-simulated fracture-induced failure modes. For this structure, the net result of including the fracture-induced collapse is a 32% reduction in the collapse margin ratio, CMR, from 2.5 for the simulated sidesway-only case, to 1.9. The conditional probability of collapse at the MCE increases from 8% to 14%. (Note: these margins and the collapse probabilities do not include the spectral shape factor, which is considered in the evaluation of acceptance criteria.)

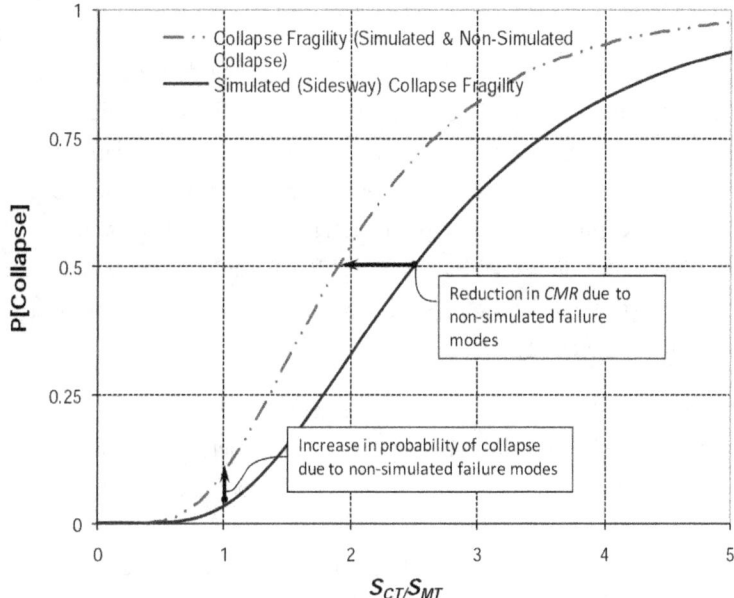

Figure 10-7   Comparison of steel special moment frame collapse fragilities for sidesway-only and combined simulated and non-simulated (sidesway and fracture-induced) collapse (not adjusted for spectral shape).

Compare to Acceptance Criteria

Finally, the combined fragility data, reflecting the likelihood of both simulated and non-simulated collapse, should be compared to the acceptance criteria in Chapter 7. For this structure, the spectral shape factor, SSF, of 1.41 is determined from Table 7-1b with $T = 0.94$ seconds and $\delta_u/\delta_{y,eff} = 7.0$, which gives an $ACMR$ of 2.7. The acceptable $ACMR$ is obtained from Table 7-3, with $\beta_{TOT} = 0.500$. The $ACMR$ is compared to an acceptable collapse margin ratio, $ACMR_{10\%}$, of 1.90 (based on the 10% probability of collapse limit), and easily satisfies the acceptance criteria.

If the Methodology were applied to a complete set of steel special moment frames, including different performance groups for high and low gravity loads, maximum and minimum seismic criteria and short-period and long period systems, each individual archetype would be compared to $ACMR_{20\%}$, and the average of each performance group would be compared to $ACMR_{10\%}$ from Table 7-3.

10.3 Collapse Evaluation of Seismically Isolated Structures

10.3.1 Introduction

Seismic isolation (commonly known as base isolation) is a technology that is intended to protect facility function and provide substantially greater damage control than conventional, fixed-base, structures for moderate and strong earthquake ground motions. For extreme ground motions, seismically isolated structures are expected to be at least as safe against collapse as their conventional counterparts. To ensure adequate performance, ASCE/SEI 7-05 requires explicit evaluation of the design of every isolated structure under Maximum Considered Earthquake (MCE) ground motions and comprehensive testing of prototype isolator units to verify design properties and demonstrate stability under MCE loads.

The provisions of ASCE/SEI 7-05 require the seismic force-resisting system of the structure above the isolation system (superstructure) to be designed for response modification factors, R_I, that are smaller than the R factors permitted for conventional structures. Reduced values of the response modification factor are intended to keep the superstructure "essentially elastic" for design earthquake ground motions. To protect against potential brittle failure for extreme ground motions, ASCE/SEI 7-05 requires the superstructure to have the same ductile capacity as that required for a conventional structure of the same type in the seismic design category of interest. The provisions of ASCE/SEI 7-05 are generally considered to be conservative with respect to design of the superstructure, although the degree of conservatism, if any, is not known.

Objectives

This study is intended to demonstrate the application of the Methodology to isolated structures, which have fundamentally different dynamic response characteristics, performance properties and collapse failure modes than those of conventional, fixed-base structures. Special issues include the following:

Period Definition. The fundamental-mode "effective" period of an isolated structure is based on secant stiffness at the response amplitude of interest, rather than "elastic" stiffness used to define the period, T, of conventional structures.

Record-to-Record (RTR) Variability. Record-to-record variability may be smaller than 0.4 because base isolated structures typically do not undergo large period elongation before collapse. Are the recommendations for reduced record-to-record (RTR) variability in Chapter 7 suitable for isolated structures?

Test Data and Modeling Uncertainty. Can the collapse margin ratio (*CMR*) of isolated structures be evaluated using the same uncertainty associated with test data and modeling of the superstructure as considered for a conventional structure with the same type of seismic-force-resisting system?

Spectral Shape Factor (SSF). Can the spectral shape factor (*SSF*) used to adjust the *CMR* of isolated structures be calculated using the same methods as those specified for conventional structures (Appendix B)?

This study is also intended to illustrate how the Methodology can be used as a tool for assessing the validity of current design requirements, in this case Chapter 17 of ASCE/SEI 7-05, and to develop improved code provisions. In order to evaluate the design requirements for isolated structures, this study specifically explores the sensitivity of collapse performance to the following key design properties of isolated structures:

Superstructure Strength. How does collapse performance vary for superstructures that have different design strength levels (e.g., superstructures designed for different effective values of the R_I factor)? Effective R_I values yielding higher and lower design base shears than the code-specified values are considered.

Superstructure Ductility. How does superstructure ductility influence collapse performance (e.g., performance of special moment frame superstructures as compared to that of ordinary moment frame superstructures)?

Moat Wall Clearance. How does collapse performance vary for isolated structures that have different amounts of clearance between the isolated structure and the moat wall?

Scope and Approach

The scope of this study is necessarily limited and relies on archetypical models available from other examples developed in this project to represent the superstructures of isolated archetypes. Specifically, superstructures are based on the 2-dimensional, archetypical models of 4-story reinforced concrete special moment frame and ordinary moment frame systems, from the example applications included in Chapter 9. Design of isolator systems in this study varies from typical isolator design in that isolator properties are designed to satisfy ASCE/SEI 7-05 design requirements between the isolation system and the superstructure, given the reinforced concrete frame superstructures developed in the Chapter 9 examples.

The archetypical models of isolated structures incorporate force-deflection properties of isolation systems typical of actual projects that use either (1) elastomeric, rubber bearings (RB), or (2) sliding, friction-pendulum (FP) bearings. These two isolation system types are designed and archetypical models of the isolated structure evaluated for maximum and minimum SDC D ground motions (SDC D_{max} and SDC D_{min}). SDC D_{max} ground motions are typical of those used for most seismic isolation projects (e.g., projects in high seismic regions of coastal California).

This study includes discussion of background information necessary for proper application of the Methodology to isolated structures and related development of isolated archetypes. Archetype configurations, nonlinear analysis techniques, and collapse performance methods are described with reference to specific differences in applying the methodology to isolated structures. Collapse evaluation results are reported for archetypes that comply with the design requirements of ASCE/SEI 7-05 (referred to herein as Code-Conforming archetypes) and for archetypes that deviate from current requirements (Non-Code-Conforming archetypes). The Non-Code-Conforming archetypes demonstrate potential use of the Methodology as a tool for code development, by evaluating collapse performance for archetype models that have weaker (or stronger) superstructures, less ductility, or different moat clearances than those specified in ASCE/SEI 7-05.

10.3.2 Isolator and Structural System Information

Archetypes of isolated structures must be designed using established design requirements, and modeling of isolated archetypes must be supported by appropriate test data (Chapter 3). Further, the quality of the design requirements and test data must be rated for establishing system collapse uncertainty (Chapter 7).

Design Requirements

Archetypes of isolated structures are designed according to the provisions of ASCE/SEI 7-05 and related superstructure design codes, including ACI 318-05, except as some specific provisions are modified, or ignored, to evaluate the effects of reduced superstructure strength, or limited ductility, or moat clearance in the Non-Code-Conforming archetypes. Chapter 17 of ASCE/SEI 7-05 requires thorough and rigorous design of the isolated structure, including explicit evaluation of the isolation system for MCE ground motions, and peer review.

For the purpose of assessing the composite uncertainty in the Methodology, the isolation system and superstructure design requirements are rated as (A) Superior, as they are thorough, detailed and vetted through the building code process.

Test Data

The requirements for test data relate both to testing of superstructure components (i.e., reinforced concrete beams, columns and connections), and testing of prototype isolator units. The test data related to reinforced concrete elements is rated (B) Good, as discussed in the Chapter 9 examples. While there is a large amount of test data on reinforced concrete components, there are still several areas where test data are not complete (i.e. tests of beams with slabs, tests to very large deformations).

Test data related to isolation units is both qualitatively and quantitatively different. For the purposes of modeling conventional structures, test data is taken from a variety of different researchers regarding components which are similar, but not identical, to the components being modeled in the structure. In contrast, the provisions ASCE/SEI 7-05 require prototype testing of isolator units for the purpose of establishing and validating the design properties of the isolation system and verifying stability for MCE response. These tests are specific to the isolation system installed in a particular building, and follow detailed requirements for force-deflection response outlined in the design requirements. As a result, there is substantially smaller uncertainty related to the test data in an isolated system than the superstructure.

A rating of (B) Good is assigned to the uncertainty assessment for test data for these systems. This rating is associated with uncertainty in test data related to superstructure modeling and potential isolator failure modes at loads and displacements greater than those required for isolator prototype testing by Section 17.8 of ASCE/SEI 7-05.

10.3.3 Modeling Isolated Structure Archetypes

This section identifies the basic configuration, systems and elements of the isolated structures of this study and provides a general overview of the index archetype models used to evaluate collapse performance of these systems. Section 10.3.4 describes specific design properties for the isolation system, moat wall clearance, and the superstructure of each model.

The seismic-force-resisting system of an isolated structure includes: (1) the isolation system; and (2) the seismic-force-resisting system of the superstructure above the isolation system. The isolation system includes individual isolator units (e.g., elastomeric or sliding bearings), structural elements that transfer seismic force between elements of the isolation system (e.g., beams just above isolators) and connections to other structural elements. Energy dissipation devices (dampers) are sometimes used to supplement damping of isolator units, but such devices are not considered in this study.

Isolated structures are typically low-rise or mid-rise buildings that have relatively stiff superstructures. This study assumes that the seismic-force-resisting system of the superstructure is a reinforced-concrete moment frame system, in order to make use of already developed models for Chapter 9 that explicitly capture sidesway collapse. A relatively short height (4 stories) is used to assure adequate stiffness of the superstructure.

Isolated structures typically have a "moat" around all or part of the perimeter of the building at the ground floor level. The moat is usually covered by architectural components (e.g., cover plates) that permit access to the building, but do not inhibit lateral earthquake displacement of the isolated structure.

Impact with the moat wall can cause collapse. Accordingly, Section 17.2.5.2 of ASCE/SEI 7-05 requires a minimum building separation to "retaining walls and other fixed obstructions" not less than the total maximum (MCE) displacement of the isolation system. The intent of this provision is to limit the likelihood of such impacts, even for strong ground motions. However, Section 17.2.4.5 of ASCE/SEI 7-05 recognizes that providing clearance for MCE ground motions may not be practical for some systems and permits isolation system design to incorporate a "displacement restraint" that would limit MCE displacement, provided certain criteria are met. These criteria include the requirement that "the structure above the isolation system is checked for stability and ductility demand of the maximum considered earthquake." This study shows how the Methodology can be used to perform this check.

Index Archetype Models

Models of isolated systems consist of a superstructure model, an isolator model, and a moat wall model. Each of these components must be capable of capturing the inelastic effects in the structure up until the point at which the structure collapses. Analysis models are two-dimensional, and neglect possible torsional effects. A schematic diagram of the index archetype model for isolated systems is illustrated in Figure 10-8.

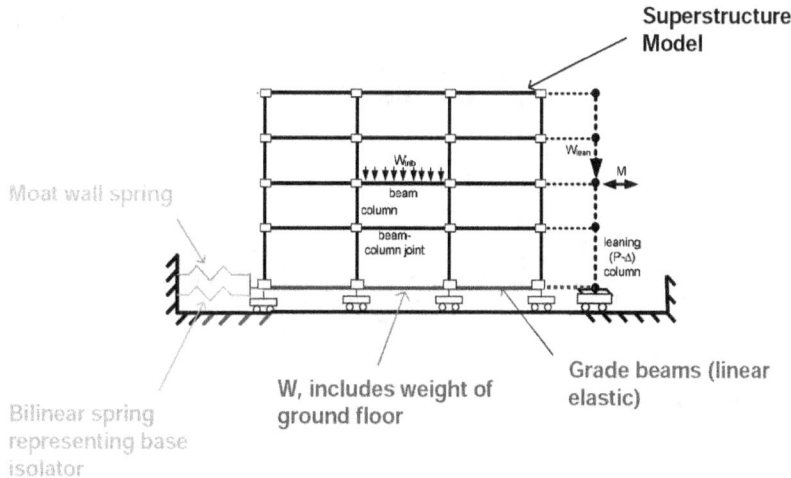

Figure 10-8   Index archetype model for isolated systems.

The superstructure is modeled using the same assumptions as described in the Chapter 9 examples. The model incorporates material nonlinearities in beams, columns and beam-column joints, as well as deterioration of strength and stiffness as the structure becomes damaged. As before, the superstructure model includes a leaning (P-delta) column to account for the effect of the seismic mass on the gravity system.

Isolator Modeling

The isolation system bearings (isolators) are modeled using a bilinear spring between the foundation and the ground-floor of the superstructure, as shown in Figure 10-8. The bilinear spring is assumed to be a non-degrading, fully hysteretic element. Some isolators exhibit significant changes in properties with repeated cycles of loading as a result of heating, but explicit modeling of such behavior is currently in its infancy and is beyond the scope of this study. Bilinear springs are commonly used in practice and provide sufficiently accurate estimates of nonlinear response when their stiffness and damping properties are selected to match those of the isolators (Kircher, 2006).

Figure 10-9 illustrates the modeled force-displacement response of nominal, upper-bound and lower-bound bilinear springs used to represent isolators in this study. In general, isolators are modeled with nominal spring properties. However, in certain cases, isolators are modeled with upper-bound and lower-bound spring properties to evaluate the effects of these properties on collapse. Upper-bound and lower-bound properties are also needed for isolation system design. Section 17.5 of ASCE/SEI 7-05 requires lower-bound properties for calculation of isolation system displacements, and upper-bound properties for calculation of design forces. The range of upper-bound and lower-bound bilinear spring properties used to model isolators is based on the prototype testing acceptance criteria of Section 17.8.4 of ASCE/SEI 7-05.

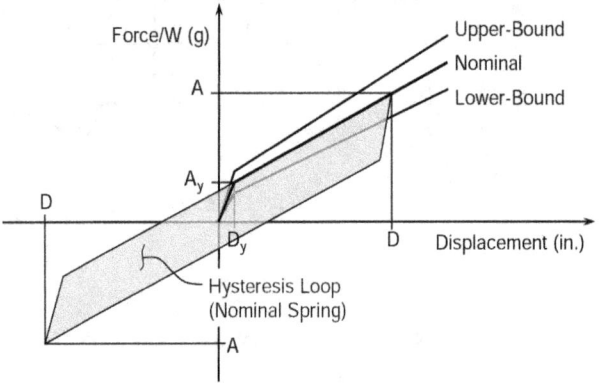

Figure 10-9   Example nominal, upper-bound and lower-bound bilinear springs and hysteretic properties used to model the isolation system.

Design of the isolation system is defined by the two control points, the "yield" point (D_y, A_y) and the post-yield point (D, A), located somewhere on the yielded portion of the curve. The resulting bilinear response is illustrated in Figure 10-9. The yield point represents the dynamic friction level of sliding bearings (e.g., FP bearings) and is closely related to the (normalized) characteristic strength of elastomeric bearings (e.g., lead-rubber or high-damping rubber bearings). The post-yield point is used simply to define the slope of yielded system (and the isolation system is assumed capable of displacing without failure beyond this point). The properties are assumed to be symmetric for positive and negative displacements.

Amplitude-dependent values of effective stiffness, k_{eff}, effective period, T_{eff} (in seconds), and effective damping, β_{eff}, of the isolation system may be calculated by the following equations:

$$k_{eff} = \frac{AW}{D} \qquad (10\text{-}1)$$

$$T_{eff} = 2\pi \sqrt{\frac{D}{gA}} \qquad (10\text{-}2)$$

$$\beta_{eff} = \frac{2}{\pi} \sqrt{\frac{A_y D - D_y A}{DA}} \qquad (10\text{-}3)$$

Equations 10-1, 10-2, and 10-3 are consistent with the definitions of effective stiffness, k_{eff}, and effective damping, β_{eff}, of Section 17.8.5 of ASCE/SEI 7-05. In Section 17.8.5, these equations are used to determine the force-deflection characteristics of the isolation system from the tests of prototype isolators.

ASCE/SEI 7-05 defines two amplitude-dependent fundamental-mode periods for isolated structures based on secant stiffness, T_D, (effective period at the design displacement), and T_M, (effective period at the MCE displacement). Values of T_D and T_M are typically close together. Isolated systems are initially stiff (see Figure 10-9), but yield early and become very flexible, so this representation of period is different from conventional fixed-base structures, which use an estimation of initial stiffness for an "elastic" period.

The MCE fundamental-mode period, T_M, as defined by ASCE/SEI 7-05, is used for evaluation of seismic collapse performance of isolated structures in this study:

$$T_M = 2\pi \sqrt{\frac{W}{k_{Mmin} g}} \qquad (10\text{-}4)$$

where:

T_M = effective period, in seconds, of the seismically isolated structure at the maximum displacement in the direction under consideration, as prescribed by Equation 17.5-4 of ASCE/SEI 7-05,

W = effective seismic weight of the structure above the isolation interface, as defined in Section 17.5.3.4 of ASCE/SEI 7-05, and

k_{Mmin} = minimum effective stiffness, in kips/in, of the isolation system at the maximum displacement in the horizontal direction under consideration, as prescribed by Equation 17.8-6 of ASCE/SEI 7-05.

Moat Wall Modeling

Nonlinear springs are used to represent the effects of impact with the moat wall when the seismic demand on the isolated system exceeds the clearance provided. The moat wall is represented by 5 symmetrical gap springs implemented in parallel as illustrated in Figure 10-10.

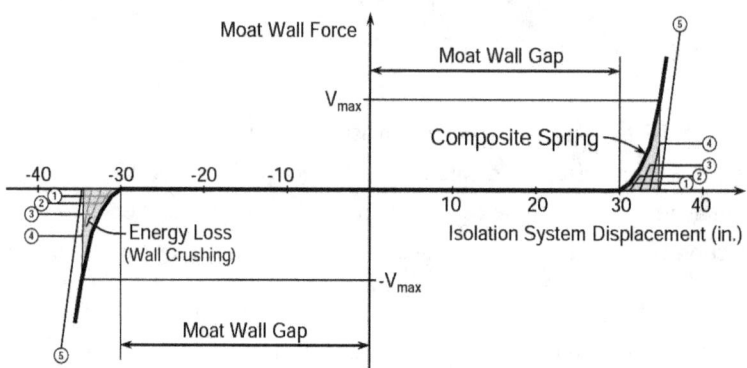

Figure 10-10   Five individual springs and effective composite spring used to model moat wall resistance for an example 30-inch moat wall gap.

Gap springs have zero force until the isolated structure reaches the moat wall, and then begin to resist further displacement of the isolated structure. The five springs engage sequentially to effect increasing stiffness and nonlinear resistance as the structure pushes into the moat wall. Gap springs are modeled as inelastic elements to account for energy loss due to localized crushing at the structure-wall interface. Gap spring properties are defined relative to the strength of the superstructure such that moat wall force is equal to the strength of the superstructure, V_{max}, at approximately 4 inches of moat wall displacement. (Note. A similar result could have been obtained using a more complicated force-displacement relationship and a single gap spring, depending on available models in software.)

Model Limitations

It should be noted that the models used in this study, and illustrated in Figure 10-9 and Figure 10-10, are relatively simplistic representations of isolated systems and neglect many complex aspects of isolator performance. Bilinear properties of the isolator neglect stiffening that may occur at large displacements, and reduction in characteristic strength that can occur for repeated cycles of loading. The isolation system model also neglects potential uplift at isolators (when overturning loads exceed gravity loads and isolator tension capacity). Local uplift of isolators is a potential failure mode, but permitted by Section 17.2.4.7 of ASCE/SEI 7-05 provided that "resulting deflections do not cause overstress or instability of isolator units or other structure elements." Softening and damping loss due to heat effects in elastomeric bearings are not considered. Despite these limitations, the models represented in Figures 10-8, 10-9, and 10-10 are expected to give reasonable predictions of dynamic response and collapse performance, as well as relative differences in performance associated with variability in design parameters.

Uncertainty due to Model Quality

Model quality is rated as (B) Good for the purpose of assessing the composite uncertainty in the performance predictions of index archetype models with a reinforced concrete special moment frame superstructure, and (C) Fair for index archetype models with a reinforced concrete ordinary moment frame superstructure. These model quality ratings are the same as those assigned to the reinforced concrete frame examples in Chapter 9.

Since design of isolated structures require a nonlinear, building-specific model of the superstructure, the model of an isolated structure represents an individual building rather than one archetype within a performance group, as in the Chapter 9 examples. In addition, the isolation system filters out some higher-mode effects that contribute to collapse variability in fixed-base structures. Therefore, it may be possible for an isolated system to earn a better model rating than the constituent superstructure. However, in this study, the simplified modeling of isolators, such as neglecting cyclic changes to bearing properties, does not warrant better model quality ratings.

10.3.4 Design Properties of Isolated Structure Archetypes

Specific design properties of index archetype models in this study include isolation system properties, moat wall clearance and superstructure properties. The archetypes are selected to probe the effects of the critical design parameters on collapse performance. The goal of considering a wide variety of archetype configurations is to assess the validity of current code provisions. As such, archetypes are developed for both "Code-Conforming" systems that comply with ASCE/SEI 7-05 requirements, and "Non-Code-Conforming" systems that deviate from ASCE/SEI 7-05 requirements in terms of superstructure strength or ductility.

Isolation System Design Properties

Design properties of the isolation system are developed using the equations and design requirements of the Equivalent Lateral Force Procedure, Section 17.5 of ASCE/SEI 7-05. These equations provide a convenient basis for design and are commonly used for preliminary design and review of isolated structures.

This study considers archetype isolation systems that are representative of systems with either elastomeric rubber bearings (RB) or sliding friction pendulum (FP) bearings. Isolation systems are designed for either SDC D_{max} or SDC D_{min} seismic criteria (Site Class D). The response characteristics of isolation systems with either RB or FP bearings are sufficiently similar for strong (SDC D_{max}) ground motions to permit modeling both systems with the same set of bilinear springs properties (i.e., a single set of "generic" properties is used to represent both systems). Such is not the case for moderate (SDC D_{min}) ground motions, and different spring properties are used to model isolation systems with RB and FP bearings.

Nominal isolation system design properties for SDC D_{max} are given in Table 10-2 for the generic (GEN) system, and the RB system and FP system for SDC D_{min}. Upper and lower-bound spring properties are also given for the generic system. Only nominal properties are shown for the RB and FP systems in SDC D_{min} (although MCE design parameters still utilize upper-bound and lower-bound properties, as required by Section 17.5 of ASCE/SEI 7-05).

Table 10-2 also provides design values of the effective period, T_M, effective damping, β_M, and total maximum displacement, D_{TM}, for each system. For each archetype isolation system, design values of the yield and the post-yield control points are selected such that the corresponding values of maximum (MCE) displacement, D_M, effective period, T_M, effective stiffness, and effective damping, β_M, meet the following criteria:

- Values of D_M, T_M and β_M comply fully with the equations and requirements of Section 17.5 of ASCE/SEI 7-05.

- Values of effective stiffness and damping are consistent with actual isolation system properties (i.e. for isolations systems with either elastomeric or sliding bearings)[2].

- Values of effective stiffness and damping in the isolator reduce response such that forces required for design of the superstructure are approximately the same as the design base shear required for a conventional fixed-base system of the same type and configuration, so that models of code-compliant systems from Chapter 9 can be used for the superstructure (without re-design).

---

[2] The effective stiffness properties of rubber bearings represent a "low modulus" rubber compound and are assumed to be the same as those of friction pendulum bearings to limit the number of models. The corresponding effective period, T_M, is somewhat atypical of rubber bearing systems which generally have an effective period less than or equal to 3 seconds.

Calculation of D_{TM} was based on the Equivalent Lateral Procedure of Section 17.5 of ASCE/SEI 7-05. Total maximum displacement includes an additional 15 percent of torsional displacement, consistent with Equation 17.5-6 of ASCE/SEI 7-05 (assuming a square configuration of the building in plan):

$$D_{TM} = 1.15 D_M \tag{10-5}$$

Table 10-2   Isolation System Design Properties

Isolator Properties		Force-Deflection Curve				MCE Design Parameters		
		Yield Point		Post-Yield		T_M (sec)	β_M (% crit.)	D_{TM} (in.)
Type	Range	D_y (in)	A_y (g)	D (in)	A (g)			
Generic Elastomeric or Sliding Systems - D_{max} Designs								
GEN	Nominal	0.5	0.05	23.3	0.225			
GEN-UB	Upper-Bound	0.5	0.06	21.1	0.23	3.47	10.5%	29.3
GEN-LB	Lower-Bound	0.5	0.04	25.5	0.22			
Rubber (RB) or Friction Pendulum (FP) Systems - D_{min} Designs								
RB	Nominal	1.5	0.04	6.5	0.081	3.18	12.5%	8.5
FP	Nominal	0.1	0.04	4.2	0.072	2.78	28%	5.7

Isolation System Clearance

The performance of isolated structures may also be dependent on the clearance of the isolated system and Table 10-3 summarizes the moat wall clearance (gap) distances used in the archetype isolation systems. For the generic (GEN) isolation system (SDC D_{max} design), five different moat wall distances are used to investigate the effects of this parameter on collapse performance.

Moat wall clearance is based on a fraction of the total maximum displacement, D_{TM}, plus a little extra displacement for fit-up tolerance. Unless dynamic analysis can justify a smaller value, $1.0D_{TM}$ is the minimum clearance permitted by Section 17.5 of ASCE/SEI 7-05. For the generic system, moat wall gap displacements of $0.6D_{TM}$ and $0.8D_{TM}$ test the consequences of restricting isolation system displacement, and clearances of $1.2D_{TM}$ and $1.4D_{TM}$ evaluate the benefits of having extra clearance.

Moat wall clearance is influenced by site conditions (e.g., sloping site), building configuration and architectural features, but economic considerations usually dictate design at or near the minimum required displacement, so that moat wall clearances between approximately $0.8D_{TM}$ and $1.0D_{TM}$ are typical (when the configuration has a moat wall). A moat wall clearance of $1.4D_{TM}$, or greater (42 inches in this study), is not common. ASCE/SEI 7-05 does not permit moat wall clearance less than $0.8D_{TM}$ unless the superstructure is explicitly evaluated for stability at MCE demand (which is not typically done).

Table 10-3   Summary of Moat Wall Clearance (Gap) Distances

Isolation System Properties			Moat Wall Gap Distance (inches)				
Type	Displacement (in.)		Approximate Fraction of Code Minimum				
	D_{IM}	Fit-up	$0.6\,D_{IM}$	$0.8\,D_{IM}$	$1.0\,D_{IM}$	$1.2\,D_{IM}$	$1.4\,D_{IM}$
Generic Elastomeric or Sliding Systems - D_{max} Designs							
GEN	29.3	0.7	18	24	30	36	42
GEN-UB	29.3	0.7			30		
GEN-LB	29.3	0.7			30		
Rubber (RB) or Friction Pendulum (FP) Systems - D_{min} Designs							
RB	8.5	0.5			9		
FP	5.7	0.3			6		

Superstructure Design Properties

Isolated structure archetypes are grouped as Code-Conforming and Non-Code-Conforming archetypes on the basis of whether the superstructure meets code requirements for strength, ductility, and detailing. Code-Conforming archetypes include systems with reinforced concrete special moment frame superstructures that conform to all the design requirements of Chapter 17 of ASCE/SEI 7-05. Non-Code-Conforming archetypes include systems with superstructures that do not conform, either in terms of design strength, such as reinforced concrete special moment frame superstructures designed for less than the minimum required base shear, or in terms of ductility, such as reinforced-concrete ordinary moment frame superstructures not permitted for use as a SDC D system. In a few cases, the Non-Code-Conforming superstructures exceed code requirements.

Table 10-4 summarizes design properties for the 3 superstructures of Code-Conforming archetypes used in this study:

(C1) – A reinforced concrete special moment frame (perimeter frame) system, designed for base shear, $V_s = 0.092\,W\,(R_I = 2.0)$.

(C2) – A reinforced concrete special moment frame (space frame) system, designed for base shear, $V_s = 0.092\,W\,(R_I = 2.0)$.

(C3) – A reinforced concrete special moment frame (space frame) system, designed for base shear, $V_s = 0.077\,W$.

The first two systems (C1, C2) are superstructures of isolated archetypes designed for SDC D_{max} seismic criteria, and the last system (C3) is the superstructure of isolated archetypes designed for SDC D_{min} seismic criteria. The base shear of the last system is governed by the limit of Section 17.5.4.3 of ASCE/SEI 7-05, which requires the base shear, V_s, not be less than 1.5 times either the "yield level" of an elastomeric system or the "breakaway" friction level of a sliding system. In this case, the base shear ($V_s = 0.077\,W$) is approximately equal to 1.5 times 0.05, the upper-bound yield level of systems designed for SDC D_{min} seismic criteria with either RB or FP bearings.

Table 10-4 Isolated Structure Design Properties for Code-Conforming Archetypes

Arch. ID	Isolated Structure Archetype Design Properties						
	Superstructure			Isolation System		Isolated Structure	
	V_s/W	Ω	V_{max}/W	Type	Gap (in.)	T_M (sec.)	S_{MI} (g)
Reinforced Concrete Special Moment Frame Perimeter Systems Evaluated at D_{max}							
C1-1	0.092	1.6	0.15	GEN	18	3.47	0.26
C1-2	0.092	1.6	0.15	GEN	24	3.47	0.26
C1-3	0.092	1.6	0.15	GEN	30	3.47	0.26
C1-4	0.092	1.6	0.15	GEN	36	3.47	0.26
C1-5	0.092	1.6	0.15	GEN	42	3.47	0.26
Reinforced Concrete Special Moment Frame Space Frame Systems Evaluated at D_{max}							
C2-1	0.092	3.3	0.30	GEN	18	3.47	0.26
C2-2	0.092	3.3	0.30	GEN	24	3.47	0.26
C2-3	0.092	3.3	0.30	GEN	30	3.47	0.26
C2-4	0.092	3.3	0.30	GEN	36	3.47	0.26
C2-5	0.092	3.3	0.30	GEN	42	3.47	0.26
C2-3U	0.092	3.3	0.30	GEN-UB	30	3.47	0.26
C2-3B	0.092	3.3	0.30	GEN-LB	30	3.47	0.26
Reinforced Concrete Special Moment Frame Space Frame Systems Evaluated at D_{min}							
C3-1	0.077	3.7	0.28	RB	9	3.18	0.09
C3-2	0.077	3.7	0.28	FP	6	2.78	0.11

The Code-Conforming archetypes include systems that meet all code requirements for the superstructures (C1, C2 and C3). Isolation system design meets all code requirements with the exception that moat wall gap distances in C1-1, C1-2, C2-1 and C2-2, would typically not meet code requirements unless special stability analyses were performed.

Table 10-5 summarizes the design properties of the Non-Code-Conforming Archetypes. These isolated structures evaluate the effects of modifying code requirements for (a) superstructure strength and (b) superstructure ductility. Variation in superstructure strength includes some structures, such as NC1, that exceed code strength requirements, while others, such as NC2, do not meet code strength requirements.

Table 10-5   Isolated Structure Design Properties for Non-Code-Conforming Archetypes

Arch. ID	Isolated Structure Archetype Design Properties						
	Superstructure			Isolation System		Isolated Structure	
	V_s/W	Ω	V_{max}/W	Gap (in.)	Type	T_M (sec.)	S_{MI} (sec.)
Reinforced Concrete Special Moment Frame Space Frame Systems Evaluated at D_{max}							
NC1-1	0.164	2.8	0.46	18	GEN	3.47	0.26
NC1-2	0.164	2.8	0.46	24	GEN	3.47	0.26
NC1-3	0.164	2.8	0.46	30	GEN	3.47	0.26
NC1-4	0.164	2.8	0.46	42	GEN	3.47	0.26
NC2-1	0.046	5.2	0.24	18	GEN	3.47	0.26
NC2-2	0.046	5.2	0.24	24	GEN	3.47	0.26
NC2-3	0.046	5.2	0.24	30	GEN	3.47	0.26
NC2-4	0.046	5.2	0.24	42	GEN	3.47	0.26
Reinforced Concrete Ordinary Moment Frame Space Frame Systems Evaluated at D_{max}							
NC3-1	0.246	1.9	0.47	30	GEN	3.47	0.26
NC3-2	0.246	1.9	0.47	42	GEN	3.47	0.26
NC4-1	0.164	1.8	0.30	30	GEN	3.47	0.26
NC4-2	0.164	1.8	0.30	42	GEN	3.47	0.26
NC5-1	0.092	1.9	0.17	30	GEN	3.47	0.26
NC5-2	0.092	1.9	0.17	42	GEN	3.47	0.26

While reinforced concrete ordinary moment frame systems from Chapter 9 are used for the purpose of investigating the effects of superstructure ductility, it should be noted that ASCE/SEI 7-05 does not permit isolated systems with a reinforced concrete ordinary moment frame superstructure in regions of high seismicity. The 2006 IBC does, however, allow isolation systems with either a steel ordinary moment frame or steel OCBF (braced frame) in regions of high seismicity when designed in accordance with AISC 341-05 and $R_I = 1.0$.

10.3.5 Nonlinear Static Analysis for Period-Based Ductility, SSFs, Record-to-Record Variability and Overstrength

The Methodology requires a nonlinear static (pushover) analysis to determine the overstrength of the archetype and to evaluate values of period-based ductility ($\mu_T = \delta_u/\delta_{y,eff}$) for determining the spectral shape factor, *SSF*, and, in certain instances, record-to-record variability, β_{RTR}. Although the overstrength parameter, Ω_0, is not used in isolation design, pushover analysis provides a useful tool for evaluating the strength of the superstructure relative to the level of lateral force in the isolation system.

Pushover Analysis of Isolated Systems

Nonlinear static analysis is performed on isolated structures with the isolation system free to displace. Pushover forces are based on a uniform pattern of lateral load emulating the approximate pattern of uniform lateral displacement of the isolated structure (at displacements up to significant yielding of the superstructure). Figure 10-11 illustrates results of nonlinear static analysis for an isolated structure and for the same superstructure on a fixed-base.[3] Approximately the same ultimate strength is obtained from the two analyses, but the isolated structure effectively shares system ductility between displacement of the isolation system and displacement of the superstructure.

It is also noted that pushover analysis of isolated systems is performed without the moat wall springs (as the sudden increase in stiffness associated with the moat wall is inconsistent with the assumptions used in developing the relationship between μ_T and *SSF* in Appendix B).

Superstructure Overstrength Properties

A static pushover analysis of each of the archetype superstructures listed in Table 10-3 and Table 10-4 was performed to determine the actual maximum strength, V_{max}, and to compare actual strength with design strength, V_s. Values of normalized design strength (V_s/W) and normalized maximum strength (V_{max}/W) are plotted in Figure 10-12.

---

[3] Nonlinear static analysis of the fixed-base structure is based on the lateral load pattern prescribed by Equation 12.8-13 of ASCE/SEI 7-05.

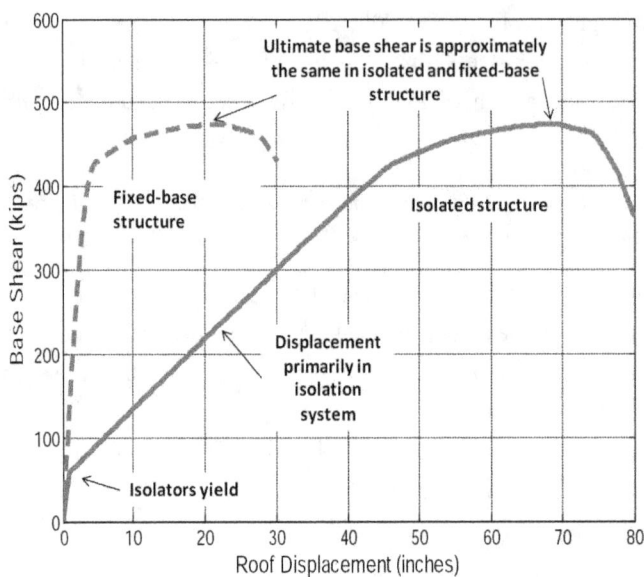

Figure 10-11   Pushover curves of a non-code-conforming isolated structure (NC1) and the same superstructure on a fixed base.

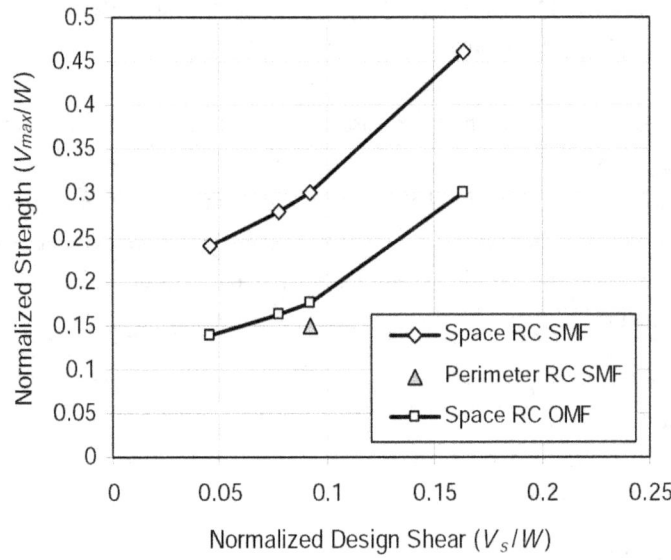

Figure 10-12   Normalized design shear (V_s/W) and maximum strength (V_{max}/W) of superstructures.

As Figure 10-12 shows, the maximum strength of the superstructure is system dependent and does not decrease in proportion to design base shear. When designed with a base shear of $0.092\,W$, the perimeter reinforced concrete special moment frame has an ultimate strength $0.15\,W$, the space reinforced concrete special moment frame has an ultimate strength of $0.30\,W$ and the space reinforced concrete ordinary moment frame has an ultimate strength of $0.17\,W$. Space frames typically have higher overstrength (relative to the design lateral load) than perimeter frames due to the higher contribution of gravity loads in space frame designs. Reinforced concrete ordinary moment frame systems typically have lower overstrength because of a lack of capacity-design provisions. The relative strength of the superstructure and isolation system can have a significant influence on the performance of isolated structures under extreme loading.

System Period-Based Ductility and Spectral Shape Factor (SSF)

Pushover analysis of the isolated system (including both isolator and superstructure) is used for determining period-based ductility, $\mu_T = \delta_u/\delta_{y,eff}$, for computation of the *SSF*. The calculation of μ_T is illustrated in Figure 10-13.

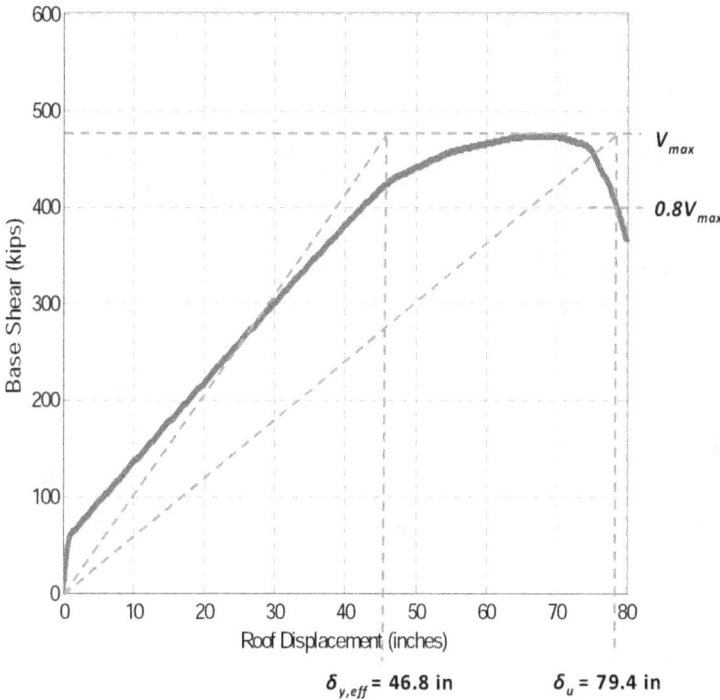

$\delta_{y,eff} = 46.8$ in $\delta_u = 79.4$ in

Figure 10-13   Illustration of calculation of period-based ductility, μ_T, for isolated system NC-1, where $\mu_T = \delta_u/\delta_{y,eff}$.

Calculation of the effective yield roof drift displacement, $\delta_{y,eff}$, for base isolated systems is given by the formula:

$$\delta_{y,eff} = \frac{V_{max}}{W}\left[\frac{g}{4\pi^2}\right]T_M^2 \qquad (10\text{-}6)$$

which is based on Equation 6-7 for values of the modal coefficient, $C_0 = 1.0$ and fundamental period, $T = T_1 = T_M$. The modal coefficient, C_0, is defined by Equation 6-8 and generally has values near unity for isolated structures with roof drift displacement dominated by lateral displacement of the isolation system displacement. In the limiting case of roof drift displacement equal to isolation system displacement (i.e., no drift in the superstructure), the ultimate roof drift displacement, δ_u, would be equal to the effective yield roof drift displacement, $\delta_{y,\text{eff}}$.

Due to inherent flexibility in isolated systems, period-based ductility is smaller for base isolated systems than for fixed base systems because base isolated systems have a larger effective yield roof drift displacement, $\delta_{y,\text{eff}}$, relative to the ultimate roof displacement, δ_u. For fixed-base special moment frames in SDC $D_{\max}$, μ_T is approximately equal to 11, whereas for isolated special moment frames, μ_T ranges between 1.4 and 2.5. The period-based ductility values obtained in this study are relatively large due to the flexibility of reinforced concrete frame superstructures. Typically, isolated buildings would not have as much displacement in the superstructure.

Record-To-Record Uncertainty (β_{RTR})

For most conventional (fixed-base) structures, the period-based ductility, μ_T is greater than or equal to 3.0, and record-to-record variability, β_{RTR}, is equal to 0.40. Isolated systems have limited period-based ductility (i.e., $\mu_T \leq 3.0$), and Equation 7-2 is used to calculate β_{RTR} for these systems. Values of β_{RTR} calculated for base isolated systems using Equation 7-2 range from 0.24 to 0.35, with an average value of 0.27.

Values of SSF and β_{RTR} for isolated systems were computed in accordance with Table 7-1 and Equation 7-2, respectively, which were derived for conventional fixed-base structures. The resulting values of SSF and β_{RTR} showed good agreement with values of spectral shape adjustment and record-to-record variability obtained directly from collapse assessment of isolated systems.

10.3.6 Collapse Evaluation Results

Evaluation Process and Acceptance Criteria

The base isolated systems listed in Tables 10-4 and 10-5 are analyzed and evaluated, as described in Chapter 5 through Chapter 7. Nonlinear dynamic analyses are used with the Far-Field record set to determine the median spectral acceleration at which the structure collapses ($\hat{S}_{CT}$). The collapse margin ratio, CMR, is computed as the ratio of the median collapse capacity, $\hat{S}_{CT}$, and the MCE demand, S_{MT}. These results are reported in Tables 10-6 and 10-7.

Note that for base isolated reinforced concrete ordinary moment frames, CMR accounts for both the sidesway simulated collapse modes and non-simulated column shear failure modes not included in simulation models. Collapse due to column shear failure is predicted when the column loses its vertical-load-carrying capacity, on the basis of component fragility functions as described in Chapter 9. Section 10.2 describes the procedure for incorporating non-simulated failure modes.

Acceptance criteria are based on the adjusted collapse margin ratio ($ACMR$) which is the CMR modified by the spectral shape factor, SSF, to account for the unique spectral shape of rare ground motions. The computed values of period-based ductility, μ_T, for base isolated systems are reported in Table 10-6 and Table 10-7. The spectral shape factor is determined from Table 7-1 as a function of μ_T and building period ($T > 1.5$ seconds for all the isolated systems).

The composite (total) uncertainty, β_{TOT}, associated with collapse must be assessed in order to compare the $ACMR$ to the acceptance criteria. Isolated archetypes with ductile (reinforced concrete special moment frame) superstructures are assigned ratings of (B) Good for modeling, (B) Good for test data, and (A) Superior for design requirements. The isolated archetypes with non-ductile (reinforced concrete ordinary moment frame) superstructures have ratings of (C) Fair for modeling, (B) Good for test data, and (A) Superior for design requirements.

For isolated structures, which have limited period-based ductility (i.e., $\mu_T <$ 3.0), record-to-record variability, β_{RTR}, is calculated using Equation 7-2. Values of record-to-record variability range from 0.2 to 0.3 for the isolated archetypes of this study. These smaller values are supported by a special study on record-to-record variability described in Appendix A.

Record-to-record variability is combined with other sources of uncertainty (i.e., modeling, test data and design requirements, respectively) to determine total composite uncertainty, β_{TOT}, in accordance with Equation 7-5. For the isolated systems considered, β_{TOT} ranges from 0.375 to 0.475. The acceptable $ACMR$ in Chapter 7 will depend on the total uncertainty for each isolated system archetype.

Since design of isolated structures is based on building-specific testing and evaluation, archetype systems are not grouped into performance groups in this study. Instead, each archetype is evaluated individually in comparison with the acceptable $ACMR$. In judging acceptability of each isolation system archetype, the computed $ACMR$ is compared to the acceptable $ACMR$

associated with a 10% probability of collapse, $ACMR_{10\%}$. The use of the 10% criteria for evaluating the performance of an individual building is consistent with recommendations contained in Appendix F.

Collapse Results for Code-Conforming Archetypes

Collapse results for the Code-Conforming base isolation system archetypes are tabulated in Table 10-6a, Table 10-6b, and Table 10-6c.

Table 10-6a Collapse Results for Code-Conforming Archetypes: Various Gap Sizes

Arch. No.	Gap Size (in.)	Computed Collapse Margin Ratio				Acceptable *ACMR*	
		CMR	μ_T	SSF	ACMR	β_{TOT}	$ACMR_{10\%}$
Perimeter Reinforced Concrete Special Moment Frame Systems Evaluated at D_{max}							
C1-1	18	1.54	1.94	1.23	1.89	0.425	1.72
C1-2	24	1.66	1.94	1.23	2.04	0.425	1.72
C1-3	30	1.67	1.94	1.23	2.05	0.425	1.72
C1-4	36	1.70	1.94	1.23	2.09	0.425	1.72
C1-5	42	1.70	1.94	1.23	2.09	0.425	1.72
Space Reinforced Concrete Special Moment Frame Systems Evaluated at D_{max}							
C2-1	18	1.92	1.57	1.18	2.27	0.400	1.67
C2-2	24	2.16	1.57	1.18	2.55	0.400	1.67
C2-3	30	2.19	1.57	1.18	2.58	0.400	1.67
C2-4	36	2.40	1.57	1.18	2.83	0.400	1.67
C2-5	42	2.52	1.57	1.18	2.97	0.400	1.67

Table 10-6b Collapse Results for Code-Conforming Archetypes: Nominal (GEN), Upper-Bound (GEN-UB) and Lower-Bound (GEN-LB) Isolator Properties

Arch. No.	Isolator Prop's.	Computed Collapse Margin Ratio				Acceptable *ACMR*	
		CMR	μ_T	SSF	ACMR	β_{TOT}	$ACMR_{10\%}$
Reinforced Concrete Special Moment Frame Space Frame Systems Evaluated at D_{max} - 30-inch Gap							
C2-3	GEN	2.19	1.57	1.18	2.58	0.400	1.67
C2-3U	GEN-UB	2.08	1.83	1.21	2.53	0.400	1.67
C2-3L	GEN-LB	2.20	1.35	1.15	2.52	0.375	1.62

Table 10-6c Collapse Results for Code-Conforming Archetypes: Minimum Seismic Criteria (SDC D$_{min}$)

Arch. No.	Isolator Props.	Computed Collapse Margin Ratio				Acceptable ACMR	
		CMR	μ_T	SSF	ACMR	β_{TOT}	ACMR$_{10\%}$
Reinforced Concrete Special Moment Frame Space Frame Systems Evaluated at D$_{max}$ - 30-inch Gap							
C3-1	RB	4.01	1.94	1.15	4.60	0.425	1.72
C3-2	FP	4.86	2.54	1.18	5.75	0.475	1.84

Comparison of computed and acceptable values of the *ACMR* reveals that the code-conforming base isolation systems all easily meet the acceptance criteria in this Methodology. The perimeter frame system (C1) has consistently smaller *ACMR*s than the space frame system (C2) due to the smaller lateral overstrength inherent in perimeter frame systems. This study indicates that base isolated systems have comparable levels of safety to code-conforming, conventional fixed-base structures.

Results summarized in Table 10-6a and plotted in Figure 10-14 illustrate the effect of the moat wall clearance distance or gap size. The smallest gap sizes (18 in. and 24 in.) would generally not be allowed by Chapter 17 of ASCE/SEI 7-05, and are not code-compliant. These results are included for comparison purposes only. It is noted, however, that even systems with moat clearances less than the code minimum are acceptable according to the collapse criteria of the Methodology. There may be some benefit to increasing the gap size even beyond 30 inches in SDC D$_{max}$. This benefit is especially apparent for the space frame systems, which have sufficient overstrength to avoid significant nonlinear behavior even when the forces in the isolator are large. These results indicate that the code criteria for base isolated systems are adequate, and may be conservative in the case of moat wall clearance criteria for structures that have sufficient overstrength.

As shown in Table 10-6b, variation in isolator properties does not have a significant effect on the computed collapse margin ratio. Variation in isolator properties is based on ASCE/SEI 7-05 Section 17.8.4.3, which specifies that there not be more than a 20% change in initial stiffness during testing of a prescribed range of prototypes. This variation is not intended to account for differences between modeled and actual behavior, such as softening and damping loss due to heating effects. When upper-bound (GEN-UB) and lower-bound (GEN-LB) isolator properties are used, *ACMR*s are very close to the results for the nominal properties (GEN).

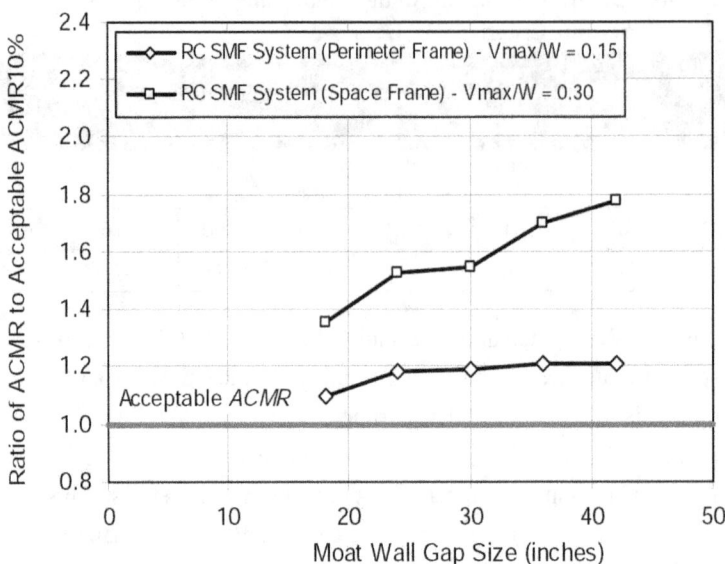

Figure 10-14   Ratios of computed *ACMR* to acceptable $ACMR_{10\%}$ for reinforced concrete special moment frame Code-Conforming isolated archetypes with various moat wall gap sizes, evaluated for SDC D_{max} seismic criteria.

Whereas Tables 10-6a and 10-6b report collapse margins evaluated for SDC D_{max} seismic criteria, Table 10-6c reports collapse margin ratios computed for RB and FP systems in SDC D_{min}. As observed in Chapter 9, the collapse margins tend to increase as the seismic criteria decreases such that *ACMR*s in Table 10-6c are large, and high seismic criteria (SDC D_{max}) govern the acceptability of this system.

Collapse Results for Non-Code-Conforming Archetypes

Collapse results for the Non-Code-Conforming isolation archetypes are summarized in Tables 10-7a and 10-7b. Computed *CMR*s include both simulated and non-simulated failure modes for the reinforced concrete ordinary moment frame structures.

In this study, Non-Code-Conforming archetypes are systems that violate or exceed certain code provisions in order to examine the effects of superstructure strength (i.e., superstructures designed for a higher or lower base shear than required according to ASCE/SEI 7-05) and superstructure ductility (i.e., superstructures designed as ordinary moment frames without the ductile detailing requirements for special moment frame systems). All results in this section are for SDC D_{max}.

Table 10-7a Collapse Results for Isolated Archetypes with Ductile Superstructures and Normalized Design Shear Values (V_s/W) Not Equal to the Code-Required Value ($V_s/W = 0.092$)

Arch. No.	Gap Size (in.)	Computed Collapse Margin Ratio				Acceptable ACMR	
		CMR	μ_T	SSF	ACMR	β_{TOT}	ACMR$_{10\%}$
reinforced concrete special moment frame Space Frame Systems Evaluated at D_{max} - V_s/W = 0.164, V_{max}/W = 0.46							
NC1-1	18	2.08	1.47	1.17	2.42	0.400	1.67
NC1-2	24	2.28	1.47	1.16	2.66	0.400	1.67
NC1-3	30	2.32	1.47	1.16	2.70	0.400	1.67
NC1-4	42	2.64	1.47	1.16	3.08	0.400	1.67
reinforced concrete special moment frame Space Frame Systems Evaluated at D_{max} - V_s/W = 0.046, V_{max}/W = 0.24							
NC2-1	18	1.62	1.76	1.21	1.95	0.400	1.67
NC2-2	24	1.93	1.76	1.21	2.33	0.400	1.67
NC2-3	30	2.03	1.76	1.21	2.45	0.400	1.67
NC2-4	42	2.4	1.76	1.21	2.89	0.400	1.67

Table 10-7b Collapse Results for Isolated Archetypes with Non-Conforming (Non-Ductile) Superstructures of Various Normalized Design Shear Values (V_s/W)

Arch. No.	Gap Size (in.)	Computed Collapse Margin Ratio				Acceptable ACMR	
		CMR	μ_T	SSF	ACMR	β_{TOT}	ACMR$_{10\%}$
reinforced concrete ordinary moment frame Space Frame Systems Evaluated at D_{max} - Vs/W = 0.246, V_{max}/W = 0.45							
NC3-1	30	1.36	1.74	1.20	1.64	0.500	1.90
NC3-2	42	1.73	1.74	1.20	2.08	0.500	1.90
reinforced concrete ordinary moment frame Space Frame Systems Evaluated at D_{max} - Vs/W = 0.164, V_{max}/W = 0.30							
NC4-1	30	1.08	1.93	1.23	1.32	0.500	1.90
NC4-2	42	1.62	1.93	1.23	1.99	0.500	1.90
reinforced concrete ordinary moment frame Space Frame Systems Evaluated at D_{max} - Vs/W = 0.092, V_{max}/W = 0.17							
NC5-1	30	1.02	1.89	1.22	1.25	0.500	1.90
NC5-2	42	1.27	1.89	1.22	1.55	0.500	1.90

The effect of superstructure strength is demonstrated by the *ACMR* results shown in Table 10-7a. Figure 10-15 shows trends in the ratio of calculated *ACMR* to acceptable *ACMR*$_{10\%}$. For a reinforced concrete special moment frame space frame (NC1-3) with a design base shear of $V_s/W = 0.164$ (above Code minimum) and gap size of 30 inches, the computed *ACMR* is 2.57, a factor of approximately 1.7 times the limit on acceptable *ACMR*. When the design base shear is $V_s/W = 0.092$ (C1-3), the minimum value required by

ASCE/SEI 7-05, the computed *ACMR* is slightly smaller, approximately 1.6 times the limiting *ACMR*. For a smaller design base shear, $V_s/W = 0.046$ (NC2-3), below Code minimum, the computed *ACMR* is 1.5 times the limiting *ACMR*. These results suggest that a large difference in the design base shear (varying between 0.046 and 0.164 times the weight of the structure) has only a modest impact on the collapse results.

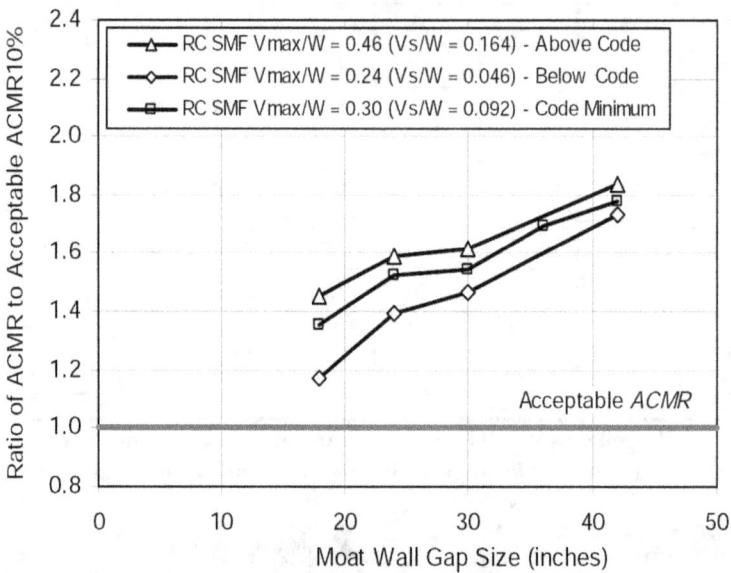

Figure 10-15   Ratios of computed *ACMR* to acceptable $ACMR_{10\%}$ for reinforced concrete special moment frame isolated archetypes with various superstructure strengths and moat wall gap sizes, evaluated for SDC D_{max} seismic criteria

The reason the design base shear does not have a greater impact on the *ACMR* is that the maximum strength of these structures is not linearly related to the design base shear (as shown in Figure 10-12). The design rules for special moment frame systems require capacity design and strong-column-weak-beam requirements such that the actual strength is much larger than the design strength, particularly for space frame systems which have significant gravity loads providing additional overstrength. As a result, the structure designed with $V_s/W = 0.046$ has a true strength of $V_{max}/W = 0.24$, accounting for its high collapse capacity. These observations indicate that the maximum strength of the structure (V_{max}/W) is a much better indicator of collapse performance of isolated structures than the base shear used for design of the superstructure (V_s/W).

Figure 10-16 shows the assessed collapse performance of reinforced concrete special moment frame and reinforced concrete ordinary moment frame isolated archetypes as a function of the strength of the superstructure

(V_{max}/W). These isolated systems have a peak MCE response of approximately 0.25 g. Reinforced concrete special moment frame archetypes that have a superstructure strength greater than 0.25 g have approximately the same $ACMR$ (as indicated by flattening of the curves with increasing strength). For lower strength values, systems with less strength have progressively lower values of $ACMR$. Due to their ductility, reinforced concrete special moment frame systems with superstructure strength that is significantly less than the MCE demand meet the acceptance criteria of the Methodology.

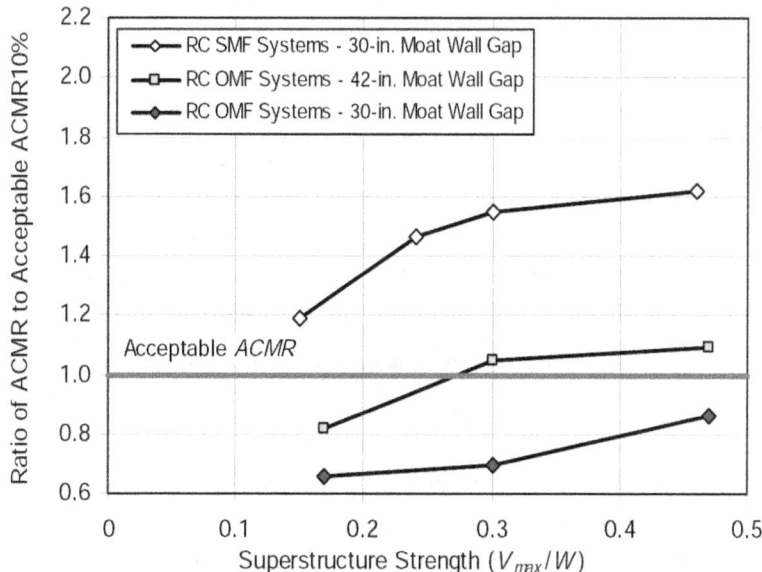

Figure 10-16   Ratios of computed $ACMR$ to acceptable $ACMR_{10\%}$ for reinforced concrete special moment frame and reinforced concrete ordinary moment frame isolated archetypes with different superstructure strengths, evaluated for SDC D_{max} seismic criteria.

The isolated archetypes in Table 10-7b are all reinforced concrete ordinary moment frame systems, which are not permitted for use in regions of high seismicity (SDC C and D) by Table 12.2-1 of ASCE/SEI 7-05. The collapse results for reinforced concrete ordinary moment frame systems are shown in Figure 10-16 as a function of superstructure strength, and Figure 10-17 as a function of moat wall gap size.

Only two of the reinforced concrete ordinary moment frame systems in Table 10-7b are found to have acceptable collapse performance. This result indicates that certain less-ductile superstructures may not provide adequate collapse safety for isolated buildings, even with large moat wall clearance. However, isolated buildings with strong superstructures (e.g., significant overstrength), and large moat wall clearances, do have acceptable collapse performance (as indicated by archetypes NC3-2 and NC4-2), even when the superstructures are less ductile.

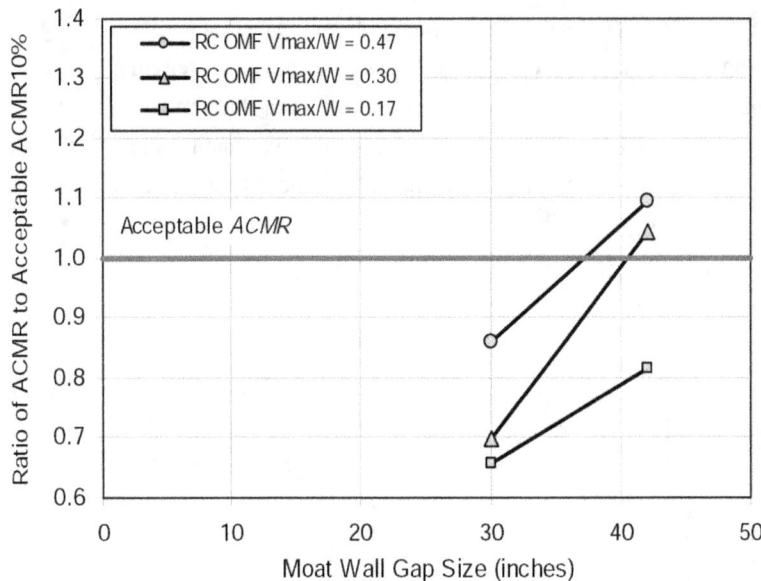

Figure 10-17   Ratios of computed $ACMR$ to acceptable $ACMR_{10\%}$ for reinforced concrete ordinary moment frame isolated archetypes with various moat wall gap sizes, evaluated for SDC D_{max}

Reinforced concrete ordinary moment frame systems that fail to meet the acceptance criteria (NC3-1, NC4-1, NC5-1 and NC5-2) have $ACMRs$ between 15% and 35% lower than the minimum acceptable value. Increasing the moat wall clearance, increasing superstructure strength, or improving the ductility of ordinary moment frame superstructures would likely improve their performance sufficiently to meet the acceptance criteria. Where large moat wall clearance is provided (e.g., 42 inches), analyses suggest that reinforced concrete ordinary moment frame systems with superstructure strength V_{max}/W greater than approximately 0.25 g would be acceptable. For smaller moat wall clearances, a significantly stronger or more ductile structure would be required (see Figure 10-16). This study also did not investigate the use of less ductile systems for lower levels of seismicity (e.g., SDC D_{min}), where superstructure strength and ductility demands are lower.

The effect of reinforced concrete ordinary moment frame superstructure ductility on collapse performance can be investigated by examining how the seismic performance of isolated reinforced concrete ordinary moment frame systems change if design rules are modified such that brittle shear failure of columns is prevented. Assuming all other design requirements remain unchanged, $ACMRs$ for these structures are recalculated assuming brittle shear failure of columns is prevented. Results are plotted in Figure 10-18.

On average, preventing column shear failure in these example archetypes increased *ACMRs* by 25%. Figure 10-18 shows that this design change is sufficient to improve the collapse performance of reinforced concrete ordinary moment frame systems with a 30-inch gap such that the calculated *ACMRs* meet (or nearly meet) the acceptability criteria of the Methdology.

These results suggest the desirability of a comprehensive study of limited ductility systems on isolators, as collapse performance depends on superstructure design rules. A more detailed examination of base isolated reinforced concrete ordinary moment frame systems may also be warranted with shear failure and post-shear failure degradation incorporated directly into the models (rather than using non-simulated failure modes).

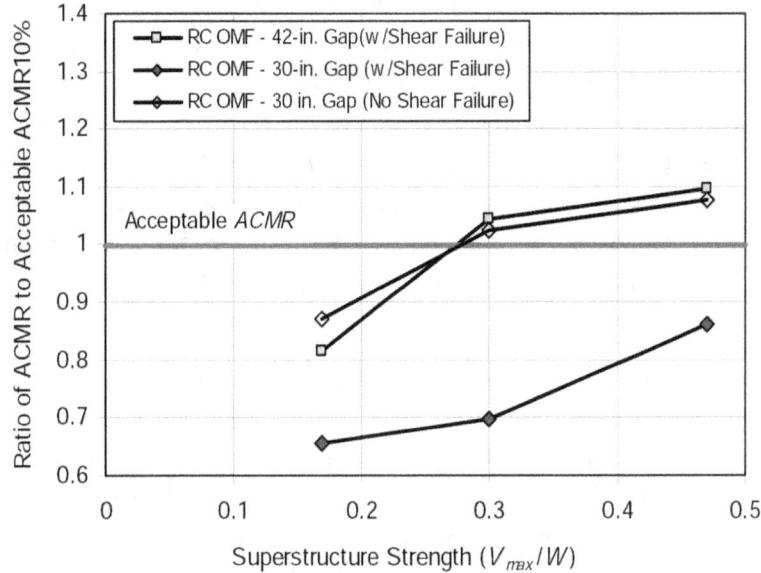

Figure 10-18   Ratios of computed *ACMR* to acceptable $ACMR_{10\%}$ for reinforced concrete ordinary moment frame isolated archetypes with various moat wall gap sizes, evaluated for SDC D_{max}, illustrating the effect of column shear failure on assessed collapse performance.

10.3.7 Summary and Conclusion

This study illustrates application of the Methodology to isolated structures, which have fundamentally different dynamic response characteristics, performance properties and collapse failure modes than those of conventional, fixed-base structures. It demonstrates that, when evaluated in accordance with the Methodology, base isolated systems provide levels of safety against collapse that are comparable to conventional, fixed-base structures. When compared to conventional structures, the major benefit of isolating structures is in reduced earthquake damage and loss, but these effects have not been quantified in this study.

This study also demonstrates the potential use of the Methodology as a tool for assessing validating and developing improvements to current design requirements, in this case requirements for isolated structures. In particular, the Methodology can be used to investigate and possibly improve on apparent conservatisms or limitations for isolated structures related to the following:

Superstructure Design Strength. Performance of isolated structures is much more closely related to the "true" maximum strength, V_{max}, than the design strength, V_s, of the superstructure due to inherent differences in the overstrength of different systems (and designs). For systems with high inherent overstrength, such as the special space frame designs, low design base shears can be used. Other systems with less overstrength may behave differently when isolated. Design base shear requirements of ASCE/SEI 7-05 may be conservative for highly ductile structures which have large amounts of inherent overstrength. Rather than relying on generic values of the R_I factor (and unknown amounts overstrength), to define design strength, the Methodology could be used to establish appropriate criteria for design based on maximum strength, verified by pushover analysis.

Less Ductile (Non-Complying) Superstructures. In a similar process, the Methodology could also be used to establish appropriate levels of true strength for less ductile (non-complying) systems not currently permitted for use with isolated buildings in high seismic regions. While the process would be the same, true strength required for design of less ductile systems would likely be significantly higher than that required for more ductile systems to achieve the same acceptable level of collapse performance.

Moat Wall Gap Clearance (Displacement Restraint System). In the case of moat wall clearance, superstructures with sufficient strength and ductility were observed to be relatively insensitive to the size of the clearance. This suggests that moat wall clearance requirements of ASCE/SEI 7-05 might be conservative for highly ductile superstructures. The Methodology could be used to design isolation systems with limited displacement capacity for less intense, more common, earthquake ground motions. If the moat wall or other displacement restraint mechanism is used to limit excessive displacement of isolators for rare (MCE) ground motions, isolators could then be smaller and more economical, and provide a more practical solution in regions of very high seismic demands.

Chapter 11

Conclusions and Recommendations

This recommended Methodology provides a rational basis for establishing global seismic performance factors (SPFs), including the response modification coefficient, R factor, the system overstrength factor, Ω_0, and the deflection amplification factor C_d of new seismic-force-resisting systems proposed for inclusion in model building codes. The Methodology also provides a more reliable basis for re-evaluation of seismic performance factors of seismic-force-resisting systems currently available in model building codes and reference standards.

This chapter describes assumptions and limitations of the Methodology, summarizes observations and conclusions resulting from its development, discusses an adaptation of the Methodology to collapse performance evaluation for an individual building, and provides recommendations for further study.

11.1 Assumptions and Limitations

The Methodology is intended to apply broadly to all buildings, recognizing that this objective may not be fully achieved for certain seismic environments and building configurations. Likewise, the Methodology has incorporated certain simplifying assumptions deemed appropriate for reliable evaluation of seismic performance. Key assumptions and potential limitations of the Methodology are summarized in the following sections.

11.1.1 Far-Field Record Set Ground Motions

The Methodology specifies the same set of ground motions (i.e., Far-Field record set) for collapse performance evaluation of all systems. Records of the Far-Field record set are unambiguously defined (including scaling) to avoid any subjectivity in the ground motions used for nonlinear dynamic analyses. The Far-Field record set is a robust sample of strong motion records from large magnitude events. Even so, these records have inherent limitations for certain buildings.

Buildings at Sites near Active Faults

Two sets of ground motion data, the Far-Field record set (records at sites at least 10 km from fault rupture) and the Near-Field record set (records at sites within 10 km of fault rupture) were developed. An internal study, documented in Appendix A, found that the collapse margin ratio, CMR, was somewhat smaller for a system designed for "near-fault" (SDC E) seismic criteria, and evaluated using the Near-Field record set, than for same system designed for SDC D seismic criteria, and evaluated using the Far-Field record set. This implies that somewhat smaller values of the response modification coefficient, R, would be appropriate for design of buildings near active faults.

Various alternatives were considered, including the development of a separate set of seismic performance factors for design of buildings near active faults. For simplicity, as well as consistency with ASCE/SEI 7-05, a single set of seismic performance factors based on collapse assessments using the Far-Field record set was chosen. In so doing, the Methodology implicitly accepts somewhat greater life safety risk for buildings located close to active faults. This is consistent with the approach in ASCE/SEI 7-05, which implicitly accepts somewhat greater life safety risk for buildings near active faults by limiting MCE ground motions to deterministic values of seismic hazard.

Although the Methodology uses the Far-Field record set to establish seismic performance factors for a new seismic-force-resisting systems, the Near-Field record set would be more appropriate for verifying life safety performance of an individual building located near an active fault.

Buildings at Sites in the Central and Eastern United States or Subject to Deep Subduction Earthquakes

The Far-Field record set is a robust sample of all strong motion records from large magnitude events recorded at sites greater than 10 km from fault a rupture. No attempt was made to limit records based on tectonic setting or fault mechanism during the record selection process. The Far-Field record set is dominated by shallow crustal earthquakes, representative of areas in the Western United States. It does not include strong motions records from deep subduction earthquakes, or from Central and Eastern United States earthquakes, since such records do not exist.

Duration of strong shaking is an important parameter in collapse analyses of degrading systems, and strong motion records were purposely selected from large magnitude events to adequately capture shaking duration. Very large magnitude earthquakes associated with deep subduction zone events would be expected to have longer durations of strong shaking, on average. Central and Eastern United States events could have different shaking characteristics also affecting the collapse margin ratio and related seismic performance factors. In spite of limitations in available deep subduction zone, or Central and Eastern United States earthquake records, actual earthquake records, rather than artificial or theoretical ground motions, were selected as the basis for collapse assessment.

Buildings with Very Long Periods

The usefulness of earthquake records at very long periods is limited by the ability of strong motion instruments to accurately record long-period vibration. Records from older instruments may only be accurate to one or two seconds, while records from the newer instruments are typically accurate to at least 10 seconds. The Far-Field and Near-Field record sets include only those records deemed accurate to a period of at least 4 seconds by the agency responsible for processing the record. Most records in these sets are accurate to a period of at least 10 seconds.

The Methodology conservatively limits the elastic period of index archetype configurations to a period not greater than 4 seconds to ensure valid evaluations of collapse performance. Records from the Far-Field and Near-Field record sets should be used with caution to evaluate collapse performance of buildings with elastic periods greater than 4 seconds (i.e., very tall buildings), since the spectral content of some records may not be valid, and the associated value of the spectral shape factor, *SSF*, may be overstated by the Methodology.

11.1.2 Influence of Secondary Systems on Collapse Performance

The Methodology evaluates collapse performance of the seismic-force-resisting system, ignoring the influence of secondary systems, such as gravity systems and nonstructural components, which are not included in the designation of the seismic-force-resisting system. Such systems can either improve or diminish the collapse performance of a system of interest.

Potential for Improved Performance

For many buildings, elements of the gravity or nonstructural systems can significantly improve collapse performance by providing additional resistance to lateral forces. This is particularly true when the secondary system has a larger lateral capacity than that of that of the primary system (FEMA, 2009).

Incorporation of elements of the gravity or nonstructural systems in index archetype configurations was considered. It was ultimately decided to limit participating elements to those defined as part of the seismic-force-resisting system. It was considered inappropriate to take the beneficial effects of elements and components that would not be subject to regulation under the earthquake design provisions. As an alternative, it was considered permissible to include elements of the gravity or nonstructural systems in the assessment, if they were also included in the definition of the seismic-force-resisting system, and subject to the criteria contained within the system design requirements.

Potential for Reduced Performance

Partial collapse of buildings has occurred when elements of the gravity or nonstructural systems are not able to sustain lateral deformations of the seismic-force-resisting system. Options for explicitly modeling and evaluating gravity system performance were considered. Ultimately it was considered to be problematic, and not completely relevant to the evaluation of seismic performance factors for a seismic-force-resisting system, which must be qualified for use with many different secondary systems. Rather, displacement compatibility requirements of ASCE/SEI 7-05 are relied upon to protect secondary systems from collapse failure.

It should be noted that current displacement compatibility requirements of Section 12.12.4 of ASCE/SEI 7-05 may not be adequate to protect against premature failure of a gravity system. They only apply to SDC D (and SDC E and F) structures, and are based on design story drift, which is substantially less than the peak inelastic story drift of ductile seismic-force-resisting systems at the point of incipient collapse.

11.1.3 Buildings with Significant Irregularities

Significant irregularities, including torsion (horizontal structural irregularity Types 1a and 1b, Table 12.3-1, ASCE/SEI 7-05) and soft/weak story (vertical structural irregularity Types 1a and 1b, Table 12.3-2, ASCE/SEI 7-05), are known contributors to building collapse. Options for explicit modeling of irregularity were considered, but internal studies showed this to be unnecessary for evaluating collapse margins and related seismic performance factors for generic seismic-force-resisting systems. Limits on the use of equivalent lateral force (ELF) analysis given in Table 12.6-1 of ASCE/SEI 7-05, and related design conservatisms of the ELF procedure (e.g., accidental torsion, P-delta effects), were considered adequate to either limit the effects of significant irregularities, or require more detailed, dynamic analysis.

It should be noted that limits on the use of the ELF procedure are more restrictive for SDC D (and SDC E and F) structures, and certain design requirements (e.g., amplification of accidental torsion) do not apply to SDC B structures, so that there may be some additional life safety risk due to irregularity inherent to systems designed for either SDC B or SDC C criteria.

11.1.4 Redundancy of the Seismic-Force-Resisting System

Section 12.3.4.2 of ASCE/SEI 7-05 requires the seismic-force-resisting system of structures assigned to Seismic Design Categories D, E, and F to be designed for seismic loads increased by the redundancy factor, ρ, where $\rho =$ 1.3, unless the configuration meets certain requirements for redundancy. Options for explicit modeling of non-redundant systems were considered, but internal studies showed this to be unnecessary for evaluating collapse margins and related seismic performance factors for generic seismic-force-resisting systems. The Methodology assumes $\rho = 1.0$ for design of structural system archetypes, since larger values of ρ would be unconservative for collapse evaluation of archetypes that generally meet redundancy requirements of Table 12.3-3 of ASCE/SEI 7-05.

11.2 Observations and Conclusions

In the development of this Methodology, selected seismic-force-resisting systems were evaluated to illustrate the application of the Methodology and verify its methods. Results of these studies provide insight into the collapse performance of buildings and appropriate values of seismic performance factors. Observations and conclusions in terms of generic findings applicable to all systems, and specific findings for certain types of seismic-force-resisting systems are described below. These findings should be considered generally representative, but not necessarily indicative of all possible trends, given limitations in the number and types of systems evaluated.

11.2.1 Generic Findings

The following generic findings and conclusions apply to seismic-force-resisting systems in general.

Systems Approach

Collapse performance (and associated seismic performance factors) must be evaluated in terms of the behavior of the overall seismic-force-resisting system, and not the behavior of individual components or elements of the system. Collapse failure modes are highly dependent on the configuration and interaction of elements within a seismic-force-resisting system. Seismic performance factors should be considered as applying to an entire seismic-force-resisting system, and not elements comprising it.

Precision of Seismic Performance Factors

In general, there is no practical difference in the collapse performance of systems designed with fractional differences in the response modification coefficient, R. For example, collapse performance of structures designed for $R = 6$ and $R = 6.5$ is essentially the same, all else being equal. There is a discernible, but modest difference in collapse performance for systems designed for moderately different values of R, for example $R = 6$ and $R = 8$. There is, however, a significant difference in collapse performance for systems designed using different multiples of R, as in $R = 3$ versus $R = 6$. Current values of R provided in Table 12.2-1 (e.g., 3, 3-1/4 and 3-1/2) reflect a degree of precision that is not supported by results of example collapse evaluations.

Spectral Content of Ground Motions

Consideration of spectral content (spectral shape) of ground motions can be very important to the evaluation of collapse performance of ductile structures. Epsilon-neutral earthquake records that are scaled to represent very rare ground motions (ground motions corresponding to large positive values of epsilon) can significantly overestimate demand on ductile structures. The Methodology incorporates a spectral shape factor, SSF, that adjusts calculated response to account for the spectral content of rare earthquake ground motions and avoid overestimation of nonlinear response.

Short-Period Buildings

Consistent with prior research, values of collapse margin ratio are consistently smaller for short-period buildings, regardless of the type of seismic-force-resisting system. Unless they have adequate design strength, short-period buildings generally do not meet collapse performance objectives of the Methodology.

These findings suggest a possible need for period-dependent seismic performance factors, such as a short-period value and a 1-second value of the response modification coefficient, R, for each system. At present, the Methodology determines a single value of each seismic performance factor, independent of period, and consistent with the design requirements of ASCE/SEI 7-05. It could, however, be modified to determine period-dependent values of each factor.

Governing Seismic Design Category

Values of collapse margin ratio for a seismic-force-resisting system designed and evaluated for SDC D are generally smaller than corresponding values of collapse margin ratio for the same seismic-force-resisting system designed and evaluated for SDC C, all else being equal.

This trend is attributed to the increasing role of gravity loads in the strength of seismic-force-resisting components as the level of seismic design decreases. The size and strength of seismic-force-resisting components supporting both seismic and gravity loads is not necessarily proportional to a decrease in seismic design loads, since gravity loads do not decrease. The role of gravity loads in the strength of the seismic-force-resisting system is directly related to system overstrength, Ω, which also tends to increase as the level of seismic design decreases.

These findings suggest that the response modification coefficient, R, will generally be governed by the Seismic Design Category with the strongest ground motions for which the system is proposed for use (e.g., SDC D for seismic-force-resisting systems that are permitted for use in all SDCs). In general, R factors based on SDC D evaluations are conservative for design of the same seismic-force-resisting system designed for SDC C or SDC B seismic criteria.

These findings also suggest a possible need for seismic load-dependent seismic performance factors. For example, a system designed for SDC C could be assigned a larger value of R than would be required for the same system designed for SDC D. At present, the Methodology determines a single value of each seismic performance factor, independent of seismic design criteria, and consistent with the design requirements of ASCE/SEI 7-05. It could, however, be modified to determine SDC-specific values of each factor.

Overstrength

Consistent with prior research, values of collapse margin ratio are strongly related to calculated values of overstrength, regardless of the type of seismic-force-resisting system. For larger values of overstrength, larger values of collapse margin ratio are observed.

Calculated values of overstrength for different index archetype models vary widely, depending on configuration and Seismic Design Category. Results suggest that current values of system overstrength, Ω_O, given in Table 12.2-1 of ASCE/SEI 7-05 are not representative of the actual overstrength present in seismic-force-resisting systems. Values in the table generally vary between 2 and 3 for all systems (except cantilevered structures), while calculated values of Ω varied from as low as 1.5 to over 6 for structural system archetypes evaluated in the development of the Methodology.

Distribution of Inelastic Response

Consistent with prior research, values of collapse margin ratio are strongly related to the distribution of inelastic response over the height of a structure, regardless of the type of seismic-force-resisting system. Values of collapse margin ratio are significantly larger for systems in which the inelastic response is more evenly distributed throughout the system.

11.2.2 Specific Findings

The following key findings and conclusions apply specifically to reinforced-concrete moment frame systems, wood light-frame systems, and base-isolated systems that were evaluated as part of the development of the Methodology.

Example Application – Reinforced Concrete Special Moment Frame Systems

Evaluation of reinforced concrete special moment frame systems found that, in general, designs based on current ASCE/SEI 7-05 requirements ($R = 8$) meet the collapse performance objectives of this Methodology. However, trends in collapse margin ratio results suggested a potential collapse deficiency for taller buildings that did not meet minimum base shear requirements consistent with requirements in the predecessor document, ASCE 7-02. This information was made available to the ASCE 7 Seismic Committee, and a special code change proposal was passed in 2007 (Supplement No. 2), amending the minimum base shear requirements of ASCE/SEI 7-05 to correct for this potential deficiency.

The root cause of the potential deficiency in collapse performance of reinforced concrete special moment frame systems is the localization of large lateral deformations in the lower stories, and the associated detrimental P-delta effects on post-yield behavior. While imposing a minimum value of design base shear eliminated the potential deficiency for these systems, other approaches could have been used, including the introduction of height limits (which would not be practical), enhancement of P-delta and strong-column-weak-beam design criteria (which would require additional study), or the development of period-dependent response modification coefficients, R (which would effectively increase the design base shear for taller buildings).

Example Application - Reinforced-Concrete Ordinary Moment Frame Systems

Example evaluation of reinforced concrete ordinary moment frame systems found that, in general, designs based on current ASCE/SEI 7-05 requirements ($R = 3$) meet the collapse performance objectives of this Methodology, considering that these systems are only permitted for use in SDC B structures.

Example Application - Wood Light-Frame Systems

Example evaluations of wood light-frame systems found that, in general, designs based on current ASCE/SEI 7-05 requirements ($R = 6$) meet the collapse performance objectives of this Methodology.

Supporting Study - Seismically-Isolated Structures

A special study of seismically-isolated buildings focused on the performance of isolated special and ordinary reinforced concrete frame systems. This study found that code-conforming isolated structures with ductile superstructures (e.g., special moment frame systems) generally meet collapse performance objectives, but that isolated structures with non-ductile superstructures (e.g., ordinary moment frame systems) may not. However, it also found that with higher strengths, superstructures of isolated systems did not need to have the full ductility capacity required of conventional buildings to achieve acceptable collapse performance. This is an important finding for the introduction of more economically detailed superstructures of isolated buildings, although more comprehensive studies would be required to develop appropriate code requirements.

Pushover evaluations of isolated structures found poor correlation between the true (maximum) strength of the superstructure and the design strength. This suggests that nonlinear static (pushover) analyses would be more appropriate for verifying adequate overstrength of the superstructure than selecting superstructure strength based on the approximate values of R_I, and would also provide a more reliable and economical basis for design.

11.3 Collapse Evaluation of Individual Buildings

Although developed as a tool to establish seismic performance factors for generic seismic-force-resisting systems, the Methodology could be readily adapted for collapse assessment of an individual building system. As such, it could be used to demonstrate adequate collapse performance of the structural system of a building designed using performance-based design methods, permitted by Section 11.1.4 of ASCE/SEI 7-05. Specific methods for collapse evaluation of an individual building system are described in Appendix F.

11.3.1 Feasibility

It is anticipated that buildings designed using performance-based methods will likely be large or important structures. Projects using performance-based design methods typically utilize detailed models for linear and nonlinear analyses of the building, and peer review is commonly required. Such projects are already set up to utilize many components of the Methodology.

11.3.2 Approach

The Methodology is based on the concept of collapse level ground motions, defined as the level of ground motions that cause median collapse (i.e., one-half of the records in the set cause collapse). For a building to meet the collapse performance objectives of this Methodology, the median collapse capacity must be an acceptable amount above the MCE ground motion demand level (i.e., the adjusted collapse margin ratio, $ACMR$, must exceed acceptable values).

By starting with an acceptable collapse probability (for MCE ground motions) and working backwards through the Methodology, values of the spectral shape factor, SSF, and collapse margin ratio, CMR, can be evaluated to determine the level of ground motions corresponding to median collapse. The Methodology can be "reverse engineered" to determine the level of ground motions for which not more than one-half of the records should cause collapse. By scaling the record set to this level, trial designs for a subject building can be evaluated. If the analytical model of the trial design survives one-half or more of the records without collapse, then the building has a collapse probability that is equal to (or less than) the acceptable collapse probability (for MCE ground motions), and meets collapse performance objective of the Methodology.

11.4 Recommendations for Further Study

The following recommendations are provided for possible future studies that would help to: (1) further improve or refine the Methodology; or (2) utilize the Methodology to investigate and develop potential improvements to the seismic provisions of ASCE/SEI 7-05.

11.4.1 Studies Related to Improving and Refining the Methodology

Comprehensive Evaluation of Existing Systems

The primary objective of this study would be to "beta test" the Methodology to further verify that this procedure will reliably and reasonably quantify building seismic performance for the various building systems that are, or will eventually become adopted by current building codes and standards organizations.

The Methodology could also be used to set minimum acceptable design criteria for standard code-approved systems and to provide guidance in the selection of appropriate design criteria for other systems when linear design methods are applied. It is possible that the Methodology could then be used to modify or eliminate those systems or requirements that cannot reliably meet, or do not relate, to these objectives.

Component Qualification

The Methodology thus far has been developed to comprehensively evaluate entire building systems, including variations in system configuration. This study would be intended to modify and simplify the procedure so that it can be used on individual building components.

The need for a simplified component methodology has been identified by construction materials industries and codes and standards organizations for the purpose of reliably and accurately quantifying seismic performance of various building components that currently are, or will eventually become available for use as seismic-force-resisting components within building systems.

Simplified Methods

Collapse simulation is a detailed, data-intensive process, which has a high degree of uncertainty. In order to comprehensively evaluate entire building systems, and all permissible variations in system configuration, the Methodology is necessarily complex. This study would investigate and test possible short cuts in the process to simplify the overall application of the Methodology for use when more comprehensive evaluations are not necessarily warranted.

Modeling of Short-Period Structures

The Methodology does not currently provide any specific guidance for modeling foundation flexibility and considering possible beneficial effects of soil-structure-interaction on the collapse performance of short-period

---

structures. This study would investigate these effects and, if justified, develop guidance for explicitly modeling of foundation flexibility and adjustment factors (similar to *SSF*) that would adjust collapse margin ratios as function of period, for evaluation of structural system archetypes assumed to be fixed on a rigid base.

11.4.2 Studies Related to Advancing Seismic Design Practice and Building Code Requirements (ASCE/SEI 7-05)

Period-Dependent and Seismic Load-Dependent R factors

The equivalent lateral force (ELF) procedure of ASCE/SEI 7-05 defines base shear in terms of a single value of R for a given seismic-force-resisting system, independent of the period of the structure or the level of seismic design criteria. Studies show that designs based on a single value of R do not necessarily have consistent collapse performance, and that such inconsistencies could be reduced or eliminated by specifying period-dependent or seismic load-dependent values of R.

Use of period-dependent and load-dependent values of R would require substantial revision to the equivalent lateral force method and related requirements in seismic design codes and standards, but could provide a basis for more efficient and economical design. The Methodology could be used to investigate period and seismic load dependency of R, and the feasibility of incorporating such factors. Potential benefits would be greater in regions of lower seismicity, where seismic load-dependent values of R are likely to be larger than those now specified for the same system in regions of high seismicity.

Consideration of Gravity and Nonstructural Systems and Components

The Methodology could be used to investigate the importance of the gravity system and certain nonstructural components to collapse performance, and investigate the feasibility of enhancing current seismic design requirements to more appropriately incorporate these systems in the seismic design process. This would include accounting for both the possible beneficial and detrimental effects of these systems on collapse performance. Comparisons of collapse results for archetypical models both with and without selected gravity system components could be made to quantify the results.

Structural Irregularities and Redundancy

The Methodology could be used to investigate the importance of structural system regularity and redundancy to collapse performance, and investigate possible changes to current seismic design requirements with regard to these characteristics. Comparisons of collapse results for archetypical models both with and without features of regularity and redundancy could be made to quantify the effects.

Structures with Isolation and Damping Systems

Due to limited U.S. experience in strong earthquakes, code development committees have necessarily and appropriately imposed certain conservative restrictions on design requirements for seismically isolated and damped structures. The Methodology could be used to evaluate the importance of current design requirements for isolated systems, and investigate the feasibility of removing unnecessary conservatism. Comparisons of collapse results of archetypical models with non-ductile superstructures of varying strength levels could be made to quantify required strength for meeting collapse performance objectives. Removal of unnecessary conservatism would reduce system cost and support greater use of these types of protective systems.

Appendix A
Ground Motion Record Sets

This appendix describes the selection of ground motion record sets for collapse assessment of building structures using nonlinear dynamic analysis (NDA) methods. It summarizes the characteristics of the Far-Field and Near-Field record sets and defines the scaling methods appropriate for collapse evaluation of building archetypes based on incremental dynamic analysis. The Methodology utilizes the Far-Field record set for nonlinear dynamic analysis and related collapse assessment of archetype models.

Both the Far-Field and Near-Field record sets have average epsilon values that are lower than expected for Maximum Considered Earthquake (MCE) ground motions, and therefore can substantially underestimate calculated collapse margin ratios without appropriate adjustment for spectral shape effects. Adjustment of results from nonlinear dynamic analysis using these record sets is described in Section A.4 (and Appendix B).

Three ground-motion-related studies are included at the end of this appendix. The first study (Section A.11) compares collapse performance for archetypes evaluated using Far-Field and Near-Field record sets, respectively, and shows that collapse margins are somewhat smaller for the Near-Field set record. The second study (Section A.12) addresses the robustness of the Far-Field record set and shows that the selection criteria of Section A.7 yield a sufficiently large number of representative records (for sites not close to fault rupture). The third study (Section A.13) investigates record-to-record variability, β_{RTR}, of collapse results and develops a simple relationship for estimating β_{RTR} as a function of period-based ductility, μ_T, for systems that have limited period elongation.

A.1   Introduction

Ground motion record sets include a set of ground motions recorded at sites located greater than or equal to 10 km from fault rupture, referred to as the "Far-Field" record set, and a set of ground motions recorded at sites less than 10 km from fault rupture, referred to as the "Near-Field" record set. The Near-Field record set includes two subsets: (1) ground motions with strong pulses, referred to as the "NF-Pulse" record subset, and (2) ground motions without such pulses, referred to as the "NF-No Pulse" record subset.

A.2 Objectives

The Methodology requires a set of records that can be used for nonlinear dynamic analysis of buildings and evaluation of the probability of collapse for Maximum Considered Earthquake (MCE) ground motions. These records meet a number of conflicting objectives, described below.

- **Code (*ASCE/SEI 7-05*) Consistent** – The records should be consistent (to the extent possible) with the ground motion requirements of Section 16.1.3.2 of ASCE/SEI 7-05 *Minimum Design Loads for Buildings and Other Structures* (ASCE, 2006a) for three-dimensional analysis of structures. In particular, "ground motions shall consist of pairs of appropriate horizontal ground motion acceleration components that shall be selected and scaled from individual recorded events."

- **Very Strong Ground Motions** – The records should represent very strong ground motions corresponding to the MCE motion. In high seismic regions where buildings are at greatest risk, few recorded ground motions are intense enough, and significant upward scaling of the records is often required.

- **Large Number of Records** – The number of records in the set should be "statistically" sufficient such that the results of collapse evaluations adequately describe both the median value and record-to-record (RTR) variability of collapse capacity.

- **Structure Type Independent** – Records should be broadly applicable to collapse evaluation of a variety of structural systems, such as systems that have different dynamic response properties or performance characteristics. Accordingly, records should not depend on period, or other building-specific properties of the structure.

- **Site Hazard Independent** – The records should be broadly applicable to collapse evaluation of structures located at different sites, such as sites with different ground motion hazard functions, site and source conditions. Accordingly, records should not depend on hazard de-aggregation, or other site- or hazard-dependent properties.

No single set of records can fully meet all of the above objectives due, in part, to inherent limitations in available data. Large magnitude events are rare, and few existing earthquake ground motion records are strong enough to collapse large fractions of modern, code-compliant buildings. In the United States, strong-motion records date to the 1933 Long Beach earthquake, with only a few records obtained from each event until the 1971 San Fernando earthquake.

Even with many instruments, strong motion instrumentation networks (e.g., Taiwan and California) provide coverage for only a small fraction of all regions of high seismicity. Considering the size of the earth and period of geologic time, the available sample of strong motion records from large-magnitude earthquakes is still quite limited (and potentially biased by records from more recent, relatively well-recorded events).

A.3 Approach

The Methodology requires a set of ground motion records for collapse assessment of archetypical models that are appropriate for incremental dynamic analysis (Vamvatsikos and Cornell, 2002), as adapted herein. Incremental dynamic analysis (IDA) makes use of multiple response history analyses for a given ground motion record of increasing intensity until collapse occurs or the model otherwise reaches a collapse limit state. This process is repeated for a set of ground motion records of sufficient number to determine median collapse and record-to-record variability.

The Methodology follows the IDA concept of increasing ground motion intensity to collapse, but applies each sequential increase in intensity, collectively, to the entire set of ground motion records. Similar to IDA, the Methodology characterizes intensity in terms of response spectral acceleration at the fundamental period, T, of the system of interest, except that intensity is defined collectively by the median spectral acceleration of the record set, S_T, rather than by different intensities for each record. Record set intensity, S_T, is the parameter used to define record set intensity when the record set is scaled to a particular level of ground motions (e.g., MCE spectral acceleration). The median value of collapse spectral acceleration, $\hat{S}_{CT}$, is the value of S_T when the record set is scaled such that one-half of the records affect collapse.

To ensure broad representation of different recorded earthquakes, sets of ground motions contain records selected from all large-magnitude events in the PEER NGA database (PEER, 2006a). The PEER NGA database is described in Section A.6. A sufficient number of the strongest ground motion records are selected from each event to permit statistical evaluation of record-to-record variability.

Record selection does not distinguish ground motion records based on either site condition or source mechanism. However, distance to fault rupture is used to develop separate Far-Field and Near-Field record sets (for examination of potential differences in collapse fragility due to near-source directivity and pulse effects).

A.4 Spectral Shape Consideration

Spectral shape, i.e., frequency content of ground motions, can significantly influence the calculation of collapse fragility and the collapse margin ratio of "ductile" structures. Baker and Cornell (2006) have shown that rare ground motions in the Western United States, such as those corresponding to the MCE, have a distinctive spectral shape that, for a given fundamental-period spectral acceleration, causes the record to be less damaging than other records of less intensity.

In essence, the shape of the spectrum of rare ground motions drops off more rapidly at periods both greater and less than the fundamental period of interest (i.e., has less energy), as compared to spectra of other (less rare) records. The amount by which spectral shape can influence the collapse ratio is a function of the "rareness" of the ground motions. For ductile structures located in coastal California, accounting for this spectral shape effect can cause a 40% to 60% increase in the collapse margin ratio (i.e., median collapse capacity).

The ground motion record sets do not directly incorporate the effect of spectral shape. Direct incorporation of spectral shape would necessarily require records to be selected based on the fundamental period of the structure, resulting in a different set of records for each structure of differing period. Rather, collapse margin ratios calculated using the ground motion record sets are adjusted for spectral shape effects based on structure deformation capacity and seismic design category, described in Chapter 7, using factors developed in Appendix B.

A.5 Maximum Considered Earthquake and Design Earthquake Demand (ASCE/SEI 7-05)

The seismic provisions of ASCE/SEI 7-05 specify ground motions and design requirements in terms of the structure's Seismic Design Category (SDC), which is a function of the level of design earthquake (DE) ground motions and the Occupancy Category of the structure. The Methodology is based on life safety and assumes all structures to be either Occupancy Category I or II (i.e., structures that do not have special functionality requirements). Seismic Design Categories for Occupancy I and II structures vary from SDC A in regions of very low seismicity (which is of least interest) to SDC E in regions of highest seismicity near active faults.

The seismic provisions of ASCE/SEI 7-05 define MCE demand in terms of mapped values of short-period spectral acceleration, S_S, and 1-second spectral acceleration, S_1, site coefficients, F_a and F_v, and a standard response spectrum shape. For seismic design of the structural system, ASCE/SEI 7-05 defines the DE demand as two-thirds of the MCE demand. Archetypical systems are designed for DE ground motions and evaluated for collapse using the corresponding set of MCE ground motions.

Mapped values of spectral acceleration vary greatly by seismic region. The Methodology defines MCE and DE ground motions by the range of spectral accelerations associated with Seismic Design Categories B, C, and D, respectively. For each SDC, maximum and minimum ground motions are based on the respective upper-bound and lower-bound values of DE spectral acceleration, as given in Table 11.6-1 of ASCE/SEI 7-05, for short-period response, and in Table 11.6-2 of ASCE/SEI 7-05, for 1-second response. MCE spectral accelerations are derived from DE spectral accelerations for site coefficients corresponding to Site Class D (stiff soil) following the requirements of Section 11.4 of ASCE/SEI 7-05.

Tables A-1A and A-1B list values of spectral acceleration, site coefficients and design parameters for maximum and minimum ground motions of Seismic Design Categories B, C, and D. Figure A-1 shows MCE response spectra for these ground motions.

Table A-1A   Summary of Mapped Values of Short-Period Spectral Accelerations, Site Coefficients, and Design Parameters Used for Collapse Evaluation of Seismic Design Categories D, C, and B Structure Archetypes

Seismic Design Category		Maximum Considered Earthquake			Design
Maximum	Minimum	S_S (g)	F_a	S_{MS} (g)	S_{DS} (g)
D		1.5	1.0	1.5	1.0
C	D	0.55	1.36	0.75	0.50
B	C	0.33	1.53	0.50	0.33
	B	0.156	1.6	0.25	0.167

Table A-1B   Summary of Mapped Values of 1-Second Spectral Accelerations, Site Coefficients, and Design Parameters Used for Collapse Evaluation of Seismic Design Categories D, C, and B Structure Archetypes

Seismic Design Category		Maximum Considered Earthquake			Design
Maximum	Minimum	S_1 (g)	F_v	S_{M1} (g)	S_{D1} (g)
D		0.60	1.50	0.90	0.60
C	D	0.132	2.28	0.30	0.20
B	C	0.083	2.4	0.20	0.133
	B	0.042	2.4	0.10	0.067

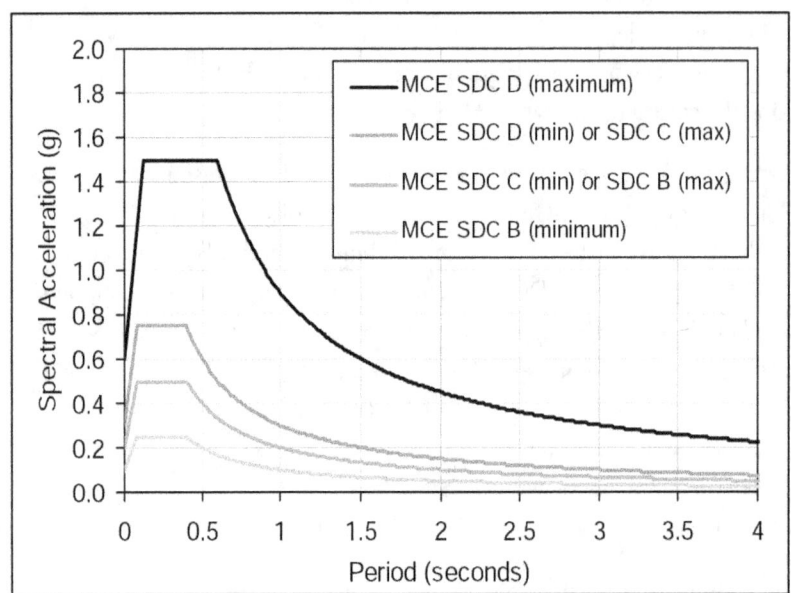

Figure A-1　Plots of MCE response spectral accelerations used for collapse evaluation of Seismic Design Categories D, C, and B structure archetypes.

The 1-second value of spectral acceleration shown in Table A-1B, $S_1 = 0.60$, is rounded upward slightly for convenience, and should be taken as $S_1 < 0.60$g, for the purpose of evaluating minimum base shear design requirements, i.e., Equation 12.8-6 of ASCE/SEI 7-05 should not be used for design of index archetypes.

Maximum values of seismic criteria for SDC D ($S_s = 1.5$ g and $S_1 = 0.60$ g) are based on the effective boundary between deterministic (near-source) and probabilistic regions of MCE ground motions, as defined in Section 21.2 of ASCE/SEI 7-05. The Methodology purposely excludes SDC D structures at deterministic (near-source) sites defined by 1-second spectral acceleration equal to or greater than 0.60 g, i.e., Section 11.6 of ASCE/SEI 7-05 defines SDC D structures as having 1-second spectral acceleration values as large as 0.75 g.

The seismic response coefficient, C_s, is used for design of the structure archetypes. This coefficient is based on the fundamental period, T, which is computed using Equation A-1. T is the code-defined period, and not the period computed using eigenvalue analysis of the structural model.

$$T = C_u T_a = C_u \quad C_t \quad h_n^x \geq 0.25 \text{ seconds} \qquad \text{(A-1)}$$

where h_n is the height, in feet, of the building above the base to the highest level of the structure, and values of the coefficient, C_u, are given in Table 12.8-1 and values of period parameters, C_t and x, are given in Table 12.8-2 of ASCE/SEI 7-05.

Period parameters, C_t and x are a function of structure type, distinguishing between steel moment frames, reinforced concrete moment frames, eccentrically braced steel frames and all other stiff structural systems. Example values of the fundamental period, T, and corresponding MCE spectral acceleration, S_{MT}, are given in Table A-2 for concrete moment resisting frame structures of various heights, ranging from 1-story to 20-stories. Table A-2 lists example values of the fundamental period and MCE spectral acceleration corresponding to maximum and minimum seismic criteria of Seismic Design Categories B, C, and D.

Table A-2   **Example values of the Fundamental Period, T, and Corresponding MCE Spectral Acceleration, S_{MT}, for Reinforced Concrete Moment Frame Structures of Various Heights**

System Properties		Seismic Design Category (SDC) Max and Min Seismic Criteria							
		SDC D$_{max}$		SDC C$_{max}$		SDC B$_{max}$		SDC B$_{min}$	
				SDC D$_{min}$		SDC C$_{min}$			
No. of Stories	Height (ft.)	T (sec.)	S_{MT} (g)	T (sec.)	S_{MT} (g)	T (sec.)	S_{MT} (g)	T (sec.)	S_{MT} (g)
1	15	0.26	1.50	0.27	0.75	0.30	0.50	0.31	0.25
2	28	0.45	1.50	0.48	0.62	0.52	0.38	0.55	0.183
4	54	0.81	1.11	0.87	0.34	0.95	0.21	0.99	0.101
8	106	1.49	0.60	1.60	0.188	1.74	0.115	1.81	0.055
12	158	2.13	0.42	2.29	0.131	2.49	0.080	2.59	0.039
20	262	3.36	0.27	3.60	0.083	3.92	0.051	4.08	0.024

Section A.8 discusses scaling of record sets and provides factors for scaling the Far Field record set to match MCE spectral acceleration, S_{MT}, for Seismic Design Categories B, C and D, respectively.

A.6   PEER NGA Database

The PEER NGA database is an extension of the earlier PEER Strong Motion Database that was first made publicly available in 1999. The PEER NGA database is composed of over 3,550 ground motion recordings that represent over 160 seismic events (including aftershock events) ranging in magnitude from M4.2 to M7.9.

Each ground motion recording includes two horizontal components of acceleration; the vertical acceleration component is also available for many records. In addition, the rotated fault-normal and fault-parallel motion components are also available. This database contains records from a wide range of domestic and international sources including the United States Geological Survey, the California Division of Mines and Geology, the Central Weather Bureau, the Earthquake Research Department of Turkey, and many others.

The PEER NGA database was chosen for use in this study for several reasons, including the large number of ground motion records, and the availability of fault-normal and fault-parallel components of ground motion. This report utilizes version 6.0 of the PEER NGA database. After this work was completed, version 7.3 of the database was released, and some ground motion parameters differ slightly in the new version, such as PGV. Even so, the variations are small and have no tangible impact on the FEMA P695 Methodology.

A.7   Record Selection Criteria

This section describes record selection criteria developed to meet Methodology objectives. Each criterion is listed, followed by a brief discussion of the intent of the rule.

- **Source Magnitude.** Large-magnitude events pose the greatest risk of building collapse due to inherently longer durations of strong shaking and larger amounts of energy released. Ground motions of smaller magnitude (M < 6.5) events can cause building damage (typically of a nonstructural nature), but are not likely to collapse new structures. Even when small magnitude events generate strong ground motions, the duration of strong shaking is relatively short and the affected area is relatively small. In contrast, large-magnitude events can generate strong, long duration, ground motions over a large region, affecting a much larger population of buildings.

- **Source Type.** Record sets include ground motions from earthquakes with either strike-slip or reverse (thrust) sources. These sources are typical of shallow crustal earthquakes in California and other Western United States locations. Few strong-motion records are available from other source mechanisms.

- **Site Conditions.** Record sets include ground motions recorded on either soft rock (Site Class C) or stiff soil (Site Class D) sites. Records on soft soil (Site Class E) or sites susceptible to ground failure (Site Class F) are not used. Relatively few strong-motion records are available for Site Class B (rock) sites.

- **Site-Source.** The 10 km source-to-site distance boundary between Near-Field and Far-Field records is arbitrary, but generally consistent with the "near fault" region of MCE design values maps in ASCE/SEI 7-05. Several different measures of this distance are available. For this project, the source-to-site distance was taken as the average of Campbell and Joyner-Boore fault distances provided in the PEER NGA database.

- **Number of Records per Event.** Strong-motion instruments are not evenly distributed across seismically active regions. Due to the number of instruments in place at the time of the earthquake, some large-magnitude events have generated many records, while others have produced only a few. To avoid potential event-based bias in record sets, not more than two records are taken from any one earthquake for a record set. The two-record limit was applied separately to the Near-Field "Pulse" and "No Pulse" record sets, respectively. When more than two records of an event pass the other selection criteria, the two records with highest peak ground velocity are selected.

- **Strongest Ground Motion Records.** The limits of greater than 0.2 g on PGA and greater than 15 cm/sec on PGV are arbitrary, but generally represent the threshold of structural damage (for new buildings) and capture a large enough sample of the strongest ground motions (recorded to date) to permit calculation of record-to-record variability.

- **Strong-Motion Instrument Capability.** Some strong-motion instruments, particularly older models, have inherent limitations on their ability to record long-period vibration accurately. Most records have a valid frequency content of at least 8 seconds, but some records do not, and records not valid to at least 4 seconds are excluded from the record sets. The record sets are considered valid for collapse evaluation of tall buildings with elastic fundamental periods up to about 4 seconds.

- **Strong-Motion Instrument Location.** Strong-motion instruments are sometimes located inside buildings (e.g., ground floor or basement) that, if large, can influence recorded motion due to soil-structure-foundation interaction. Instead, instruments located in free-field location or on ground floor of a small building should be used.

A.8   Scaling Method

Scaling of ground motion records is a necessary element of nonlinear dynamic analysis, since few, if any, available unscaled records are strong enough to collapse modern buildings. The scaling process of the Methodology involves two elements: normalization and scaling.

Normalization of Records. Individual records of a given set are normalized by their respective peak ground velocities. In essence, some records are factored upwards (and some factored downwards), while maintaining the same overall ground motion strength of the record set. Normalization by peak ground velocity is a simple way to remove unwarranted variability between records due to inherent differences in event magnitude, distance to source, source type and site conditions, while still maintaining the inherent aleatory (i.e., record-to-record) variability necessary for accurately predicting collapse fragility.

Normalization is done with respect to the value of peak ground velocity computed in the PEER NGA database (PGV_{PEER}), which is the geometric mean of PGV of the two horizontal components considering different record orientations. The geometric mean (or geomean) is the square root of the product of the two horizontal components and is a common parameter used to characterize ground motions.

The following formulas define the normalization factor, NM_i, and calculation of normalized horizontal components of the ith record, respectively:

$$NM_i = \text{Median}(PGV_{PEER,i})/PGV_{PEER,i} \qquad \text{(A-2)}$$

$$NTH_{1,i} = NM_i \times TH_{1,i}$$
$$NTH_{2,i} = NM_i \times TH_{2,i} \qquad \text{(A-3)}$$

where:

NM_i = Normalization factor of both horizontal components of the ith record (of the set of interest),

$PGV_{PEER,i}$ = Peak ground velocity of the ith record (PEER NGA database),

$\text{Median}(PGV_{PEER,i})$ = Median of $PGV_{PEER,i}$ values of records in the set,

$NTH_{1,i}$ = Normalized i^{th} record, horizontal component 1,

$NTH_{2,i}$ = Normalized i^{th} record, horizontal component 2,

$TH_{1,i}$ = Record i, horizontal component 1 (PEER NGA database), and

$TH_{2,i}$ = Record i, horizontal component 2 (PEER NGA database).

Records and their corresponding values of PGV$_{PEER}$ are taken directly from the PEER NGA database. For Near-Field records, the components rotated to the fault normal (FN) and fault parallel (FP) directions are utilized. Horizontal components of Far-Field records are not rotated based on the orientation with respect to the fault and their as-recorded orientation is used. Horizontal components of each record are normalized by the same normalization factor, NM_i, to maintain the relative, as-recorded strength of the two components (which can be important for analysis of three-dimensional archetype models.

Table A-3 provides median values of 5%-damped spectral acceleration, $\hat{S}_{NRT}$, of normalized Far-Field and Near-Field record sets, respectively. The Far-Field record set is described in Section A.9 and the Near-Field record set is described in Section A.10, and Tables A-4D and A-6D summarize normalization factors for each record of the Far-Field and Near-Field ground motion sets, respectively.

Scaling of Record Sets. For collapse evaluation, the Methodology requires the set of normalized ground motion records to be collectively scaled upward (or downward) to the point that causes 50 percent of the ground motions to collapse the archetype analysis model being evaluated. This approach is used to determine the median collapse capacity, $\hat{S}_{CT}$, of the model. Once this has been established, the Methodology requires that the collapse margin ratio, CMR, be computed between median collapse capacity, $\hat{S}_{CT}$, and MCE demand, S_{MT}.

In the same manner, record sets can be scaled to a specific level of ground motions e.g., MCE spectral acceleration. Record normalization and record set scaling to match a particular level of ground motions parallels the ground motion scaling requirements of Section 16.1.3.2 of ASCE/SEI 7-05, with the notable exception that the median value of the scaled record set need only match the MCE demand at the fundamental period, T, rather than over the range of periods required by ASCE/SEI 7-05.

Table A-3 provides scaling factors for anchoring the normalized Far-Field record set to the MCE demand level of interest. Scaling factors (for anchoring the normalized Far-Field record set to MCE demand) depend on the fundamental period of the building, T, which is shown in the left column of the table. Figure A-2 illustrates the median spectrum of the normalized Far-Field record set anchored to various levels of MCE demand at a period of 1 second.

Table A-3 Median 5%-Damped Spectral Accelerations of Normalized Far-Field and Near-Field Record Sets and Scaling Factors for Anchoring the Normalized Far-Field Record Set to MCE Spectral Demand[1].

Period $T = C_u T_a$ (sec.)	Median Value of Normalized Record Set $\hat{S}_{NRT}$ (g)		Scaling Factors for Anchoring Far-Field Record Set to MCE Spectral Demand			
	Near-Field Set	Far-Field Set	SDC D$_{max}$	SDC C$_{max}$ / SDC D$_{min}$	SDC B$_{max}$ / SDC C$_{min}$	SDC B$_{min}$
0.25	0.936	0.779	1.93	0.96	0.64	0.32
0.3	1.020	0.775	1.94	0.97	0.65	0.32
0.35	0.939	0.761	1.97	0.99	0.66	0.33
0.4	0.901	0.748	2.00	1.00	0.67	0.33
0.45	0.886	0.749	2.00	0.89	0.59	0.30
0.5	0.855	0.736	2.04	0.82	0.54	0.27
0.6	0.833	0.602	2.49	0.83	0.55	0.28
0.7	0.805	0.537	2.40	0.80	0.53	0.27
0.8	0.739	0.449	2.50	0.83	0.56	0.28
0.9	0.633	0.399	2.50	0.83	0.56	0.28
1.0	0.571	0.348	2.59	0.86	0.58	0.29
1.2	0.476	0.301	2.49	0.83	0.55	0.28
1.4	0.404	0.256	2.51	0.84	0.56	0.28
1.6	0.356	0.208	2.70	0.90	0.60	0.30
1.8	0.319	0.168	2.98	0.99	0.66	0.33
2.0	0.284	0.148	3.05	1.02	0.68	0.34
2.2	0.258	0.133	3.08	1.03	0.68	0.34
2.4	0.230	0.118	3.18	1.06	0.71	0.35
2.6	0.210	0.106	3.28	1.09	0.73	0.36
2.8	0.190	0.091	3.53	1.18	0.79	0.39
3.0	0.172	0.080	3.75	1.25	0.83	0.42
3.5	0.132	0.063	4.10	1.37	0.91	0.46
4.0	0.104	0.052	4.29	1.43	0.95	0.48
4.5	0.086	0.046	4.34	1.45	0.96	0.48
5.0	0.072	0.041	4.43	1.48	0.98	0.49

1. Spectral acceleration values and scaling factors may be based on linear interpolation for periods not listed in the table.

In Figure A-2, the median spectrum of the normalized Far-Field record set is factored by 2.59 to match the maximum MCE spectral acceleration of SDC D$_{max}$ at a period of 1 second, and factored by 0.86 to match the minimum MCE spectral acceleration of SDC D$_{min}$ at a period of 1 second. The figure shows that the median spectrum of the Far-Field record set tends to match

MCE spectral acceleration at periods near the "anchor" period, but can substantially deviate from MCE spectra at other periods. Such deviations are unavoidable due to the approximate shape of code-defined spectra.

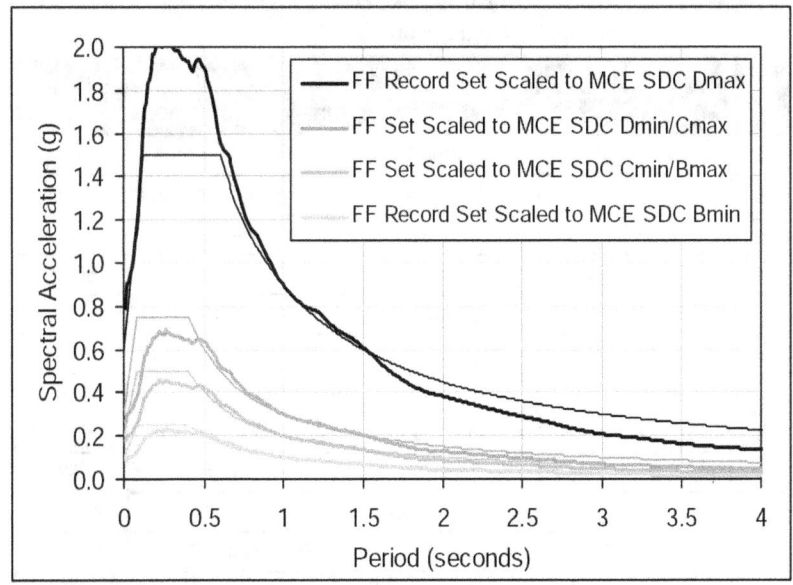

Figure A-2   Example anchoring of median spectrum of the Far-Field record set to MCE spectral acceleration at 1 second for maximum and minimum seismic criteria of Seismic Design Categories B, C and D.

A.9   Far-Field Record Set

The Far-Field record set includes twenty-two records (44 individual components) selected from the PEER NGA database using the criteria from Section A.7 of this appendix.

For each record, Table A-4A summarizes the magnitude, year, and name of the event, as well as the name and owner of the station. The twenty-two records are taken from 14 events that occurred between 1971 and 1999. Of the 14 events, eight were California earthquakes and six were from five different foreign countries. Event magnitudes range from M6.5 to M7.6 with an average magnitude of M7.0 for the Far-Field record set.

For each record, Table A-4B summarizes site and source characteristics, epicentral distances, and various other measures of site-source distance. Site characteristics include shear wave velocity and the corresponding *NEHRP* Site Class. Sixteen sites are classified as Site Class D (stiff soil sites) and the remaining six are classified as Site Class C (very stiff soil sites). Fifteen records are from events of predominantly strike-slip faulting and the remaining seven records are from events of predominantly thrust (or reverse) faulting.

Table A-4A   Summary of Earthquake Event and Recording Station Data for the Far-Field Record Set

ID No.	Earthquake			Recording Station	
	M	Year	Name	Name	Owner
1	6.7	1994	Northridge	Beverly Hills - Mulhol	USC
2	6.7	1994	Northridge	Canyon Country-WLC	USC
3	7.1	1999	Duzce, Turkey	Bolu	ERD
4	7.1	1999	Hector Mine	Hector	SCSN
5	6.5	1979	Imperial Valley	Delta	UNAMUCSD
6	6.5	1979	Imperial Valley	El Centro Array #11	USGS
7	6.9	1995	Kobe, Japan	Nishi-Akashi	CUE
8	6.9	1995	Kobe, Japan	Shin-Osaka	CUE
9	7.5	1999	Kocaeli, Turkey	Duzce	ERD
10	7.5	1999	Kocaeli, Turkey	Arcelik	KOERI
11	7.3	1992	Landers	Yermo Fire Station	CDMG
12	7.3	1992	Landers	Coolwater	SCE
13	6.9	1989	Loma Prieta	Capitola	CDMG
14	6.9	1989	Loma Prieta	Gilroy Array #3	CDMG
15	7.4	1990	Manjil, Iran	Abbar	BHRC
16	6.5	1987	Superstition Hills	El Centro Imp. Co.	CDMG
17	6.5	1987	Superstition Hills	Poe Road (temp)	USGS
18	7.0	1992	Cape Mendocino	Rio Dell Overpass	CDMG
19	7.6	1999	Chi-Chi, Taiwan	CHY101	CWB
20	7.6	1999	Chi-Chi, Taiwan	TCU045	CWB
21	6.6	1971	San Fernando	LA - Hollywood Stor	CDMG
22	6.5	1976	Friuli, Italy	Tolmezzo	--

Site-source distances are given for the closest distance to fault rupture, Campbell R distance, and Joyner-Boore horizontal distance to the surface projection of the rupture. Based on the average of Campbell and Boore-Joyner fault distances, the minimum site-source distance is 11.1 km, the maximum distance is 26.4 km and the average distance is 16.4 km for the Far-Field record set.

Table A-4B   Summary of Site and Source Data for the Far-Field Record Set

ID No.	Site Data		Source (Fault Type)	Site-Source Distance (km)			
	NEHRP Class	Vs_30 (m/sec)		Epicentral	Closest to Plane	Campbell	Joyner-Boore
1	D	356	Thrust	13.3	17.2	17.2	9.4
2	D	309	Thrust	26.5	12.4	12.4	11.4
3	D	326	Strike-slip	41.3	12	12.4	12
4	C	685	Strike-slip	26.5	11.7	12	10.4
5	D	275	Strike-slip	33.7	22	22.5	22
6	D	196	Strike-slip	29.4	12.5	13.5	12.5
7	C	609	Strike-slip	8.7	7.1	25.2	7.1
8	D	256	Strike-slip	46	19.2	28.5	19.1
9	D	276	Strike-slip	98.2	15.4	15.4	13.6
10	C	523	Strike-slip	53.7	13.5	13.5	10.6
11	D	354	Strike-slip	86	23.6	23.8	23.6
12	D	271	Strike-slip	82.1	19.7	20	19.7
13	D	289	Strike-slip	9.8	15.2	35.5	8.7
14	D	350	Strike-slip	31.4	12.8	12.8	12.2
15	C	724	Strike-slip	40.4	12.6	13	12.6
16	D	192	Strike-slip	35.8	18.2	18.5	18.2
17	D	208	Strike-slip	11.2	11.2	11.7	11.2
18	D	312	Thrust	22.7	14.3	14.3	7.9
19	D	259	Thrust	32	10	15.5	10
20	C	705	Thrust	77.5	26	26.8	26
21	D	316	Thrust	39.5	22.8	25.9	22.8
22	C	425	Thrust	20.2	15.8	15.8	15

For each record, Table A-4C summarizes key record information from the PEER NGA database. This record information includes the record sequence number and file names of the two horizontal components, as well as the lowest frequency (longest period) for which frequency content is considered fully reliable.

Maximum values of as-recorded peak ground acceleration, PGA_{max}, and peak ground velocity, PGV_{max}, are also given for each record. The term "maximum" implies that the larger value of the two components is reported. Peak ground acceleration values vary from 0.21 g to 0.82 g with an average PGA_{max} of 0.43 g. Peak ground velocity values vary from 19 cm/second to 115 cm/second with an average PGV_{max} of 46 cm/second.

Table A-4C   Summary of PEER NGA Database Information and Parameters of Recorded Ground Motions for the Far-Field Record Set

ID No.	PEER-NGA Record Information				Recorded Motions	
	Record Seq. No.	Lowest Freq (Hz.)	File Names - Horizontal Records		PGA_{max} (g)	PGV_{max} (cm/s.)
			Component 1	Component 2		
1	953	0.25	NORTHR/MUL009	NORTHR/MUL279	0.52	63
2	960	0.13	NORTHR/LOS000	NORTHR/LOS270	0.48	45
3	1602	0.06	DUZCE/BOL000	DUZCE/BOL090	0.82	62
4	1787	0.04	HECTOR/HEC000	HECTOR/HEC090	0.34	42
5	169	0.06	IMPVALL/H-DLT262	IMPVALL/H-DLT352	0.35	33
6	174	0.25	IMPVALL/H-E11140	IMPVALL/H-E11230	0.38	42
7	1111	0.13	KOBE/NIS000	KOBE/NIS090	0.51	37
8	1116	0.13	KOBE/SHI000	KOBE/SHI090	0.24	38
9	1158	0.24	KOCAELI/DZC180	KOCAELI/DZC270	0.36	59
10	1148	0.09	KOCAELI/ARC000	KOCAELI/ARC090	0.22	40
11	900	0.07	LANDERS/YER270	LANDERS/YER360	0.24	52
12	848	0.13	LANDERS/CLW-LN	LANDERS/CLW-TR	0.42	42
13	752	0.13	LOMAP/CAP000	LOMAP/CAP090	0.53	35
14	767	0.13	LOMAP/G03000	LOMAP/G03090	0.56	45
15	1633	0.13	MANJIL/ABBAR--L	MANJIL/ABBAR--T	0.51	54
16	721	0.13	SUPERST/B-ICC000	SUPERST/B-ICC090	0.36	46
17	725	0.25	SUPERST/B-POE270	SUPERST/B-POE360	0.45	36
18	829	0.07	CAPEMEND/RIO270	CAPEMEND/RIO360	0.55	44
19	1244	0.05	CHICHI/CHY101-E	CHICHI/CHY101-N	0.44	115
20	1485	0.05	CHICHI/TCU045-E	CHICHI/TCU045-N	0.51	39
21	68	0.25	SFERN/PEL090	SFERN/PEL180	0.21	19
22	125	0.13	FRIULI/A-TMZ000	FRIULI/A-TMZ270	0.35	31

For each record, Table A-4D summarizes the 1-second spectral acceleration of both horizontal components, peak ground velocity reported in the PEER NGA database, PGV_{PEER}, normalization factors, NM_I, and values of PGA_{max} and PGV_{max} after normalization by PGV_{PEER}.

Table A-4D   Summary of Factors Used to Normalize Recorded Ground Motions, and Parameters of Normalized Ground Motions for the Far-Field Record Set

| ID No. | As-Recorded Parameters | | | Normaliz-ation Factor | Normalized Motions | |
| | 1-Sec. Spec. Acc. (g) | | PGVPEER (cm/s.) | | PGA_{max} (g) | PGV_{max} (cm/s.) |
	Comp. 1	Comp. 2				
1	1.02	0.94	57.2	0.65	0.34	41
2	0.38	0.63	44.8	0.83	0.40	38
3	0.72	1.16	59.2	0.63	0.52	39
4	0.35	0.37	34.1	1.09	0.37	46
5	0.26	0.48	28.4	1.31	0.46	43
6	0.24	0.23	36.7	1.01	0.39	43
7	0.31	0.29	36.0	1.03	0.53	39
8	0.33	0.23	33.9	1.10	0.26	42
9	0.43	0.61	54.1	0.69	0.25	41
10	0.11	0.11	27.4	1.36	0.30	54
11	0.50	0.33	37.7	0.99	0.24	51
12	0.20	0.36	32.4	1.15	0.48	49
13	0.46	0.28	34.2	1.09	0.58	38
14	0.27	0.38	42.3	0.88	0.49	39
15	0.35	0.54	47.3	0.79	0.40	43
16	0.31	0.25	42.8	0.87	0.31	40
17	0.33	0.34	31.7	1.17	0.53	42
18	0.54	0.39	45.4	0.82	0.45	36
19	0.49	0.95	90.7	0.41	0.18	47
20	0.30	0.43	38.8	0.96	0.49	38
21	0.25	0.15	17.8	2.10	0.44	40
22	0.25	0.30	25.9	1.44	0.50	44

Normalization factors vary from 0.41 to 2.10. After normalization, peak ground acceleration values vary from 0.18 g to 0.58 g with an average PGA_{max} of 0.40 g. Peak ground velocity values vary from 36 cm/second to 54 cm/second with an average PGV_{max} of 42 cm/second. Table A-5 shows that normalization of the records (by PGV_{PEER}) has reduced the dispersion in PGV_{max} to a level consistent with that of PGA_{max} without appreciably affecting average values of PGA_{max} or PGV_{max} for the record set.

| Table A-5 | Far-Field Record Set (as-Recorded and After Normalization): Comparison of Maximum, Minimum and Average Values of Peak Ground Acceleration (PGA$_{max}$) and Peak Ground Velocity (PGV$_{max}$), Respectively |

Parameter Value	PGA$_{max}$ (g)		PGV$_{max}$ (g)	
	As-Recorded	Normalized	As-Recorded	Normalized
Maximum	0.82	0.58	115	54
Minimum	0.21	0.18	19	36
Max/Min Ratio	3.9	3.2	6.1	1.5
Average	0.43	0.43	46	42

Figure A-3 shows response spectra of individual records of the normalized Far-Field record set, as well as the median and one and two standard deviation spectra of the set in log format. Figure A-4 shows median and one-standard deviation spectra, as well as a plot of the standard deviation (natural log of spectral acceleration) of the normalized Far-Field record set in linear form.

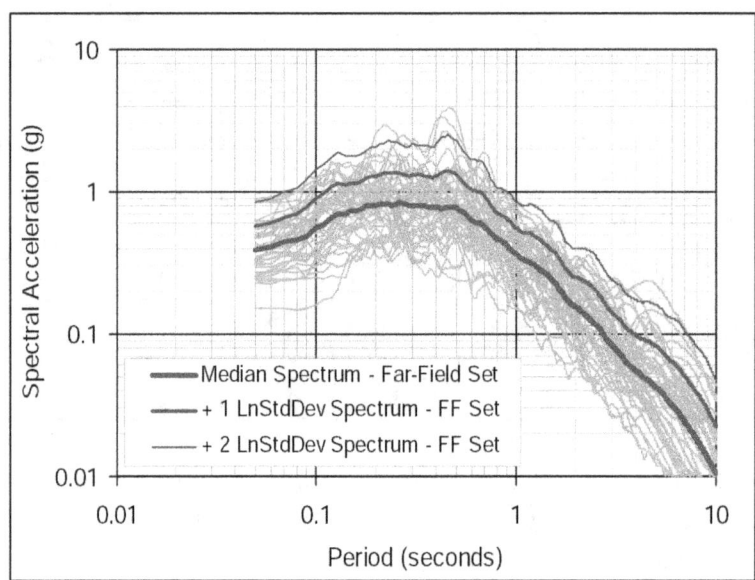

Figure A-3   Response spectra of the forty-four individual components of the normalized Far-Field record set and median, one and two standard deviation response spectra of the total record set.

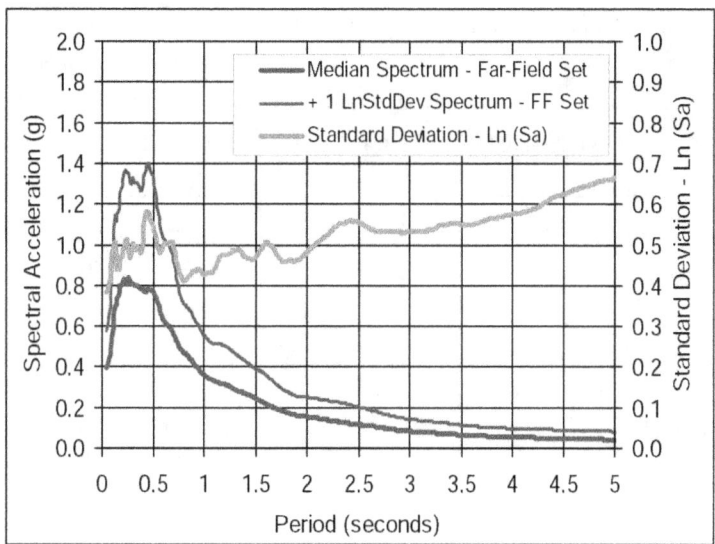

Figure A-4   Median and one standard deviation response spectra of the normalized Far-Field record set, and plot of the standard deviation (natural log) of response spectral acceleration.

The median spectrum of the normalized Far-Field record set has frequency content consistent with that of a large magnitude (M7) event recorded at 15 km from fault rupture. The median spectral acceleration at short periods is about 0.8 g and median spectral acceleration at 1-second is about 0.35 g. The domain of "constant acceleration" transitions to the domain of "constant velocity" at about 0.5 second, consistent with soft rock and stiff soil site response. Record-to-record variability (standard deviation of the logarithm of spectral acceleration) ranges from about 0.5 at short periods to about 0.6 at long periods, consistent with the values of dispersion from common ground motion attenuation relations (e.g., see Figure 8, Campbell and Borzorgnia, 2003).

Recent research has shown that spectral shape is an important aspect of collapse capacity prediction (Baker and Cornell, 2006) and that this is related to a parameter called epsilon, ε. Epsilon is defined as the number of standard deviations between the observed spectral value and the median prediction from an attenuation function such that it depends on both the period and the attenuation function used. Figure A-5 is a plot of the median value of ε calculated for the forty-four components of the Far-Field record set. These values are based on the attenuation relationship developed by Abrahamson and Silva (1997).

Median values of ε are generally small and near zero at all periods beyond 1 second, indicating that the Far-Field record set is approximately "ε-neutral."

Accordingly, collapse margin ratios (*CMRs*) based on incremental dynamic analysis using these records do not account for the effects of spectral shape, discussed in Section A.4, and are later increased to account for these effects, using the factors described in Section 7.4 and developed in Appendix B.

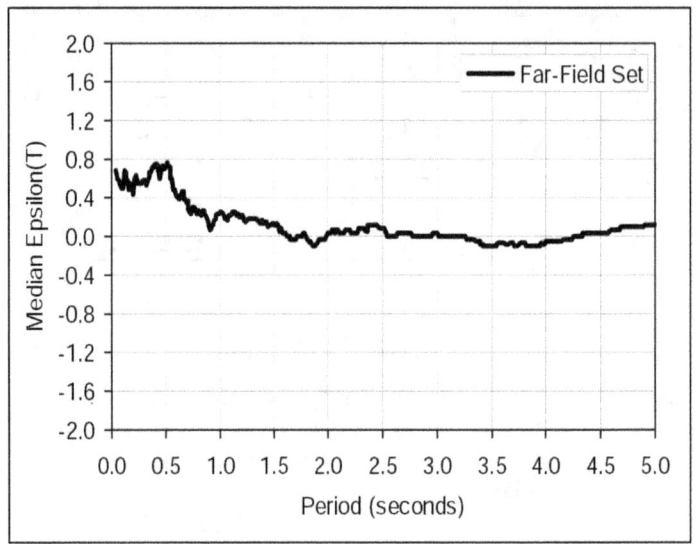

Figure A-5   Plot of median value of epsilon for the Far-Field record set.

A.10 Near-Field Record Set

The Near-Field record set includes twenty-eight records (56 individual components) selected from the PEER NGA database using the criteria from Section A.7 of this appendix. Fourteen records have pulses (Pulse subset) and fourteen records do not have pulses (No-Pulse subset), as judged by wavelet analysis classification of the records (Baker, 2007).

For each record, Table A-6A summarizes the magnitude, year and name of earthquake events and the name and owner of the station. The twenty-eight records are taken from 14 events that occurred between 1976 and 2002. Of the 14 events, seven were United States earthquakes (six in California) and seven were from five different foreign countries. Event magnitudes range from M6.5 to M7.9 with an average magnitude of M7.0.

Table A-6A   Summary of Earthquake Event and Recording Station Data for the Near-Field Record Set

| ID No. | Earthquake | | | Recording Station | |
	M	Year	Name	Name	Owner
Pulse Records Subset					
1	6.5	1979	Imperial Valley-06	El Centro Array #6	CDMG
2	6.5	1979	Imperial Valley-06	El Centro Array #7	USGS
3	6.9	1980	Irpinia, Italy-01	Sturno	ENEL
4	6.5	1987	Superstition Hills-02	Parachute Test Site	USGS
5	6.9	1989	Loma Prieta	Saratoga - Aloha	CDMG
6	6.7	1992	Erzican, Turkey	Erzincan	--
7	7.0	1992	Cape Mendocino	Petrolia	CDMG
8	7.3	1992	Landers	Lucerne	SCE
9	6.7	1994	Northridge-01	Rinaldi Receiving Sta	DWP
10	6.7	1994	Northridge-01	Sylmar - Olive View	CDMG
11	7.5	1999	Kocaeli, Turkey	Izmit	ERD
12	7.6	1999	Chi-Chi, Taiwan	TCU065	CWB
13	7.6	1999	Chi-Chi, Taiwan	TCU102	CWB
14	7.1	1999	Duzce, Turkey	Duzce	ERD
No Pulse Records Subset					
15	6.8	6.8	Gazli, USSR	Karakyr	--
16	6.5	1979	Imperial Valley-06	Bonds Corner	USGS
17	6.5	1979	Imperial Valley-06	Chihuahua	UNAMUCSD
18	6.8	1985	Nahanni, Canada	Site 1	--
19	6.8	1985	Nahanni, Canada	Site 2	--
20	6.9	1989	Loma Prieta	BRAN	UCSC
21	6.9	1989	Loma Prieta	Corralitos	CDMG
22	7.0	1992	Cape Mendocino	Cape Mendocino	CDMG
23	6.7	1994	Northridge-01	LA - Sepulveda VA	USGS/VA
24	6.7	1994	Northridge-01	Northridge - Saticoy	USC
25	7.5	1999	Kocaeli, Turkey	Yarimca	KOERI
26	7.6	1999	Chi-Chi, Taiwan	TCU067	CWB
27	7.6	1999	Chi-Chi, Taiwan	TCU084	CWB
28	7.9	2002	Denali, Alaska	TAPS Pump Sta. #10	CWB

For each record, Table A-6B summarizes site and source characteristics, epicentral distances, and various measures of site-source distance. Site characteristics include shear wave velocity and the corresponding NEHRP Site Class. Eleven sites are classified as Site Class D (stiff soil sites), fifteen are classified as Site Class C (very stiff soil sites), and the remaining two are classified as Site Class B (rock sites). Fourteen records are from events of predominantly strike-slip faulting and the remaining fourteen records are from events of predominantly thrust (or reverse) faulting.

Based on the average of Campbell and Boore-Joyner fault distances, the minimum site-source distance is 1.7 km, the maximum distance is 8.8 km, and the average distance is 4.2 km.

Table A-6B   Summary of Site and Source Data for the Near-Field Record Set

ID No.	Site Data		Source (Fault Type)	Site-Source Distance (km)			
	NEHRP Class	Vs_30 (m/sec)		Epicentral	Closest to Plane	Campbell	Joyner-Boore
Pulse Records Subset							
1	D	203	Strike-slip	27.5	1.4	3.5	0.0
2	D	211	Strike-slip	27.6	0.6	3.6	0.6
3	B	1000	Normal	30.4	10.8	10.8	6.8
4	D	349	Strike-slip	16.0	1.0	3.5	1.0
5	C	371	Strike-slip	27.2	8.5	8.5	7.6
6	D	275	Strike-slip	9.0	4.4	4.4	0.0
7	C	713	Thrust	4.5	8.2	8.2	0.0
8	C	685	Strike-slip	44.0	2.2	3.7	2.2
9	D	282	Thrust	10.9	6.5	6.5	0.0
10	C	441	Thrust	16.8	5.3	5.3	1.7
11	B	811	Strike-slip	5.3	7.2	7.4	3.6
12	D	306	Thrust	26.7	0.6	6.7	0.6
13	C	714	Thrust	45.6	1.5	7.7	1.5
14	D	276	Strike-slip	1.6	6.6	6.6	0.0
No Pulse Records Subset							
15	C	660	Thrust	12.8	5.5	5.5	3.9
16	D	223	Strike-slip	6.2	2.7	4.0	0.5
17	D	275	Strike-slip	18.9	7.3	8.4	7.3
18	C	660	Thrust	6.8	9.6	9.6	2.5
19	C	660	Thrust	6.5	4.9	4.9	0.0
20	C	376	Strike-slip	9.0	10.7	10.7	3.9
21	C	462	Strike-slip	7.2	3.9	3.9	0.2
22	C	514	Thrust	10.4	7.0	7.0	0.0
23	C	380	Thrust	8.5	8.4	8.4	0.0
24	D	281	Thrust	3.4	12.1	12.1	0.0
25	D	297	Strike-slip	19.3	4.8	5.3	1.4
26	C	434	Thrust	28.7	0.6	6.5	0.6
27	C	553	Thrust	8.9	11.2	11.2	0.0
28	C	553	Strike-slip	7.0	8.9	8.9	0.0

For each record, Table A-6C summarizes key record information from the PEER NGA database. This record information includes the record sequence number and the file names of the two horizontal components as well as the lowest frequency (longest period) for which frequency content is considered fully reliable.

Maximum values of as-recorded peak ground acceleration, PGA_{max}, and peak ground velocity, PGV_{max}, are given for each record. The term "maximum" implies the larger peak ground velocity of the two components is reported. Peak ground acceleration values range from 0.22 g to 1.43 g with an average PGA_{max} of 0.60 g. Peak ground velocity values range from 30 cm/second to 167 cm/second with an average PGV_{max} of 84 cm/second.

Table A-6C   Summary of PEER NGA Database Information and Parameters of Recorded Ground Motions for the Near-Field Record Set

ID No.	PEER-NGA Record Information				Recorded Motions	
	Record Seq. No.	Lowest Freq (Hz.)	File Names - Horizontal Records		PGA_{max} (g)	PGV_{max} (cm/s.)
			FN Component	FP Component		
Pulse Records Subset						
1	181	0.13	IMPVALL/H-E06_233	IMPVALL/H-E06_323	0.44	111.9
2	182	0.13	IMPVALL/H-E07_233	IMPVALL/H-E07_323	0.46	108.9
3	292	0.16	ITALY/A-STU_223	ITALY/A-STU_313	0.31	45.5
4	723	0.15	SUPERST/B-PTS_037	SUPERST/B-PTS_127	0.42	106.8
5	802	0.13	LOMAP/STG_038	LOMAP/STG_128	0.38	55.6
6	821	0.13	ERZIKAN/ERZ_032	ERZIKAN/ERZ_122	0.49	95.5
7	828	0.07	CAPEMEND/PET_260	CAPEMEND/PET_350	0.63	82.1
8	879	0.10	LANDERS/LCN_239	LANDERS/LCN_329	0.79	140.3
9	1063	0.11	NORTHR/RRS_032	NORTHR/RRS_122	0.87	167.3
10	1086	0.12	NORTHR/SYL_032	NORTHR/SYL_122	0.73	122.8
11	1165	0.13	KOCAELI/IZT_180	KOCAELI/IZT_270	0.22	29.8
12	1503	0.08	CHICHI/TCU065_272	CHICHI/TCU065_002	0.82	127.7
13	1529	0.06	CHICHI/TCU102_278	CHICHI/TCU102_008	0.29	106.6
14	1605	0.10	DUZCE/DZC_172	DUZCE/DZC_262	0.52	79.3
No Pulse Records Subset						
15	126	0.06	GAZLI/GAZ_177	GAZLI/GAZ_267	0.71	71.2
16	160	0.13	IMPVALL/H-BCR_233	IMPVALL/H-BCR_323	0.76	44.3
17	165	0.06	IMPVALL/H-CHI_233	IMPVALL/H-CHI_323	0.28	30.5
18	495	0.06	NAHANNI/S1_070	NAHANNI/S1_160	1.18	43.9
19	496	0.13	NAHANNI/S2_070	NAHANNI/S2_160	0.45	34.7
20	741	0.13	LOMAP/BRN_038	LOMAP/BRN_128	0.64	55.9
21	753	0.25	LOMAP/CLS_038	LOMAP/CLS_128	0.51	45.5
22	825	0.07	CAPEMEND/CPM_260	CAPEMEND/CPM_350	1.43	119.5
23	1004	0.12	NORTHR/0637_032	NORTHR/0637_122	0.73	70.1
24	1048	0.13	NORTHR/STC_032	NORTHR/STC_122	0.42	53.2
25	1176	0.09	KOCAELI/YPT_180	KOCAELI/YPT_270	0.31	73.0
26	1504	0.04	CHICHI/TCU067_285	CHICHI/TCU067_015	0.56	91.8
27	1517	0.25	CHICHI/TCU084_271	CHICHI/TCU084_001	1.16	115.1
28	2114	0.03	DENALI/ps10_199	DENALI/ps10_289	0.33	126.4

For each record, Table A-6D summarizes the 1-second spectral acceleration (both horizontal components), peak ground velocity reported in the PEER database, PGV_{PEER} (i.e., single value of peak ground velocity based on the geometric mean of rotated components), normalization factors (NM_i), and values of PGA_{max} and PGV_{max} after normalization by PGV_{PEER}. Normalization is done separately for Pulse subset and No-Pulse subsets.

Table A-6D Summary of Factors Used to Normalize Recorded Ground Motions, and Parameters of Normalized Ground Motions for the Near-Field Record Set

ID No.	As-Recorded Parameters			Normaliz-ation Factor	Normalized Motions	
	1-Sec.Spec. Acc. (g)		PGV_{PEER} (cm/s.)		PGA_{max} (g)	PGV_{max} (cm/s.)
	FN Comp.	FP Comp.				
Pulse Records Subset						
1	0.43	0.60	83.9	0.90	0.40	100.1
2	0.66	0.64	78.3	0.96	0.44	104.4
3	0.25	0.41	43.7	1.72	0.53	78.2
4	0.97	0.51	71.9	1.04	0.44	111.6
5	0.47	0.32	46.1	1.63	0.62	90.6
6	0.98	0.37	68.8	1.09	0.53	104.2
7	0.92	0.70	69.6	1.08	0.68	88.6
8	0.43	0.34	97.2	0.77	0.62	108.4
9	1.96	0.47	109.3	0.69	0.59	114.9
10	0.89	0.65	94.4	0.80	0.58	97.7
11	0.29	0.28	26.9	2.79	0.62	83.2
12	1.33	1.10	101.6	0.74	0.60	94.4
13	0.60	0.58	87.5	0.86	0.25	91.5
14	0.54	0.73	69.6	1.08	0.56	85.6
No Pulse Records Subset						
15	0.81	0.42	65.0	0.86	0.61	61.4
16	0.44	0.44	49.8	1.13	0.86	49.8
17	0.41	0.37	28.2	1.99	0.56	60.8
18	0.53	0.29	44.1	1.27	1.50	55.9
19	0.16	0.29	28.7	1.95	0.87	67.8
20	0.55	0.45	49.0	1.15	0.73	64.0
21	0.53	0.50	47.9	1.17	0.60	53.3
22	0.42	0.73	84.4	0.66	0.95	79.4
23	0.62	1.00	72.6	0.77	0.56	54.2
24	0.81	0.40	47.7	1.18	0.50	62.6
25	0.38	0.35	62.4	0.90	0.28	65.6
26	0.75	0.75	72.3	0.78	0.44	71.3
27	2.54	0.86	90.3	0.62	0.72	71.5
28	0.69	0.82	98.5	0.57	0.19	72.0

Normalization factors vary from 0.57 to 2.79. After normalization, peak ground acceleration values range from 0.19 g to 1.50 g with an average PGA_{max} of 0.60 g. Peak ground velocity values range from 50 cm/second to 115 cm/second with an average PGV_{max} of 80 cm/second. Table A-7 shows that normalization of the records (by PGV_{PEER}) has substantially reduced the dispersion in PGV_{max} without greatly affecting average values of PGA_{max} or PGV_{max}, or the dispersion in PGA_{max}.

Table A-7   Near-Field Record Set (As-Recorded and After Normalization): Comparison of Maximum, Minimum and Average Values of Peak Ground Acceleration (PGA_{max}) and Peak Ground Velocity (PGV_{max}), Respectively

Parameter Value	PGA_{max} (g)		PGV_{max} (g)	
	As-Recorded	Normalized	As-Recorded	Normalized
Maximum	1.43	1.50	167	115
Minimum	0.22	0.19	30	50
Max/Min Ratio	6.5	7.9	5.6	2.3
Average	0.60	0.60	84	80

Figure A-6 shows response spectra of individual records of the normalized Near-Field record set, as well as the median and one and two standard deviation spectra of the set in log format. Figure A-7 shows median and one-standard deviation spectra, as well as a plot of the standard deviation (natural log of spectral acceleration) of the normalized Near-Field record set in linear format.

The median spectrum of the normalized Near-Field record set has frequency content consistent with a large magnitude (M7) event recorded relatively close (5 km) to the fault rupture. The median spectral acceleration at short periods is about 1.0 g and median spectral acceleration at 1-second is about 0.6 g. Record-to-record variability (standard deviation of the logarithm of spectral acceleration) ranges from about 0.4-0.5 at short periods to about 0.6 at long periods, generally consistent (except at very short periods) with the values of dispersion from common ground motion (attenuation) relations (e.g., see Figure 8, Campbell and Borzorgnia, 2003).

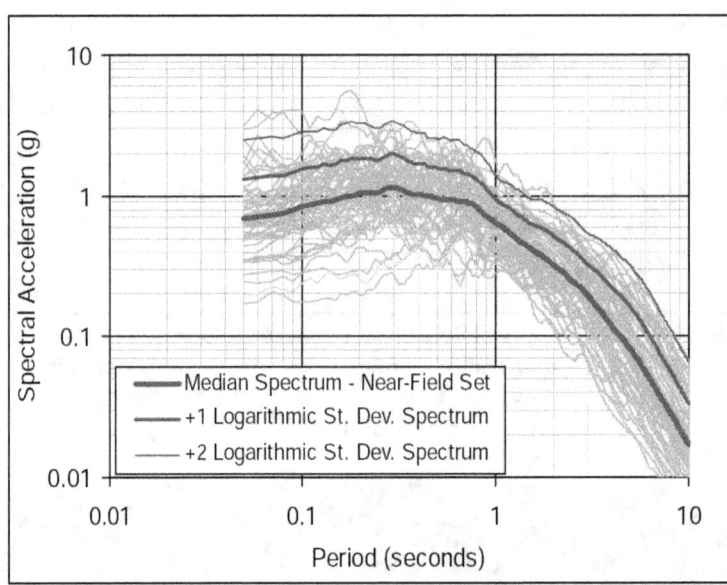

Figure A-6   Response spectra of the fifty-six individual components of the normalized Near-Field record set, and median, one and two standard deviation response spectra of the total record.

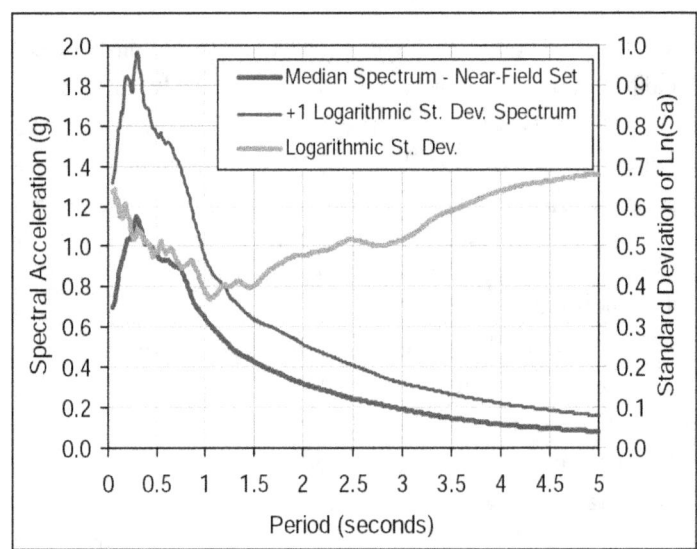

Figure A-7   Median and one standard deviation response spectra of the normalized Near-Field record set, and plot of the standard deviation (natural log) of response spectral acceleration.

Figure A-8 is a plot of the median value of epsilon calculated for the fifty-six components of the Near-Field record set. For comparison, Figure A-8 also shows the median value of ε calculated for the Far-Field record (from Figure A-5). These values are based on the attenuation relationship developed by

Abrahamson and Silva (1997). Median values of ε are generally small and near zero at all periods, with mildly positive ε values occurring at periods of 2.5 to 4.5 seconds, indicating that the Near-Field record set is approximately "epsilon neutral." Accordingly, collapse margin ratios (*CMR*s) based on incremental dynamic analysis using these records do not account for the effects of spectral shape, discussed in Section A.4, and are later increased to account for these effects, using the factors of Section 7.4 and Appendix B.

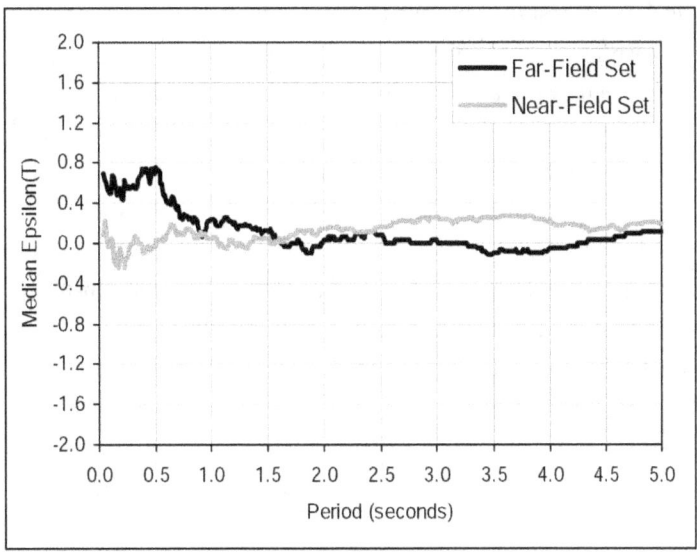

Figure A-8   Median value of epsilon for the Near-Field record set (and plot of median value of epsilon for the Far-Field record set for comparison).

A.11   Comparison of Far-Field and Near-Field Record Sets

This section compares median response spectra of the Far-Field and Near-Field record sets and collapse margin ratios (*CMR*s) calculated using these record sets. Comparisons of collapse margin ratios are made to determine if margins are, in general, substantially less for SDC E structures subjected to near-fault seismic demands than for SDC D structures subjected to SDC D_{max} seismic demands. These comparisons necessarily consider that higher seismic loads are required for design of SDC E structures (than for SDC D structures).

The three record sets used in this section to evaluate collapse margin are: (1) Far-Field record set (full set of 44 records); (2) Near-Field record set (full set of 56 records); and (3) Near-Field subset of pulse records in the fault normal (FN) direction (14 Near-Field records that are oriented in the fault normal direction and which have pulses). All records are normalized and scaled as described in Section A.8. These sets permit assessment of the effect of differences in frequency content and response characteristics of records on collapse margin. For example, are collapse margins similar for Far-Field and Near-Field record sets (when both sets are anchored to the same level of SDC E seismic criteria)? Similarly, are collapse margins substantially lower for Near-Field FN-Pulse records than for the full Near-Field record set?

Figure A-9 shows unscaled median response spectra for Near-Field and Far-Field record sets, respectively, and the ratio of these spectra. The ratio of median spectra varies from about 1.2 at short periods to about 1.6 at a period of 1-second (and over 2.0 at periods beyond 2 seconds) consistent with mapped values of ground motion required by ASCE/SEI 7-05, and other codes, for structural design near active sources. For example, the near-source coefficients of the 1997 UBC (ICBO, 1997), summarized in Table A-8, increase seismic design loads by 1.2 in acceleration domain and 1.6 in velocity domain for structures located 5 km of an active fault capable of generating large magnitude earthquakes. Corresponding increases in MCE seismic criteria are used for evaluation of collapse margin for SDC E seismic demands.

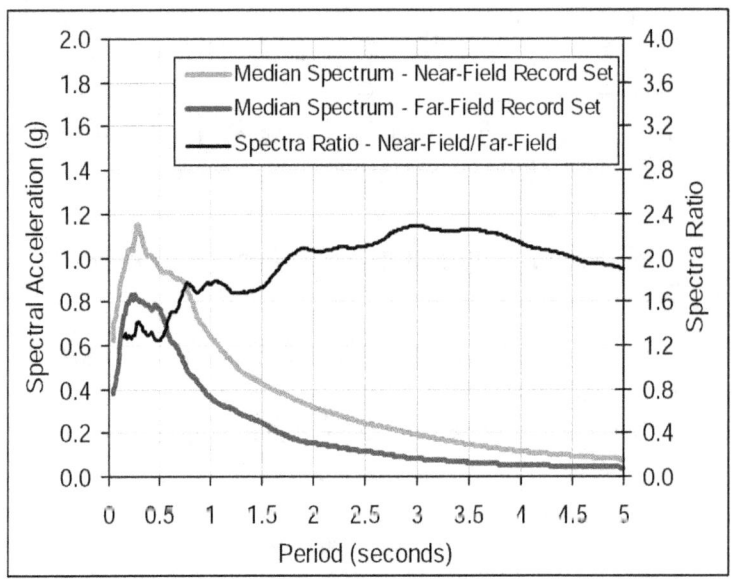

Figure A-9   Median response spectra of normalized Near-Field and Far-Field record sets, and the ratio of these spectra.

Table A-8   Near-Source Coefficients of the 1997 UBC (from Tables 16-S and 16-T, ICBO, 1997)

Spectral Domain	Closest Distance to Fault			
	≤2 km	5 km	10 km	≥15 km
Acceleration	1.5	1.2	1.0	1.0
Velocity	2.0	1.6	1.2	1.0

Figure A-10 compares unscaled median response spectra of the Far-Field record set, the full Near-Field record set, and the fault normal records of the Near-Field Pulse subset. At short periods, the three response spectra are similar, but they diverge significantly at long periods. As expected, at long periods, the median response spectrum of fault normal records of the Near-Field Pulse subset shows substantially greater seismic demand than both the Near-Field and Far-Field record sets.

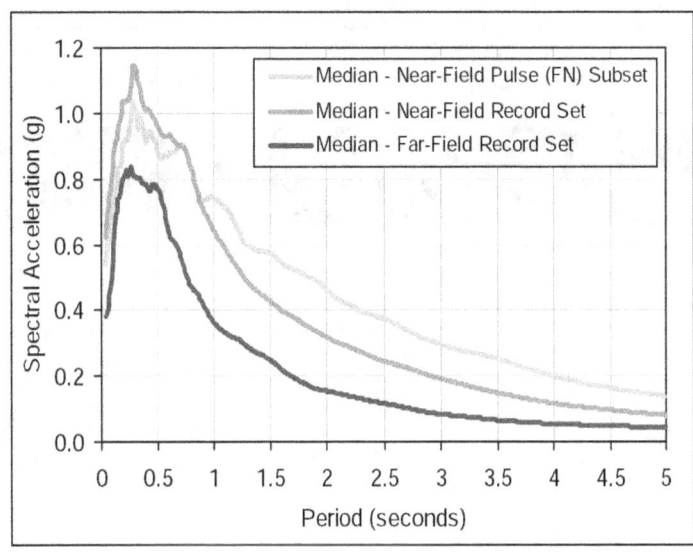

Figure A-10   Median response spectra of the normalized Near-Field record set, the Far-Field record set, and fault normal records of the Near-Field Pulse subset.

Potential differences in collapse margin ratio due to differences in the frequency content of the three record sets are investigated using archetypes of reinforced-concrete special moment-frame structures (i.e., archetypes of Chapter 9, perimeter frame and 30-foot bay configuration). Three archetype heights are considered: 1-story, 4-story and 20-story. In the case of the 20-story archetype, two designs are prepared. One design, 20-Story–NL, is without lower-bound limits on design base shear (i.e., design ignores Equation 12.8-6 of ASCE/SEI 7-05). The other design, 20-Story-02, has lower-bound base shear limits, including also the base shear limit of Equation 9.5.5.2.1-3 of ASCE 7-02 (ASCE, 2003).

Each reinforced concrete special moment frame archetype is designed in accordance with ASCE/SEI 7-05 criteria for both SDC D_{max} and SDC E design requirements (i.e. one design is prepared for SDC D_{max} and another for SDC E), respectively, with certain exceptions for 20-story archetypes. Note that the SDC E demands were not previously defined in Table A-1, since SDC E is not used in the basic assessment Methodology of this document; SDC E is only used for this comparison, so the SDC E demands are given in Table A-9.

Table A-9 summarizes fundamental periods and seismic design coefficients for each of the eight archetypes. As shown, the seismic coefficients of archetypes designed for SDC E requirements are 20% to 60% greater than those of archetypes designed for SDC D_{max} requirements.

Table A-9   Summary of Key Reinforced-Concrete Special Moment Frame Archetype Properties and Seismic Coefficients Used to Evaluate the Collapse Margin Ratio (*CMR*)

Building Archetype			Seismic Coefficient, C_s		
Height - ID	T (sec.)	C_s Limits	SDC D_{max}	SDC E	Ratio
1-Story	0.26	NA	0.125	0.15	1.2
4-Story	0.81	NA	0.092	0.149	1.6
20-Story - 02	3.36	ASCE/SEI 7-02[1]	0.044[1]	0.053[2]	1.2
20-Story - NL	3.36	None[3]	0.022	0.036	1.6

1. Archetype design includes lower-bound limit of ASCE 7-02, Equation 9.5.5.2.1-3.
2. Archetype design slightly less than C_s = 0.06 limit of ASCE/SEI 7-05, Equation 12.8-6.
3. Archetype design ignores lower-bound limit of ASCE/SEI 7-05, Equation 12.8-6.

Tables A-10A and A-10B summarize collapse margin ratios (*CMR*'s) for the eight archetypes of Table A-9. In both tables, *CMR*s of the four archetypes designed and evaluated for SDC D_{max} (far-field) sites are compared with *CMR*'s of the same four archetypes designed and evaluated for SDC E (near-field) sites.

Table A-10A compares two *CMR* values computed for SDC E sites, against the baseline *CMR* values (for SDC D_{max} design assessed using the far-field ground motion set). The SDC E *CMR* values are computed both (a) using the Far-Field ground motion set, and (b) using the Near-Field ground motion set. Both sets are anchored to SDC E demands (from Table A-9) according to section A.8. Note that the near-field record set should be used when assessing performance at a SDC E (near-field) site; the far-field set is only used for comparison and to help explain the observed differences between the computed CMR values for SDC E and SDC D_{max} sites.

Table A-10A Summary of Selected Collapse Margin Ratios (CMRs) for Reinforced-Concrete Special Moment Frame Archetypes – Comparison of CMRs for Far-Field and Near-Field Record Sets

Building Archetype	Far-Field Baseline (Sdc D_{max})	Near-Field Designs Evaluated For Sdc E Seismic Demand (Mce)			
		Near-Field Record Set		Far-Field Record Set	
Height - ID	CMR	CMR	CMR/ Baseline	CMR	CMR/ Baseline
1-Story	1.26	0.86	68%	1.03	82%
4-Story	1.98	1.32	67%	1.55	78%
20-Story - 02	1.62	1.34	83%	1.39	86%
20-Story - NL	0.82	1.01	123%	0.91	111%

Table A-10A shows the CMR values for SDC E sites, computed using both the Far-Field and Near-Field record sets. When the Near-Field record set is used for the SDC E evaluation, the CMR is an average of 30% lower as compared to SDC D_{max} (with the exception of the 20-Story-NL archetype, which will be discussed later).

This 30% difference is caused by two aspects: (1) the higher seismic demand of SDC E (it has been shown in Chapter 9 that CMR values are typically lower for buildings designed and assessed for higher seismic demands) and (2) the use of the Near-Field record set instead of the Far-Field set. To clearly separate these two effects, columns two and five show the CMR values for SDC D_{max} and SDC E, respectively, both evaluated using the Far-Field record set. This shows that the CMR values are an average of 20% lower for SDC E, simply due to SDC E having higher seismic demand. This shows that the use of Near-Field records, rather than Far-Field records, only leads to an average 10% reduction in the CMR.

Table A-10B shows how low the CMR could become if one used only the subset of Near-Field motions which are fault-normal and have pulses. However, this is not recommended for performance evaluation. For comparison, this table also replicates the results from the full Near-Field set. This table shows that the resulting CMR values are an average of 45% lower than the baseline SDC D_{max} case, and 15% lower than the values computed when the full Near-Field set is utilized.

Table A-10B Summary of Selected Collapse Margin Ratios (*CMRs*) for Reinforced-Concrete Special Moment Frame Archetypes - Comparison of *CMRs* for the Near-Field Record Set and the Near-Field Pulse FN Record Subset

Building Archetype	Far-Field Baseline (SDC D_{max})	Near-Field Designs Evaluated for SDC E Seismic Demand (MCE)			
		Near-Field Record Set		NF - Pulse FN Subset	
Height - ID	CMR	CMR	CMR/ Baseline	CMR	CMR/ Baseline
1-Story	1.26	0.86	68%	0.67	53%
4-Story	1.98	1.32	67%	0.99	50%
20-Story - 02	1.62	1.34	83%	0.92	57%
20-Story - NL	0.82	1.01	123%	0.70	85%

The 20-Story-NL archetype is the only building that does not follow the trends described in the preceding discussion. One possible reason for this is that the minimum base shear limit was not imposed in this design, which caused the design strengths to become very low (C_s = 0.022 g to 0.036 g), and the collapse capacity of extremely weak structures becomes more sensitive to changes in design strength. Another possible contributing factor is that the fundamental period, T_1, is longer for this design, and the spectral shapes change slightly for periods above 3.5 seconds (see Figure A-9).

In summary, as compared to Far-Field motions, when a structure is subjected to the full set of Near-Field records, the *CMR* is typically 10% lower. However, when a structure is subjected to the subset of FN pulse-type records, the *CMR* is decreased by about 25%. This shows that the increase in the seismic response coefficient, C_s, required for design of structures located near faults does not appear sufficient to result in performance comparable to that of the same system (i.e., same *R* factor) located further away from fault rupture.

Tables A-10A and A-10B also show that without lower-bound limits on design base shear, collapse margins for the 20-Story-NL archetype are very low; but with lower-bound limits, collapse margins for the 20-Story-02 archetype are approximately the same as the 4-Story archetype. While the margins for the 20-Story-NL archetype are very low, it should be noted that a 60% increase in SDC E design strength affects comparable or better performance than the baseline (SDC D) archetype. In fact, a 60% increase in design base shear is able to cause about the same collapse margin for the 20-Story-NL (SDC E) archetype, when evaluated with fault normal components of the Near-Field pulse records, as that of the 20-Story-NL archetype (SDC D) evaluated with Far-Field records (Table A-10B).

A.12 Robustness of Far-Field Record Set

The purpose of this Methodology is to assess the collapse safety of newly proposed structural systems. To this end, the record selection criteria of section A.7 result in an appropriate set of strong ground motions that are representative of those that may cause collapse of a new building (to the extent possible, given the limited number of recordings available).

This section evaluates the "robustness" of the ground motion selection criteria to demonstrate that the final collapse capacity predictions are not highly sensitive to small changes to the selection criteria. Specifically, the peak ground acceleration (PGA) > 0.2g and peak ground velocity (PGV) > 15cm/s limits are evaluated, as well as the requirement that the two highest PGV records be selected when there are many candidate records for an event. The other selection criteria are not investigated here because they are less subjective, or in the case of magnitudes larger than M6.5, small changes to this requirement are expected to have minimal impact on the collapse capacity predictions.

A.12.1 Approach to Evaluating Robustness

To investigate the effects of the PGA and PGV selection criteria, a set of 192 records is selected based on liberal criteria (PGA > 0.05g and PGV > 3cm/s). To reduce the number of ground motions, and still maintain the selection criteria, 20 pairs of motions are randomly selected from each event. This affected the records from Chi-Chi (261 candidate records), Loma Prieta (53 records), Northridge (70 records), and Landers (24 candidate records).

To evaluate the quantitative differences between ground motion sets, each set is used to predict the median collapse capacity (the *CMR* and *ACMR*) of three reinforced concrete special moment frame building archetypes. To cover the range of structural periods, the following building archetypes are utilized in these comparisons: 1-story reinforced concrete special moment frame ($T = 0.26$ s), 4-story reinforced concrete special moment frame ($T = 0.81$ s), and a 12-story reinforced concrete special moment frame ($T = 2.13$ s). These are three archetypes are taken from the reinforced concrete special moment frame example of Section 9.2. Each archetype is a perimeter-frame systems designed for SDC D_{max} seismic criteria. Table A.11 summarizes design properties and collapse margin results for these archetypes.

Table A-11 Summary of Design Properties and Collapse Margins of the Three Reinforced Concrete Special Moment Frame Building Archetypes Used to Evaluate Far-Field Record Set Robustness

Archetype Design ID No.	Design Configuration and Properties				Collapse Margin	
	No. of Stories	Framing Type	Seismic SDC	Period T (sec.)	CMR	ACMR
2061[1]	1	P	D_{max}	0.26	1.96	2.61
1003[1]	4	P	D_{max}	0.81	1.61	2.27
1013[2]	12	P	D_{max}	2.13	1.45	2.33

1. Design data and collapse margins taken from Tables 9-2, 9-3 and 9-8.
2. Design data and collapse margins taken from Tables 9-9 and 9-10.

A.12.2 Effects of PGA Selection Criteria Alone

Table A-12 shows the effects that the minimum PGA requirement has on the computed values of *CMR*. This table includes results for PGA > 0.05g (192 records), PGA > 0.10g (105 records), PGA > 0.15g (62 records), and PGA > 0.20g (36 records). These results clearly show that the *CMR* values increase as the minimum PGA limit is increased. Comparing the PGA limit of 0.05g and 0.20g, the *CMR* increases 31% for the 1-story building, 19% for the 4-story building, and 10% for the 12-story building, with an overall average value of 20%. The PGA limit, unsurprisingly, has a more significant effect on shorter-period structures.

Table A-12 Effects of the PGA Selection Criterion on the Computed *CMR* Values for Three Reinforced Concrete Special Moment Frame Buildings

Selection Criteria and Number of Selected Records				
Min PGA (g)	0.05	0.10	0.15	0.20
Min PGV (cm/s)	3.0	3.0	3.0	3.0
No. of Records	192	105	62	36
Arch. ID No.	**CMR**			
2061	1.76	1.97	2.24	2.31
1003	1.26	1.36	1.39	1.51
1013	1.41	1.46	1.49	1.56
Mean Value	**1.48**	**1.60**	**1.71**	**1.79**

However, the *CMR* is not the appropriate parameter for use in a complete comparison of the ground motion sets. The Adjusted Collapse Margin Ratio (ACMR) value should be used because it is the complete collapse capacity result after accounting for the effects of spectral shape. In order to compute the *ACMR* values, we must compute the Spectral Shape Factor values, as discussed in Appendix B. The following discussion provides the rationale for approximate *SSF* values that can be used to compute approximate values of *ACMR* for this comparison.

Appendix B will show that the *SSF* value is, in part, based on the mean ε values of the ground motion set. Figure A-11 shows these mean ε values for the Far-Field ground motion set, and it can be reasoned that the mean ε values of the PGA > 0.20 g set (36 records) should be similar due to the similarities in the selection criteria of the two sets. In contrast, the PGA > 0.05 g (192 records) set is more representative of all possible records, so the mean ε values is expected to be approximately zero at all periods. This observation suggests that the high PGA limit used for the Far-Field set is causing the observed mean $\varepsilon = 0.6$ at short periods.

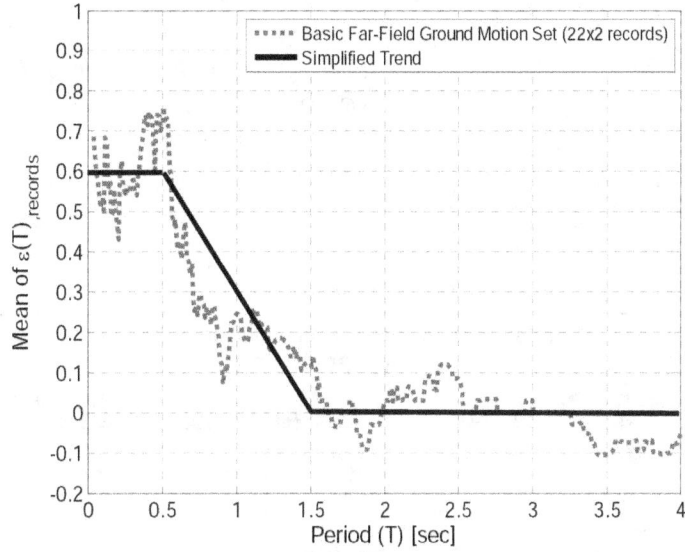

Figure A-11   Mean ε values, $\overline{\varepsilon}(T)_{,records}$, for the Far-Field ground motion set (duplicated from Figure B-3 of App. B).

Using the above rationale, Table A-13 presents the approximate *SSF* and *ACMR* values for the sets selected with PGA > 0.05 g (192 records) and PGA > 0.20 g (36 records). This shows that the differences in the mean ε values counteract much of the trend seen with the *CMR* values. For the three buildings considered here, the *ACMR* value increases by an average of 8% when the ground motions are selected based on PGA > 0.20 g rather than a more liberal selection using PGA > 0.05 g.

Table A-13 Effects of the PGA Selection Criterion on the Approximate *ACMR* Values for Three Reinforced Concrete Special Moment Frame Buildings

Selection Criteria and Number of Selected Records				
Min PGA (g)	0.05	0.20	0.05	0.20
Min PGV (cm/s)	3.0	3.0	3.0	3.0
No. of Records	192	36	192	36
Arch. ID No.	approx. *SSF*		approx. *ACMR*	
2061	1.61	1.33	2.83	3.07
1003	1.61	1.41	2.04	2.12
1013	1.61	1.61	2.27	2.51
Mean Value	1.61	1.45	2.38	2.57

A.12.3 Effects of PGV Selection Criteria Alone

Similar to previous PGA comparisons, Table A-14 shows the effect that the minimum PGV requirement has on the CMR. This shows that the impacts of the PGV requirement vary for the three buildings considered. Comparing the PGV limit of 3 cm/s and 15 cm/s, the CMR decreases 14% for the 1-story building, decreases 2% for the 4-story building, and increases 5% for the 12-story building, with an overall average decrease of 5%.

Table A-14 Effects of the PGV Selection Criterion on the Computed *CMR* Values for Three RC SMF Buildings

Selection Criteria and Number of Selected Records				
Min PGA (g)	0.05	0.05	0.05	0.05
Min PGV (cm/s)	3.0	5.0	10.0	15.0
No. of Records	192	186	130	85
Arch. ID No.	approx. *SSF*		approx. *ACMR*	
2061	1.76	1.73	1.59	1.50
1003	1.26	1.26	1.20	1.24
1013	1.41	1.42	1.47	1.48
Mean Value	1.48	1.47	1.42	1.41

For evaluating the PGV selection criterion, comparing the *CMR* value is sufficient (comparison of *ACMR* is not needed) because the PGV value is expected to affect the mean ε values at moderate or long periods. Since Figure A-11 showed that even the Far-Field set (selected for high PGV > 15 cm/s) has $\varepsilon = 0$ at moderate/long periods, it is expected that the other larger ground motion sets should also have $\varepsilon = 0$ at those periods. Therefore, the *SSF* values should approximately equal for each of the record sets shown in Table A-14.

A.12.4 Effects of both PGA and PGV Selection Criteria Simultaneously, as well as Selection of Two Records from Each Event

The previous sections investigated the minimum PGA and PGV selection criteria *individually*, and showed that they have somewhat counteracting effects on the resulting collapse capacity predictions (*CMR* and *ACMR*). To create the Far-Field record set, these two criteria are imposed *simultaneously*, and then two records are selected from each event. To select the two records from each event, the records with highest PGV are used.

Table A-15 shows results of this progression to obtain the final Far-Field ground motion set (final Far-Field set results shown in bold-italic). The three ground motion sets used to produces the collapse results shown in this table were selected as follows:

Set 1. Set selected using liberal selection criteria, with PGA > 0.05 g and PGV > 3 cm/s, and 20 records selected randomly from each event (192 records).

Set 2. Set selected from the above set of 192, but with strict imposed selection criteria of PGA > 0.20g and PGV > 15 cm/s (32 records).

Set 3. Set selected using the strict selection criteria, of PGA > 0.20 g and PGV > 15 cm/s, and two records with the highest PGV selected from each event (22 records). This is the final Far-Field record set used in the Methodology. Note that due to slight differences in the scaling method, the *CMR* values do not match exactly with those reported elsewhere in this report, so relative comparisons should be made of values within this table.

The results in Table A-15 follow directly from the trends shown previously for PGA and PGV criteria alone. Comparing Set 1 and Set 2, the *CMR* value increased by an average of 18% (*CMR* = 1.48 versus 1.74), which is driven mostly by the effects of the PGA selection criterion and the resulting differences in spectral shape. The selection of two high-PGV records from each event make the difference reduce to 12% when comparing Set 1 to Set 3 (*CMR* = 1.48 versus 1.65).

| Table A-15 | Effects of the PGA and PGV Selection Criteria on the Computed *CMR* and *ACMR* Values, for Three Reinforced Concrete Special Moment Frame Buildings, as well as The Effects of Selecting Two Records from Each Event |

Selection Criteria and Number of Selected Records									
Record Set No.	1	2	3	1	2	3	1	2	3
Min PGA (g)	0.05	0.20	0.20	0.05	0.20	0.20	0.05	0.20	0.20
Min PGV (cm/s)	3.0	15.0	15.0	3.0	15.0	15.0	3.0	15.0	15.0
Record Selection[1]	Rand	Rand	High	Rand	Rand	High	Rand	Rand	High
No. of Records	192	32	22	192	32	22	192	32	22
Arch. ID No.	CMR			approx. SSF			approx. ACMR		
2061	1.76	2.06	1.94	1.61	1.33	1.33	2.83	2.73	2.58
1003	1.26	1.53	1.43	1.61	1.41	1.41	2.04	2.16	2.01
1013	1.41	1.62	1.59	1.61	1.61	1.61	2.27	2.61	2.56
Mean Value	1.48	1.74	1.65	1.61	1.45	1.45	2.38	2.50	2.38

1. "Rand" indicates random selection from candidate records and "High" indicates biased selection from candidate records based on highest PGV values (Sec. A.7).

As with the previous comparisons of the PGA selection criterion, the differences in the *SSF* value account for most of the above differences in *CMR*, and cause the differences in *ACMR* to be much smaller. Comparing Set 1 and Set 2, *ACMR* increased by only an average of 5% (*CMR* = 2.38 versus 2.50). The selection of the two high-PGV records from each event eliminate the difference when comparing the Set 1 to Set 3 (*CMR* = 2.38 versus 2.38). Note that this average difference of 0% is rather coincidental, and the individual differences are non-zero, being -9%, -1%, and +13%, for the three buildings considered. Even the individual 1% to 13% differences in median collapse capacity are very small when comparing two sets of ground motions that were selected based on such different selection criteria.

A.12.5 Summary of the Robustness of the Far-Field Set

ACMR collapse capacity results are not highly sensitive to the PGA and PGV selection criteria or to the approach used to select the two motions from each event. More specifically, when comparing the Far-Field set (Set 3) to a set of records selected with much looser PGA and PGV criteria, the *ACMR* values vary between -9% to +13% for the three buildings considered, with an average difference of 0% which is partially coincidence when using such a small sample size. Relatively speaking, these differences of 0% to 13% are exceptionally small when comparing two ground motion sets selected using such differing criteria. This shows that the Far-Field ground motion set can be considered robust with respect to these selection criteria.

A: Ground Motion Record Sets

A.13 Assessment of Record-to-Record Variability in Collapse Fragility

According to the Methodology, the ground motion record set is scaled as described in Section A.8 in order to evaluate the median collapse capacity, $\hat{S}_{CT}$. To assess the collapse probability and accordingly the *ACMR* acceptance criteria, an estimate of the record-to-record variability in collapse capacity (β_{RTR}) is also needed. β_{RTR} is the lognormal standard deviation associated with uncertainty in response of an archetype structure due to differences in frequency content and other characteristics of ground motion records. Through the course of this project, it has become evident that β_{RTR} values are relatively consistent across different types of structures, so a fixed value of $\beta_{RTR} = 0.40$ is used in the Methodology for most structures. This section validates the choice of the 0.40 value, but shows that for structures that have limited period elongation before collapse, $\beta_{RTR} < 0.40$ is appropriate and provides guidance on how to determine β_{RTR} for these structures.

Record-to-record variability in the collapse fragility for a particular structure can be computed from the results of incremental dynamic analysis of the ground motion record set. The value of β_{RTR} obtained will depend on the scaling and normalization of the records. Two normalization and scaling methods are considered:

1. The PGV normalization method developed in this Methodology and described in Section A.8 normalizes the records by PGV and then scales the normalized record set as a group, in order to determine the median collapse capacity. An illustration of the scaled Far-Field record set is shown in Figure A-3.

2. An alternative method, termed S_a-component scaling, scales records according to the component spectral acceleration of each record individually evaluated at the period of the building. When this method is utilized, all records have exactly the same spectral acceleration at the period of interest. This is in contrast to the PGV normalization method, which maintains variability at all periods. This method is commonly used in research applications and is conceptually appealing because each individual record individually matches the target S_a level. In addition, this scaling method avoids double counting of uncertainties in prediction of the mean annual frequency of collapse, a frequently used metric of seismic performance.

The advantage of the PGV normalization method is that the entire record set can be scaled by the same scale factor, shown in Table A-3. This significantly reduces the complexity of the calculations for the user of this

Methodology. S_a-component scaling requires computing different scale factors for each record at each different period of interest.

Differences between the PGV normalization method and the S_a-component scaling method are illustrated in Figures A-12 and A-13. Figure A-12 compares the median of the record sets according to the two scaling methods. The median spectral acceleration of the records at all periods is virtually the same under the two scaling methods. Also, the median of the set under the S_a-component method does not depend on the period at which the scaling occurs. For this reason, it can be shown that the computed median collapse capacity will be the same regardless of which scaling procedure is used (see Appendix 6B of Haselton and Deierlein, 2007).

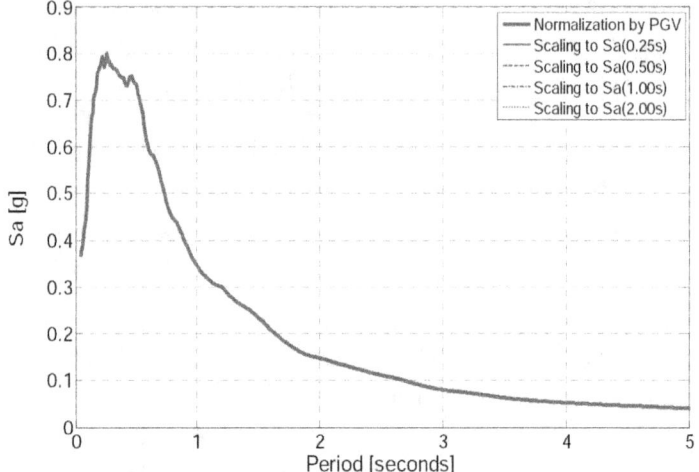

Figure A-12   Comparison of median spectra under the FEMA P695 normalization/scaling method and the S_a-component scaling method (scaled to four different periods: 0.25 s, 0.50 s, 1.0 s, and 1.2 s). The five curves on this plot are virtually indistinguishable.

Figure A-13 compares the logarithmic standard deviation of the spectral ordinates, $\sigma_{LN,Sa}$, for different scaling methods. The PGV normalization method gives relatively constant values of the $\sigma_{LN,Sa}$, regardless of the period of interest. For the S_a-component method, $\sigma_{LN,Sa}$ is zero at the period of interest, since all records have the same spectral ordinate. Away from the scaling period, $\sigma_{ln,Sa}$ values obtained from S_a-component scaling are similar to those obtained in the FEMA P695 approach.[1] Values of $\sigma_{LN,Sa}$ obtained from

---

[1] The $\sigma_{LN,Sa}$ values obtained by S_a-component scaling are notably larger at long periods away from the scaling period. The PGV normalization avoids this high uncertainty. Regardless, $\sigma_{LN,Sa}$ at long periods are not expected to significantly

S_a-component scaling indicate that the β_{RTR} for brittle structures, which collapse at periods close to the period of the building, T, is expected to be smaller than the β_{RTR} for structures that undergo significant period elongations before collapse. For brittle systems, S_a-component scaling avoids double-counting of record-to-record variability that occurs in the PGV normalization method due to the enforced variation in the record set at all periods.

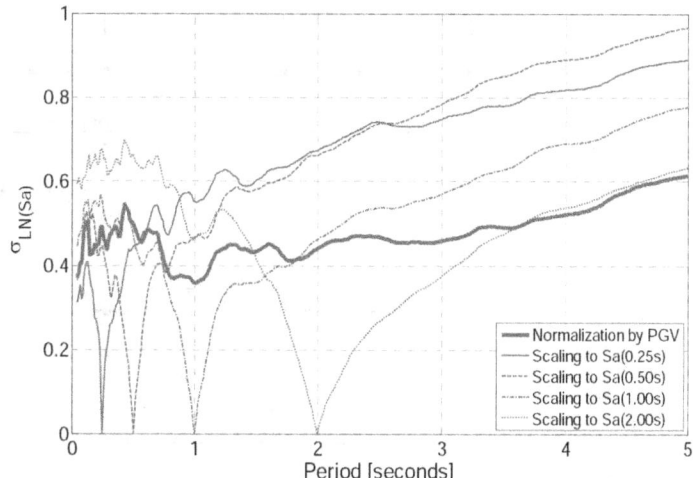

Figure A-13   Comparison of logarithmic standard deviation of spectral ordinates, $\sigma_{LN,Sa}$, for FEMA P695 scaling method and S_a-component scaling method at different periods.

To investigate the relationship between β_{RTR} and the building period elongation prior to collapse, β_{RTR} values were computed using the S_a-component scaling method for the reinforced concrete special moment frames and ordinary moment frames discussed in Chapter 9. Note that some ordinary moment frame data include hypothetical brittle non-simulated failure modes in order to get data points for very brittle systems. Period elongation is quantified by the period-based ductility parameter, μ_T, defined as the ratio of ultimate roof drift to the yield roof drift: $\mu_T = \delta_u/\delta_{y,eff}$ (see Appendix B and Chapter 7 for more details; period elongation prior to collapse would approximately be the square-root of this ratio).

The relationship between β_{RTR} and period-based ductility, μ_T, is shown in Figure A-14. The reinforced concrete special moment frames have an average period-based ductility, $\mu_T = 9$, and average $\beta_{RTR} = 0.4$ in the collapse fragility. Because these structures undergo significant period elongation

---

impact predicted structural response because the periods with high $\sigma_{ln,Sa}$ are far away from the scaling period (the elastic period of the building).

before collapse, they respond away from the pinched region of the spectrum. For reinforced concrete ordinary moment frames, which undergo less period elongation prior to collapse, Figure A-14 reveals smaller average β_{RTR} values, and shows that the β_{RTR} value clearly reduces as the building becomes less ductile (for $\mu_T < 3$).

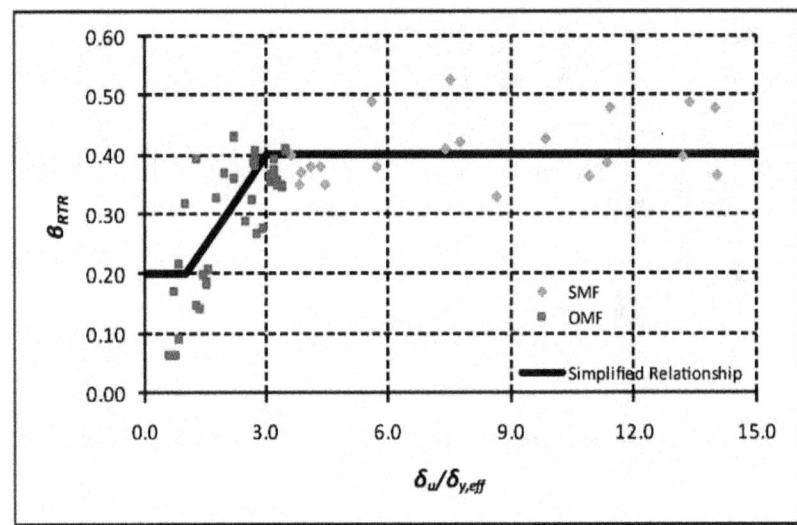

Figure A-14   Relationship between period-based ductility ($\mu_T = (\delta_u/\delta_{y,eff})$ and record-to-record variability (β_{RTR}) from collapse data for reinforced concrete frames. The FEMA P695 simplified relationship for predicting β_{RTR} is superimposed.

Figure A-14 also superimposes the FEMA P695 simplified relationship for predicting β_{RTR}. Most structures are relatively ductile and have $\mu_T > 3$. For these structures β_{RTR} should be taken as 0.40 and the totally system collapse uncertainty is given by Table 7-2. For structures with $\mu_T \leq 3$, β_{RTR} can be reduced according to Equation A-4:

$$\beta_{RTR} = 0.1 + 0.1\mu_T \leq 0.4 \qquad (A\text{-}4)$$

where β_{RTR} must be greater than or equal to 0.2. If Equation A-4 is used, the total system collapse uncertainty can be computed according to Equation 7-4, repeated below:

$$\beta_{TOT} = \sqrt{\beta^2_{RTR} + \beta^2_{DR} + \beta^2_{TD} + \beta^2_{MDL}} \qquad (A\text{-}5)$$

Values of β_{TOT} obtained should be rounded to nearest 0.05 for use in Table 7-3.

This reduction in β_{RTR} is expected to apply only to a limited number of structural systems that are either very brittle or unusual structures that have limited period elongation prior to collapse, such as base-isolated structures.

A.14 Summary and Conclusion

This appendix describes the Far-Field and Near-Field record sets and the scaling methods appropriate for collapse evaluation of building archetypes using IDA.

Both the Far-Field and Near-Field record sets have average ε values that are lower than expected for MCE motions, and therefore can substantially underestimate the *CMR* without appropriate adjustment for spectral shape effects. The adjustment method is discussed in Section 7.4 and Appendix B.

The Far-Field record set is generally appropriate for collapse evaluation of buildings, but can slightly overestimate the *CMR* of buildings at sites close to fault rupture (e.g., distances less than 10 km).

The Near-Field record set is generally appropriate for collapse evaluation of buildings at sites close to fault rupture (i.e., distances less than 10 km). Note that the Near-Field record set is not specifically required as part of the basic assessment Methodology. Even so, the Near-Field record set was developed for comparative purposes and is documented here to both substantiate the earlier comparisons and for use in other studies.

Appendix B
Adjustment of Collapse Capacity Considering Effects of Spectral Shape

This appendix describes the background and development of simplified spectral shape factors that depend on the fundamental period of the building, as well as the expected elongation of structural period as the structure collapses. These factors are used to adjust the collapse capacity to account for the frequency content (spectral shape) of the ground motion record set.

B.1   Introduction

A challenge associated with analytical prediction of structural collapse is the selection of ground motions for use in dynamic analysis. A characteristic of ground motions that can affect collapse capacity is the spectral shape. For rare ground motions in California, such as Maximum Considered Earthquake (MCE) ground motions, the spectral shape is much different than the shape of a structural design spectrum contained in ASCE/SEI 7-05 (ASCE, 2006a) or a uniform hazard spectrum (Baker, 2005; Baker and Cornell, 2006).

Figure B-1 shows the acceleration spectrum of a Loma Prieta ground motion[1] (PEER, 2006a). This motion has a MCE intensity at a period of 1.0 second, which is $S_a(1.0 \text{ sec}) = 0.9$ g for this example. This spectrum is labeled as "2% in 50 year S_a" which is the same as the MCE for this site. This figure also shows the intensity predicted by the Boore et al. (1997) attenuation prediction, consistent with the event and site associated with this ground motion. These predicted spectra include the median spectrum and the plus/minus one and two standard deviation spectra, assuming that S_a values are lognormally distributed.

---

[1] This motion is from the Saratoga station and is owned by the California Department of Mines and Geology. For this illustration, this spectrum was scaled by a factor of +1.4, in order to make the $S_a(1 \text{ s})$ demand the same as the MCE demand. For the purposes of this example, please consider this spectrum to be *unscaled*, since later values (e.g., ε) are computed using unscaled spectra.

In Figure B-1, this MCE motion has an unusual spectral shape with a "peak" from 0.6 to 1.8 seconds that is much different from the shape of a uniform hazard spectrum. This peak occurs around the period for which the motion is said to have an MCE intensity, and at this period the observed $S_a(1\ s)$ is much higher (0.9 g) than the mean expected $S_a(1\ s)$ from the attenuation function (0.3 g). This peaked shape makes intuitive sense because it seems unlikely that a ground motion with a much larger than expected spectral acceleration (meaning much higher than the mean expected) at one period would have similarly large spectral accelerations at all other periods.

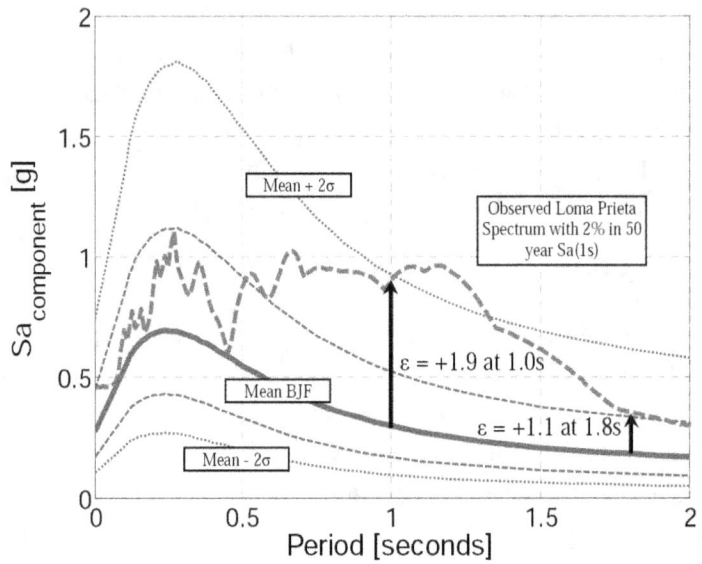

Figure B-1   Comparison of an observed spectrum with spectra predicted by Boore, Joyner, and Fumal (1997); after Haselton and Baker (2006).

Epsilon, ε, is defined as the number of logarithmic standard deviations between the observed spectral value and the median prediction from an attenuation function. At a period of 1.0 second, the spectral value is 1.9 logarithmic standard deviations above the predicted mean spectral value, so this record is said to have "$\varepsilon = 1.9$ at 1.0 second." Similarly, this record has $\varepsilon = 1.1$ at 1.8 seconds. Thus, the ε value is a function of the ground motion record, the period of interest, and the attenuation function used for ground motion prediction.

Trends shown in Figure B-1 are general to sites in coastal California where ε values ranging from 1.0 to 2.0 are typically expected for the MCE (or 2% in 50 year) ground motion level. These positive ε values come from the fact that the return period of the *ground motion* (i.e., 2,475 years for a 2% in 50 year motion) is much longer than the return period of the *event* that causes the ground motion (i.e., 150-500 years for typical events in California). Record selection for structural analyses at such sites should reflect the expected ε for the site and the ground motion hazard level of interest.

It should be noted that the expected ε value is both hazard-level and site dependent. For example, for 50% in 5 year ground motions in coastal California, ε values ranging from 0.5 to -2.0 are expected (Haselton et al., 2007, chapter 4). In the eastern United States, ε values ranging from 0.25 to 1.0 are expected for a 2% in 50 year motion. Negative ε values for a 50% in 5 year motion stems from the fact that the return period of the *ground motion* (i.e., 10 years) is much shorter than the typical return period of the *event* that causes the ground motion (e.g., 150 to 500 years). The Eastern United States has low positive ε values because seismic events are less frequent than in California, but the return periods are still shorter than the return period of a 2% in 50 year motion (i.e., 2,475 years).

Collapse capacity is defined as the $S_a(T_1)$ value that causes dynamic sidesway collapse (termed S_{CT1}). Research has shown that collapse capacity is higher for motions with a peaked spectral shape relative to motions without a peaked spectral shape. This is especially true when the peak of the spectrum is near the fundamental period of the building (T_1), and ground motions are scaled based on $S_a(T_1)$ (Haselton and Baker, 2006; Baker, 2006; Baker, 2005; Goulet et al., 2006; Zareian, 2006). Spectral accelerations at periods other than T_1 are often important to the collapse response of a building. For example, the period elongation as the building responds inelastically makes spectral values for period greater than T_1 to become important to collapse response. In addition, higher mode effects make periods less than T_1 to also become important to collapse response. Positive ε peaked spectra typically have lower spectral demands at periods away from T_1.

Past studies have shown that if $\varepsilon(T_1) = 0$ ground motions are used when $\varepsilon(T_1) = 1.5$ to 2.0 ground motions are appropriate, the median collapse capacity is under-predicted by a factor of 1.3 to 1.8 for relatively ductile structures. In cases where it is expected that the collapse-level ground motions will have high positive $\varepsilon(T_1)$ values, such as with modern buildings in high seismic areas of California, properly accounting for these values is critical.

The most direct approach to account for spectral shape is to select ground motions that have the appropriate $\varepsilon(T_1)$ expected for the site and hazard level of interest. This approach is difficult when assessing the collapse capacities of many buildings with differing T_1, because it would require a unique ground motion set for each building. To address this issue, this Appendix develops a simplified method which involves the use of a general set of ground motion records (selected independent of ε values), and a correction to the median collapse capacity estimates to account for spectral shape. In this process, the spectral shape is quantified by the $\varepsilon(T_1)$ value expected for the site and hazard level of interest.

B.2 Previous Research on Simplified Methods to Account for Spectral Shape (Epsilon)

Several recent studies have focused on how spectral shape (ε) affects collapse capacity and pre-collapse structural responses (Baker, 2006; Baker, 2005; Goulet et al., 2006; Haselton and Baker, 2006; Zareian, 2006). This Appendix does not attempt to present a full literature review of this past work.

The purpose of this Appendix is to develop a simplified method to account for spectral shape, ε. One recent study by Haselton and Deierlein (2007, Chapter 3) developed such a method, and the following is an overview of their work.

To develop the simplified method, Haselton and Deierlein first predicted the collapse capacities (in terms of S_{CT1}) of 65 modern reinforced concrete special moment frame buildings. For the collapse assessment of each building, 80 ground motions were utilized, with the goal of finding a relationship between the collapse capacity, S_{CT1}, and $\varepsilon(T_1)$. For illustration purposes, Figure B-2 shows an example of representative findings for a single building[2]. This figure shows the results of linear regression analysis that is used to define the relationship: $LN[S_{CT1}] = \beta_0 + \beta_1 \varepsilon$. The value of β_0 indicates the average collapse capacity when $\varepsilon = 0$, and the value of β_1 indicates how sensitive the collapse capacity, S_{CT1}, is to changes in the ε value. To achieve the goal of a simplified correction method, the β_1 value is a required ingredient. For this specific building, $\beta_1 = 0.315$.

It is observed from Figure B-2 that there is a great deal of scatter in the data. Even so, the p-value for the regression is 1.3×10^{-6}, which shows that the trend between collapse capacity and $\varepsilon(T_1)$ is statistically significant (note that the p-value must only be less than 0.01-0.05 for the trend to be statistically defensible, and this value is orders of magnitude smaller). Additionally, similar statistically significant trends are observed for the 118 buildings considered in this study (discussed later in section B.3.3), so this adds

---

[2] This figure is representative of findings by Haselton and Deierlein, but this specific figure is taken from collapse analyses of a 5-story wood light-frame building, which was investigated as part of this current study (model No. 15; $R = 6$, wood-only), and is used in a later section of this Appendix.

confidence for utilizing these observed trends in an adjustment method to account for proper spectral shape, ε.

As a side note, this observed scatter in the data (Figure B-2) also suggests that using a spectral shape correction method will increase the overall uncertainty in the median collapse capacity estimate. This is true, but accounting for the trend with ε also decreases the record-to-record variability. To approximately account for both of these effects in a simple manner, we will (a) use a record-to-record variability of 0.4, which neglects the reduction in record-to-record variability associated with ε-selected records, and (b) neglect the additional uncertainty associated with the variability in the Figure B-2 regression line.

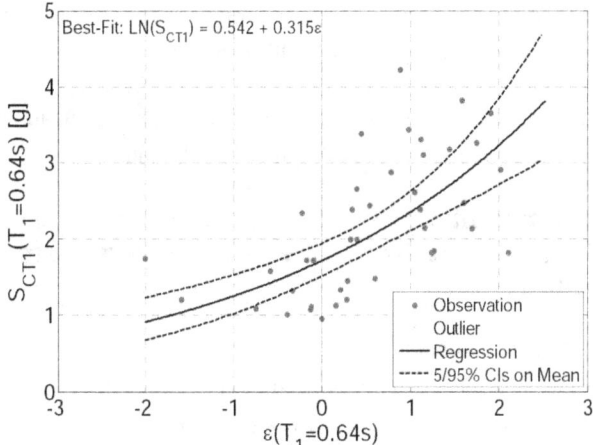

Figure B-2   Relationship between collapse spectral acceleration, S_{CT1}, and $\varepsilon(T_1)$ for a single 5-story wood light-frame building (No. 15; $R = 6$ design, wood-only model). This includes linear regression analysis results which relate $LN[S_{CT1}(T_1)]$ to $\varepsilon(T_1)$, along with confidence intervals (CIs) for the best-fit line. For this example, $\beta_0 = 0.524$ and $\beta_1 = 0.315$.

The study by Haselton and Deierlein (2007) found an average value of $\beta_1 = 0.29$ to be exceptionally consistent for modern reinforced concrete special moment frame buildings with various heights (1-story to 20-stories) and various designs, such as perimeter frame, space frame, and various bay widths.

The relationship between β_1 and inelastic building deformation capacity was also investigated. For buildings with larger inelastic deformation capacity, the effective period elongates more prior to structural collapse. This causes spectral values at period above T_1 to have larger impact on collapse response, and subsequently causes the spectral shape of the ground motion to become more important (thus increasing β_1). To investigate the impact of inelastic building deformation capacity, a set of 26 1967-era RC frame buildings was investigated in the same manner as the previous set of reinforced concrete special moment frame buildings. This revealed an average value of $\beta_1 = 0.18$, which is 35% lower than that of the reinforced concrete special moment frame buildings; this confirms that the ground motion spectral shape, ε, is less important for buildings with lower deformation capacity. To further add to the comparison, a set of 20 reinforced concrete ordinary moment frame buildings was also considered, which showed an average value of $\beta_1 = 0.19$, consistent with the finding for the 1967-era reinforced concrete frame buildings.

Using a subset of the above data, Haselton and Deierlein created a simplified equation that predicts β_1 based on the deformation capacity of the building (judged from static pushover) and the building height.

This Appendix expands on the work by Haselton and Deierlein, and proceeds to create an adapted version of their simplified methodology.

B.3 Development of a Simplified Method to Adjust Collapse Capacity for Effects of Spectral Shape (Epsilon)

The simplified method developed in this Appendix allows one to correct the collapse capacity distribution without needing to compute the $\varepsilon(T_1)$ values of the ground motion records, and without needing to perform a regression analysis. It involves using the Far-Field record set (Appendix A) for structural collapse analyses, and then applying an adjustment factor to the median collapse capacity ($\hat{S}_{CT}$).

The simplified correction factor depends on the following:

1. Differences between:
   a. the $\varepsilon(T)$ values of the ground motions used in the structural analyses (i.e., the Far-Field record set), and
   b. the $\varepsilon(T)$ value expected for the site and ground motion hazard level of interest.

2. How drastically the ground motion ε values affect the building collapse capacity; quantified by β_1, as described in section B.2. The ground motion ε values will have a greater effect on buildings with larger inelastic deformation capacity, which are those that have more extensive period elongation prior to collapse.

B.3.1 Epsilon Values for the Ground Motions in the Far-Field Set

To adjust collapse capacity predictions for spectral shape, the epsilon, $\varepsilon(T)$, values for the ground motion set used for collapse simulation are needed. Figure B-3 shows the mean $\varepsilon(T)$ values for the Far-Field record set, computed using the Abrahamson and Silva attenuation function (1997). The mean $\varepsilon(T)$ values computed using the Boore et al. (1997) attenuation function are similar but are not shown. For the Far-Field record set, mean $\varepsilon(T)$ values are approximately 0.6 for periods less than 0.5 seconds, and are nearly 0.0 for periods greater than 1.5 seconds.

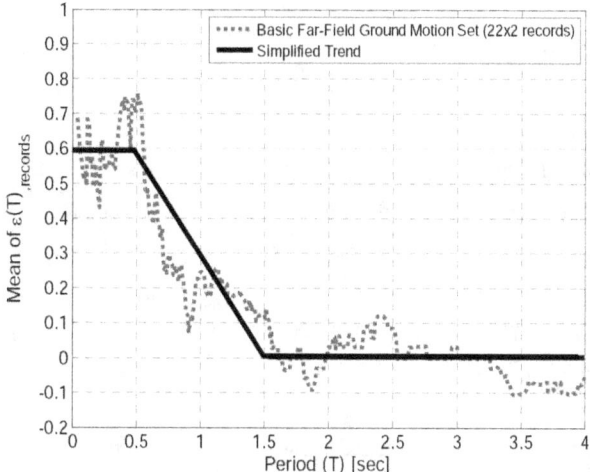

Figure B-3   Mean ε values for the Basic Far-Field ground motion set $\overline{\varepsilon}(T)_{,records}$.

The simplified trend from Figure B-3 can be express using Equation B-1:

$$\overline{\varepsilon}(T)_{,records} = (0.6)(1.5 - T)$$

(B-1)

where $0.0 \leq \overline{\varepsilon}(T)_{,records} \leq 0.6$.

Note that for consistency with the Methodology, the period used for $\overline{\varepsilon}(T)_{,records}$ is the code-defined fundamental period ($T = C_u T_a$), and not the fundamental period computed from eigenvalue analysis (T_l).

B.3.2 Target Epsilon Values

The expected epsilon, $\overline{\varepsilon}_0$, value (used in this Appendix as the "target ε") depends on both the site and hazard level of interest. A target ε is needed in the process of adjusting the median collapse capacity to account for spectral shape.

To quantify the target ε for various seismic design categories around the United States, data from the United States Geological Survey (USGS) is utilized. The USGS conducted the seismic hazard analysis for the United States and used dissagregation to determine the mean expected ε values ($\bar{\varepsilon}_0$) for various periods and hazard levels of interest (Harmsen et al., 2003; Harmsen 2001). Luco, Harmsen, and Frankel of the USGS provided electronic $\bar{\varepsilon}_0$ data for use in this study.

Figure B-4 shows a map of expected $\bar{\varepsilon}_0$ (1s) for a 2% in 50 year motion in Western United States. Values of $\bar{\varepsilon}_0$ (1s) = 0.50 to 1.25 are typical in areas other than the seismic regions of California. The values are higher in most of California, since the earthquake events have shorter return periods, with typical values being $\bar{\varepsilon}_0$ (1s) = 1.25 to 1.75, and some values ranging upward to 3.0.

Figure B-5 shows a map of expected $\bar{\varepsilon}_0$ (1s) for a 2% in 50 year motion in Eastern United States. It shows typical values of $\bar{\varepsilon}_0$ (1s) = 0.75 to 1.0, with some values reaching up to 1.25. Expected $\bar{\varepsilon}_0$ (1s) values fall below 0.75 for the New Madrid Fault Zone, portions of the eastern coast, most of Florida, southern Texas, and areas in the north-west portion of the map.

To illustrate the effects of period, Figure B-5b shows $\bar{\varepsilon}_0$ (0.2s) instead of $\bar{\varepsilon}_0$ (1s). This shows that typical $\bar{\varepsilon}_0$ (0.2s) are slightly lower and more variable, having a typical range of 0.25 to 1.0. This is in contrast to the typical range of 0.75 to 1.0 for $\bar{\varepsilon}_0$ (1s).

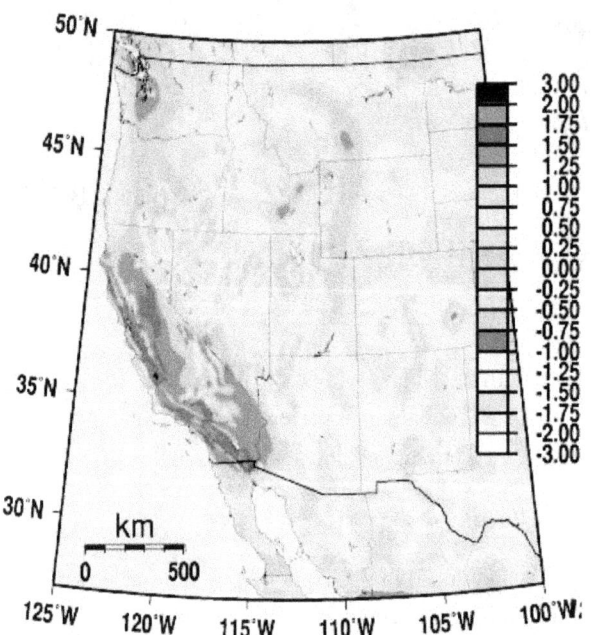

Figure B-4   Predicted $\bar{\varepsilon}_0$ values from dissagregation of ground motion hazard for Western United States. The values are for a 1.0 second period and the 2% exceedance in 50 year motion (Harmsen et al., 2003).

B: Adjustment of Collapse Capacity Considering
Effects of Spectral Shape

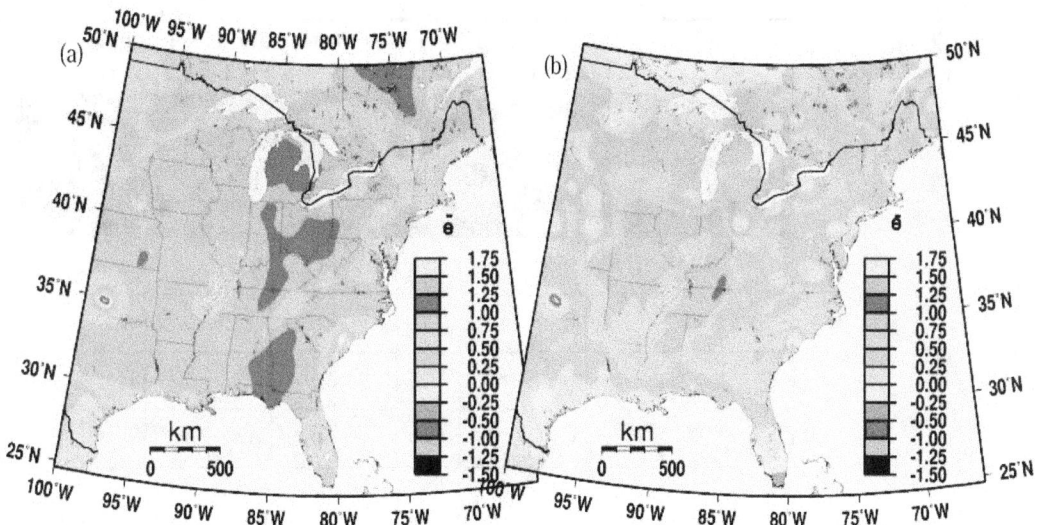

Figure B-5   Mean predicted $\bar{\varepsilon}_0$ values from dissagregation of ground
motion hazard for Eastern United States. The values are for (a)
1.0 second and (b) 0.2 second periods and the 2% exceedance
in 50 year motion (Harmsen et al., 2003).

To better quantify the $\bar{\varepsilon}_0$ values presented in the previous figures, Table B-1
and Table B-2 show the average $\bar{\varepsilon}_0$ and average spectral acceleration values
for Seismic Design Categories B, C, and D. These data are given for four
levels of ground motion: the motion with 10% exceedance in 50 years, 2% in
50 years, 1% in 50 years, and 0.5% in 50 years. These tables also show the
number of zip code data points included in each SDC.

Table B-1   Tabulated $\bar{\varepsilon}_0$ Values for Various Seismic Design Categories

| Seismic Design Category | Average ε Values | | | | | | | | Number of Zip Code Data Points |
| | $\varepsilon_0(0.2s)$ | | | | $\varepsilon_0(1.0s)$ | | | | |
	$\varepsilon_{10/50}$	$\varepsilon_{2/50}$	$\varepsilon_{1/50}$	$\varepsilon_{0.5/50}$	$\varepsilon_{10/50}$	$\varepsilon_{2/50}$	$\varepsilon_{1/50}$	$\varepsilon_{0.5/50}$	
SDC B	0.14	0.42	0.49	0.55	0.31	0.80	0.94	1.04	20,142
SDC C	0.11	0.51	0.63	0.75	0.23	0.74	0.88	1.00	7,456
SDC D	0.25	0.88	1.09	1.27	0.33	0.99	1.21	1.39	6,461
SDC D, $0.35 < S_1 < 0.599g$	0.32	0.97	1.21	1.41	0.39	1.01	1.24	1.45	1,305

Table B-2   Tabulated Spectral Demands for Various Seismic Design Categories

| Seismic Design Category | Average Sa Values | | | | | | | | Number of Zip Code Data Points |
| | Sa(0.2s) [g] | | | | Sa(1.0s) [g] | | | | |
	$Sa_{10/50}$	$Sa_{2/50}$	$Sa_{1/50}$	$Sa_{0.5/50}$	$Sa_{10/50}$	$Sa_{2/50}$	$Sa_{1/50}$	$Sa_{0.5/50}$	
SDC B	0.06	0.18	0.26	0.39	0.02	0.06	0.08	0.11	20,142
SDC C	0.11	0.31	0.46	0.66	0.04	0.10	0.14	0.19	7,456
SDC D	0.50	1.05	1.35	1.68	0.18	0.38	0.49	0.62	6,461
SDC D, $0.35 < S_1 < 0.599g$	0.61	1.31	1.69	2.09	0.21	0.46	0.61	0.77	1,305

Seismic Design Category D is treated differently from the other categories, since it is the category with highest spectral demand and often controls the collapse performance assessment, as was shown in the Chapter 9 examples. Since the higher spectral demands often control the performance, sites in SDC D with higher values of S_1 (i.e., $0.35 < S_1 < 0.599$) are used to define the target ε.

It should be noted that the $\bar{\varepsilon}_0$ values in Table B-1 are a bit lower than some may expect for SDC D sites in seismic zones of California. This comes from the values in Table B-1 being *averages* for all SDC D sites in the United States. The SDC D sites are located in seismic regions of California, as well as in the Eastern United States. Seismic sources in these two regions have widely differing return periods, which causes $\bar{\varepsilon}_0$ values to vary, with the values in California being larger. Average 2% in 50 year and 0.5% in 50 year $\bar{\varepsilon}_0$ (1.0s) values are listed below for selected California cities. These values are averages over all SDC D zip codes in a given city, and are comparable to 0.99 and 1.39 values from Table B-1 which are for the entire United States. Since these values are averages over the city, the values at each specific site in the city may be higher or lower than these values.

- 1.5 and 1.9 in San Francisco (average over 16 zip codes)

- 1.7 and 2.1 in Oakland (average over 10 zip codes)

- 1.6 and 2.0 in Berkeley (average over 3 zip codes)

- 1.6 and 2.1 in San Jose (average over 29 zip codes)

- 1.3 and 1.7 in Los Angeles (average over 58 zip codes)

- 2.0 and 2.2 in Riverside (average over 8 zip codes)

B: Adjustment of Collapse Capacity Considering Effects of Spectral Shape

Table B-1 presented both $\bar{\varepsilon}_0$ (0.2 s) and $\bar{\varepsilon}_0$ (1.0 s) values. The $\bar{\varepsilon}_0$ (1.0 s) values are used to develop the target ε values, since most building structures have periods closer to 1.0 second, or greater than 1.0 second. Buildings with periods near 0.2 seconds are relatively rare.

To complete the determination of target ε values for each Seismic Design Category, the proper ground motion hazard level must be established. Since the spectral shape, ε, adjustment will be used to modify the *median* collapse capacity the appropriate hazard level should be near this median. Table 7-3 shows that for typical uncertainty levels, the median collapse capacity must be roughly twice the MCE, for a structure to pass a 10% conditional collapse probability acceptance criterion. Therefore, the ground motion hazard level that should be used in establishing the target ε should have spectral acceleration demand that is twice (or more) of the 2% in 50 year demand (since the 2% in 50 year motion and MCE are considered to be identical for Seismic Design Categories B, C, and D). Table B-1 shows that the 0.5% in 50 year demand approximately meets this criterion, and is still conservative for SDC D, so the 0.5% in 50 year $\bar{\varepsilon}_0$ (1.0s) is used as the target ε. Based on this, the target ε value for Seismic Design Categories B and C is 1.0, and the target ε value for SDC D is 1.5.

SDC E must be treated differently, because the 2% in 50 year motion and the MCE can differ widely in near-fault regions. At sites close to a fault, the MCE motion has a shorter return period than the 2% in 50 year motion (i.e., 2,475 years). This is the result of the methodology used to construct the 1997 NEHRP maps, as explained in Appendix B of FEMA 369 (FEMA, 2001). In the near-field, the MCE is set to be 50% larger than the median predicted motion; this is approximately one logarithmic standard deviation greater than the median, so by definition $\varepsilon = 1.0$. To approximately account for the fact that the median collapse capacity is larger than the MCE motion, a target $\varepsilon = 1.2$ is used for SDC E.

B.3.3 Impact of Spectral Shape (ε) on Median Collapse Capacity

For buildings with larger inelastic deformation capacity, the effective period elongates more significantly before structural collapse, causing the spectral values at periods greater than T_1 to have more drastic impacts on collapse response. This subsequently causes the spectral shape, ε, of the ground motions to become more important. The β_1 value (defined in section B.2) is used to quantify how drastically the spectral shape, ε, affects the collapse capacity, so the β_1 is larger for buildings with larger deformation capacity. The purpose of this section is to create a predictive equation to estimate the proper β_1 value for any building. This section presents data for 118 buildings, and then shows how the data were used to create the predictive equation for β_1.

Quantification of Building Period Elongation Prior to Collapse

Fundamentally, the spectral shape, ε, is important to the building collapse capacity because the building period elongates prior to collapse; this makes the building response become affected also by the spectral values at periods greater than the initial period T_1. In creating a predictive equation for β_1, it would be ideal to base the prediction directly on the expected amount of period elongation, such as a ratio of the near-collapse period to the initial undamaged period. However, the period near collapse is an ill-defined parameter, so this method instead uses a surrogate for period elongation.

To begin creating this surrogate for period elongation, the building deformation capacity is first quantified using a pushover analysis, in accordance with the guidelines of Section 6.3. Figure B-6 shows an idealized pushover curve and shows that, according to Section 6.3, the ultimate roof displacement (δ_u) is defined as the roof displacement associated with 20% loss of base shear strength. For illustration, Figure B-6 also shows the effective yield roof displacement ($\delta_{y,eff}$), but this term is actually computed from the building period, as will be described later in this section.

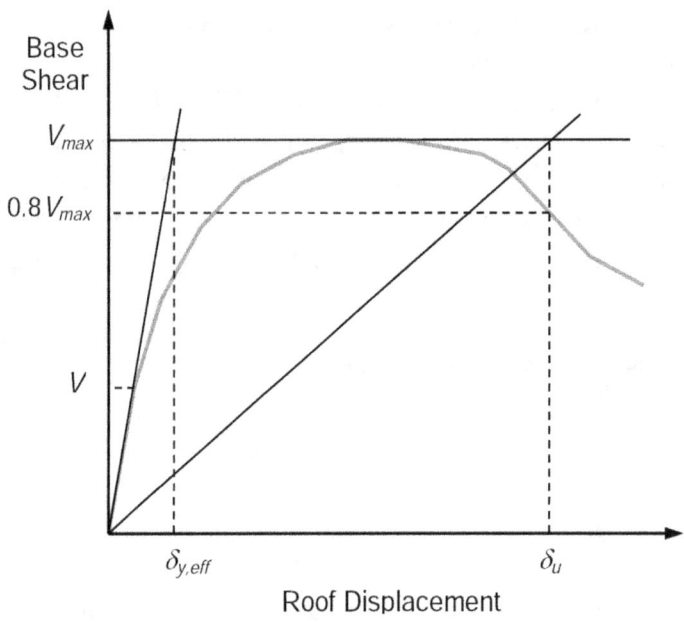

Figure B-6   Idealized nonlinear static pushover curve (from Section 6.3).

The ultimate building deformation capacity is divided by the yield deformation to compute the period-based ductility parameter, $\mu_T = \delta_u / \delta_{y,eff}$.

that is used as the surrogate for period elongation. The square root of μ_T can be approximately thought of as the ratio of the near-collapse period to the initial undamaged period. However, this is not precisely the case because the deformed shape of the building changes as the building is damaged. Even so, μ_T is an acceptable surrogate to approximately quantify the period elongation prior to collapse.

To compute the effective yield roof displacement $(\delta_{y,eff})$, the initial undamaged fundamental period of the building, T_1, is first computed using eigenvalue analysis for the undamaged structural model. This period is then converted into a roof displacement using the guidelines provided in Section 3.3.3.3.2 of ASCE/SEI 41-06 (ASCE, 2006b). To provide more stability to the Methodology, the code defined period of $T = C_u T_a$ must instead be used for cases when $T > T_1$. This requirement is imposed to ensure that analysts do not over-predict the value of β_1 by using a structural model that has an initial stiffness that is unreasonably large. The independent peer review panel should be careful to scrutinize the initial stiffness assumptions of the structural model.

Equation B-2 shows the ASCE/SEI 41-06 equation for effective yield roof displacement $(\delta_{y,eff})$, with the proper modifications for computing yield displacement (e.g., coefficients based on elastic response, and the use of yield base shear for S_a), and with the previously described period requirement. This equation is presented in Section 6.3 as Equation 6-7.

$$\delta_{y,eff} = C_0 \frac{V_{max}}{W} \left[\frac{g}{4\pi^2} \right] (\max(T, T_1))^2 \tag{B-2}$$

where C_0 relates the SDOF displacement to the roof displacement, computed according to ASCE/SEI 41-06 Section 3.3.3.3; V_{max}/W is the maximum base shear normalized by building weight, g is the gravity constant; T is the code-defined fundamental period (i.e., $C_u T_a$); and T_1 is the undamaged fundamental period of the structural model computed using eigenvalue analysis.

In summary, when creating the predictive equation for β_1, the period-based ductility parameter, μ_T, will be used as a surrogate for period elongation prior to collapse.

Predictive Equation for β_1: Database of Buildings Used as the Basis for Creating the Equation

In order to create a predictive equation for β_1, the first step is to understand what the β_1 values should be for various structural models that have various deformation capacities. To this end, the project team assembled a large database of structural models and then, for each model, completed the collapse analyses and performed the regression to compute the β_I value (as previously outlined in Section B.2). This section summarizes the set of structural models utilized, provides the important properties for each building, and report the β_I values computed using regression for each model. The results from these structural models are utilized to develop the predictive equation for β_I, as discussed in the next section.

These five building sets include a total of 118 buildings, which are described in the following list. Partial documentation of the results for these models is provided in tables that follow this list.

- 30 code-conforming reinforced concrete special moment frame buildings from Haselton and Deierlein (2007, Chapter 6). These buildings range from 1-20 stories, and are representative of currently designed buildings (ASCE/SEI 7-02 and ACI 318-02) in high seismic regions of California. Eighteen of these buildings are included in an example in section 9.2 of this report, and the full results for these 18 buildings are included in Table B-3 below. More complete documentation is provided in Haselton and Deierlein (2007). The average β_I for this set of buildings is 0.29 and the average μ_T is 11.5.

- 30 reinforced concrete special moment frame buildings that were designed and analyzed as part of a design sensitivity study completed by Haselton and Deierlein (2007, Chapter 7). This set of building designs includes variations in design strength (R value), design strong-column weak beam ratio, and design drift limits. These buildings are useful for this study because they include subsets of buildings that are identical, except for a single design change (such as the strong-column weak beam ratio, which affects the building deformation capacity); this allows trends to be seen more clearly. These data are not fully documented in this Appendix, though some of these data are used later in Figure B-7; the reader is referred to Chapter 7 of Haselton and Deierlein (2007) for full documentation.

    - 16 code-conforming reinforced concrete ordinary moment frame buildings, from Section 9.3 of this report. These buildings range from 2-12 stories, and are representative of currently designed buildings in the eastern United States. The full results for these 20 buildings are included in Table B-4 below. The average β_I for this set of buildings is 0.19, and the average μ_T is 3.4. As expected, this set of buildings shows lower β_I values being associated with buildings having lower deformation capacity.

- 26 non-ductile reinforced concrete frame buildings from Liel (2008). These buildings range from 2-12 stories, and are representative of existing 1967-era buildings in high seismic regions of California. The full results for these 26 buildings are included in Table B-5 below. The average β_1 for this set of buildings is 0.18, and the average μ_T is 2.9, consistent with the observation made from the set of reinforced concrete ordinary moment frame buildings, that lower β_1 values are to be expected for buildings with lower deformation capacity.

- 16 wood light-frame buildings, from Section 9.4 of this report. These buildings range from 1-5 stories, and are representative of currently designed buildings in high-seismic regions of the United States (SDC D). The full results for these 16 buildings are included in Table B-6 below. The average β_1 for this set of buildings is 0.33 and the average μ_T is 7.8. It is also observed that there is a modest difference between the β_1 values for the buildings designed for SDC D_{max} versus SDC D_{min}.

The following tables (Table B-3 through Table B-6) provide documentation of the results obtained using the above models. These tables include the β_1 values for each model, obtained from regression analysis, as well as the model properties that are important for prediction of β_1.

Table B-3 Documentation of Building Information and β_1 Regression Results for the Set of Reinforced Concrete Special Moment Frame Buildings

Arch. ID	Design Configuration		Building Information							β_1 from Regr.
	No. of Stories	Cs [g]	Ω	T_1 [s]	T [s]	$\delta_{y,eff}/h_r$	δ_u/h_r	μ_1		β_1
Maximum Seismic (D_{max}) and Low Gravity (Perimeter Frame) Designs, 20' Bay Width										
2069	1	0.125	1.6	0.71	0.26	0.0055	0.077	14.0		0.27
2064	2	0.125	1.8	0.66	0.45	0.0034	0.067	19.6		0.26
1003	4	0.092	1.6	1.12	0.81	0.0035	0.038	10.9		0.27
1011	8	0.050	1.6	1.71	1.49	0.0023	0.023	9.8		0.31
1013	12	0.044	1.7	2.01	2.13	0.0023	0.026	11.4		0.29
1020	20	0.044	1.6	2.63	3.36	0.0032	0.018	5.6		0.26
Maximum Seismic (D_{max}) and High Gravity (Space Frame) Designs, 20' Bay Width										
2061	1	0.125	4.0	0.42	0.26	0.0048	0.077	16.1		0.39
1001	2	0.125	3.5	0.63	0.45	0.0061	0.085	14.0		0.26
1008	4	0.092	2.7	0.94	0.81	0.0041	0.047	11.3		0.26
1012	8	0.050	2.3	1.80	1.49	0.0037	0.028	7.5		0.32
1014	12	0.044	2.1	2.14	2.13	0.0028	0.022	7.7		0.25
1021	20	0.044	2.0	2.36	3.36	0.0040	0.023	5.7		0.30
Comparison of Results - SDC D_{max} & D_{min} Seismic Design Conditions, 20' Bay Width										
4011	8	0.017	1.8	3.00	1.60	0.0028	0.010	3.6		0.25
4013	12	0.017	1.8	3.35	2.28	0.0023	0.010	4.3		0.20
4020	20	0.017	1.8	4.08	3.60	0.0021	0.008	3.9		0.15
4021	20	0.017	2.8	4.03	3.60	0.0031	0.012	3.8		0.20
Comparison of Results - 20-Foot and 30-Foot Bay Width Designs (SDC D_{max})										
1009	4	0.092	1.6	1.16	0.81	0.0037	0.050	13.4		0.32
1010	4	0.092	3.3	0.86	0.81	0.0042	0.056	13.2		0.27

B: Adjustment of Collapse Capacity Considering Effects of Spectral Shape

Table B-4 Documentation of Building Information and β_1 Regression Results for the Set of Reinforced Concrete Ordinary Moment Frame Buildings

Arch. ID	Design Configuration		Building Information							β_1 from Regr.
	No. of Stories	C_S [g]	Ω	T_1 [s]	T [s]	$\delta_{y,eff}/h_r$	δ_u/h_r	μ_T	β_1
Minimum Seismic, SDC B$_{min}$, Low Gravity (Perimeter Frame) Designs										
9101	2	0.041	2.0	1.56	0.45	0.0059	0.022	3.7	0.25
9103	4	0.023	1.8	2.81	0.81	0.0048	0.014	3.0	0.17
9105	8	0.012	2.6	4.58	1.49	0.0026	0.008	3.0	0.07
9107	12	0.010	2.3	5.8	2.13	0.0027	0.007	2.5	0.10
Minimum Seismic, SDC B$_{min}$, High Gravity (Space Frame) Designs										
9102	2	0.041	6.6	0.85	0.45	0.0067	0.020	3.0	0.15
9104	4	0.023	5.3	1.49	0.81	0.0050	0.010	2.1	0.27
9106	8	0.012	6.0	2.53	1.49	0.0045	0.014	3.0	0.22
9108	12	0.010	6.0	2.85	2.13	0.0034	0.031	9.1	0.21
Maximum Seismic, SDC B$_{max}$, Low Gravity (Perimeter Frame) Designs										
9201	2	0.087	1.6	1.23	0.45	0.0069	0.024	3.5	0.28
9203	4	0.048	1.6	1.93	0.81	0.0047	0.018	3.8	0.24
9205	8	0.026	1.5	3.39	1.49	0.0033	0.009	2.8	0.12
9207	12	0.018	1.7	4.43	2.13	0.0028	0.009	3.0	0.17
Maximum Seismic, SDC B$_{max}$, High Gravity (Space Frame) Designs										
9202	2	0.087	2.9	0.81	0.45	0.0058	0.019	3.3	0.09
9204	4	0.048	3.0	1.36	0.81	0.0050	0.011	2.2	0.27
9206	8	0.026	3.1	2.35	1.49	0.0045	0.014	3.0	0.19
9208	12	0.018	3.8	2.85	2.13	--	--	--	0.16

Table B-5 Documentation of Building Information and β_1 Regression Results for the Set of 1967-era Reinforced Concrete Buildings

| Arch. ID | Design Configuration | | C_S [g] | Ω | T_1 [s] | T [s] | $\delta_{y,eff}/h_r$ | δ_u/h_r | μ_1 | β_1 from Regr. |
	No. Stor.	Perim./ Space								β_1
3001	2	Space	0.086	1.9	1.08	0.45	0.0067	0,019	2.9	0.16
3002	2	Perim.	0.086	1.6	1.04	0.45	0.0052	0.035	6.7	0.22
3003	4	Perim.	0.068	1.2	1.96	0.81	0.0059	0.013	2.2	0.18
3004	4		0.068	1.3	1.98	0.81	0.0065	0.016	2.4	0.20
3009	4		0.068	1.5	1.98	0.81	0.0075	0.016	2.1	0.15
3010	4	Space	0.068	1.4	1.98	0.81	0.0070	0.015	2.1	0.15
3012	4		0.068	1.4	1.98	0.81	0.0070	0.016	2.3	0.19
3032	4		0.068	1.6	1.98	0.81	0.0081	0.018	2.2	0.19
3015	8	Perim.	0.054	1.1	2.36	1.49	0.0033	0.007	2.1	0.16
3034	8		0.054	1.3	1.84	1.49	0.0024	0.009	3.8	0.22
3016	8		0.054	1.6	2.20	1.49	0.0042	0.011	2.6	0.18
3017	8		0.054	1.6	2.17	1.49	0.0041	0.011	2.7	0.19
3018	8	Space	0.054	1.6	2.20	1.49	0.0042	0.012	2.9	0.19
3019	8		0.054	1.6	2.20	1.49	0.0042	0.011	2.6	0.16
3020	8		0.054	1.6	2.20	1.49	0.0042	0.011	2.6	0.16
3021	8		0.054	1.6	2.20	1.49	0.0042	0.011	2.6	0.19
3022	12	Perim.	0.047	1.1	2.75	2.13	0.0026	0.005	1.9	0.10
3035	12		0.047	1.3	2.23	2.13	0.0020	0.006	2.9	0.19
3023	12		0.047	1.9	2.26	2.13	0.0031	0.010	3.3	0.16
3024	12		0.047	2.0	2.19	2.13	0.0030	0.010	3.3	0.19
3026	12		0.047	2.0	2.26	2.13	0.0032	0.010	3.1	0.18
3027	12	Space	0.047	1.9	2.26	2.13	0.0031	0.010	3.3	0.19
3028	12		0.047	2.0	2.26	2.13	0.0032	0.007	2.2	0.16
3029	12		0.047	1.9	2.26	2.13	0.0031	0.012	3.9	0.18
3031	12		0.047	1.9	2.26	2.13	0.0031	0.012	3.9	0.15
3033	12		0.047	2.2	2.26	2.13	0.0035	0.012	3.4	0.22

B: **Adjustment of Collapse Capacity Considering Effects of Spectral Shape**

Table B-6 Documentation of Building Information and β_1 Regression Results for the Set of Wood Light-frame Buildings

Arch. ID	Design Configuration		Building Information							β_1 from Regression
	No. of Stories	Cs [g]	Ω	T_1 [s]	T [s]	$\delta_{y,eff}/h_r$	δ_u/h_r	μ_T	β_1
High Seismic (SDC D_{max}) - Low Aspect Ratios - R = 6										
1	1	0.167	1.6	0.51	0.25	0.0055	0.045	8.3	0.29
5	2	0.167	2.0	0.52	0.26	0.0045	0.045	10.1	0.31
9	3	0.167	1.6	0.61	0.36	0.0032	0.033	10.3	0.29
High Seismic (SDC D_{max}) - High Aspect Ratios - R = 6										
2	1	0.167	3.1	0.38	0.25	0.0060	0.045	7.4	0.27
6	2	0.167	2.8	0.46	0.26	0.0048	0.040	8.3	0.35
10	3	0.167	2.8	0.47	0.36	0.0034	0.041	12.1	0.32
13	4	0.167	2.7	0.52	0.45	0.0031	0.023	7.2	0.30
15	5	0.167	2.4	0.64	0.53	0.0034	0.022	6.4	0.32
Low Seismic (SDC D_{min}) - Low Aspect Ratios - R = 6										
11	3	0.063	1.6	1.10	0.41	0.0039	0.025	6.4	0.41
Low Seismic (SDC D_{min}) - High Aspect Ratios - R = 6										
3	1	0.063	2.7	0.65	0.25	0.0059	0.047	7.9	0.42
4	1	0.063	4.1	0.53	0.25	0.0059	0.045	7.7	0.29
7	2	0.063	3.1	0.74	0.30	0.0052	0.038	7.3	0.35
8	2	0.063	2.5	0.80	0.30	0.0050	0.033	6.6	0.34
12	3	0.063	3.1	0.83	0.41	0.0043	0.030	7.0	0.36
14	4	0.063	2.6	0.99	0.51	0.0041	0.025	6.1	0.31
16	5	0.063	2.4	1.12	0.60	0.0041	0.021	5.2	0.34

Predictive Equation for β_1: Development and Evaluation of Equation

Using the database of results for the 118 buildings, the goal is to create an equation that can be used to predict the value of β_1, without needing to perform a regression analysis. In an earlier section, it was already decided that β_1 will be predicted using the period-based ductility parameter, $\mu_T = \delta_u/\delta_{y,eff}$, as a surrogate for the period elongation that the building undergoes prior to collapse. Therefore, the goal of this section is to determine the proper form of this equation.

To most clearly see the relationship between β_1 and μ_T, Figure B-7 shows the trends using three sets of reinforced concrete special moment frame

B: Adjustment of Collapse Capacity Considering Effects of Spectral Shape

buildings, each with constant height. These three sets of buildings are described as follows:

- Four-story perimeter frames, with the design R value varied from 4 to 12.

- Four-story space frames, with the design strong-column weak-beam ratio varied from 0.4 to 3.0.

- Twelve-story space frames, with the design R value varied from 4 to 12.

These data show a clear trend between period-based ductility (μ_T) and β_1 for observed μ_T values up to 8-9, and suggest that the effects are saturated for larger values of μ_T. For these buildings, the β_1 value saturates at 0.32 for two of the sets, and 0.26 for the third set. These observations suggest that the predictive equation for β_1 should saturate at a μ_T value somewhere between 8 and 11, and saturate at a value of β_1 that is between 0.26 and 0.32.

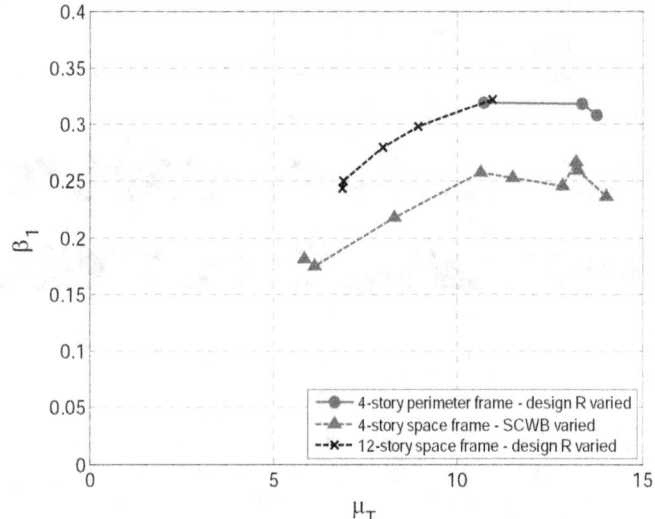

Figure B-7   Relationship between β_1 and period-based ductility, μ_T, for three sets of reinforced concrete special moment frame buildings of constant height within each set.

Considering both the reinforced concrete special moment frame data (Figure B-7 and Table B-3) and the wood light-frame data (Table B-6), it was decided that the predictive equation for β_1 should saturate at a value of $\beta_1 = 0.32$ for $\mu_T \geq 8$. This will accurately reflect the average values from the wood data, accurately reflect the saturation point for the reinforced concrete special moment frame data, and slightly over-predict the average β_1 values for the reinforced concrete special moment frame data.

Now that the saturation point has been established for the β_1 predictive equation, the equation can be created by choosing an appropriate functional

B: Adjustment of Collapse Capacity Considering
Effects of Spectral Shape

form, and then using the full set of data from the 118 buildings, with standard linear regression analysis to predict $\log(\beta_1)$ (Chatterjee et al., 2000). Several function forms were evaluated, and a power form was chosen, in order to best match the trends in the data, and enforce that $\beta_1 = 0.0$ when $\mu_T = 1.0$. Small corrections were then applied to the equation to enforce the desired saturation point of $\beta_1 = 0.32$ at $\mu_T = 8$, while still accurately fitting the data for lower values of μ_T.

The final relationship between β_1 and μ_T is shown in Equation B-3:

$$\hat{\beta}_1 = (0.14)(\mu_T - 1)^{0.42} \tag{B-3}$$

where the limit is enforced of $\mu_T \le 8.0$.

Predictive Equation for β_1: Comparisons of Predictions and Observations

In order to evaluate the prediction accuracy of Equation B-3, Figure B-8 compares the predicted and observed values of β_1. Figure B-8a shows the data points for all of the 118 buildings used to create the equation, and Figure B-8b shows the average values for each of the building subsets. Overall, this shows that the equation accurately predicts the expected values of β_1. More specifically, this shows that the equation accurately predicts the β_1 values for non-ductile buildings, accurately predicts the values for wood buildings in SDC D_{max}, slightly under-predicts the values for wood buildings in SDC D_{min}, but still accurately predicts the values for wood buildings on average (with overall average $\mu_T = 7.4$ and $\beta_1 = 0.32$), and slightly over-predicts the values for ductile reinforced concrete special moment frame buildings.

B.4 Final Simplified Factors to Adjust Median Collapse Capacity for the Effects of Spectral Shape

Following the rationale above, the spectral shape factor, *SSF*, can be computed using Equation B-4.

$$SSF = \exp\left[\beta_1 \left(\bar{\varepsilon}_o(T) - \bar{\varepsilon}(T)_{.records} \right) \right] \tag{B-4}$$

where β_1 depends on building inelastic deformation capacity (Equation B-3); $\bar{\varepsilon}_o$ depends on SDC and is equal to 1.0 for SDC B/C, 1.5 for SDC D, and 1.2 for SDC E (Section B.3.2); and $\varepsilon(T)_{.records}$ is for the Far-Field record set (Section B.3.1).

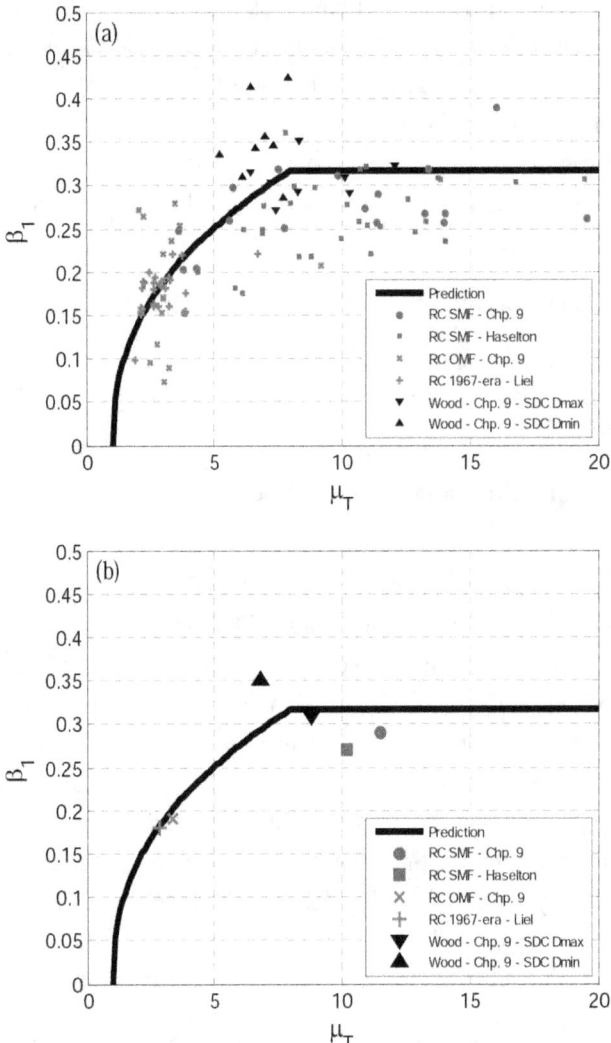

Figure B-8   Comparison of predicted and observed values of β_1, for all 118 buildings used to create the predictive equation. Figure (a) shows all the data, and Figure (b) shows the average values for the individual subsets of building types.

Table B-7 through Table B-9 present values of spectral shape factor, *SSF*, for various levels of building period-based ductility, μ_T, and various building periods, using Equation B-4. Table B-7 presents values for Seismic Design Categories B and C, Table B-8 presents values for SDC D, and Table B-9 presents values for SDC E.

To compute the adjusted collapse margin ratio, multiply the *SSF* value by the collapse margin ratio that was predicted using the Far-Field record set, as shown in Equation B-5:

$$ACMR = SSF * CMR \tag{B-5}$$

Table B-7   Spectral Shape Factors for Seismic Design Categories B, C, and D_{min}

T (sec.)	Period-Based Ductility, μ_T							
	1.0	1.1	1.5	2	3	4	6	≥ 8
≤ 0.5	1.00	1.02	1.04	1.06	1.08	1.09	1.12	1.14
0.6	1.00	1.02	1.05	1.07	1.09	1.11	1.13	1.16
0.7	1.00	1.03	1.06	1.08	1.10	1.12	1.15	1.18
0.8	1.00	1.03	1.06	1.08	1.11	1.14	1.17	1.20
0.9	1.00	1.03	1.07	1.09	1.13	1.15	1.19	1.22
1.0	1.00	1.04	1.08	1.10	1.14	1.17	1.21	1.25
1.1	1.00	1.04	1.08	1.11	1.15	1.18	1.23	1.27
1.2	1.00	1.04	1.09	1.12	1.17	1.20	1.25	1.30
1.3	1.00	1.05	1.10	1.13	1.18	1.22	1.27	1.32
1.4	1.00	1.05	1.10	1.14	1.19	1.23	1.30	1.35
≥ 1.5	1.00	1.05	1.11	1.15	1.21	1.25	1.32	1.37

Table B-8   Spectral Shape Factors for Seismic Design Category D_{max}

T (sec.)	Period-Based Ductility, μ_T							
	1.0	1.1	1.5	2	3	4	6	≥ 8
≤ 0.5	1.00	1.05	1.10	1.13	1.18	1.22	1.28	1.33
0.6	1.00	1.05	1.11	1.14	1.20	1.24	1.30	1.36
0.7	1.00	1.06	1.11	1.15	1.21	1.25	1.32	1.38
0.8	1.00	1.06	1.12	1.16	1.22	1.27	1.35	1.41
0.9	1.00	1.06	1.13	1.17	1.24	1.29	1.37	1.44
1.0	1.00	1.07	1.13	1.18	1.25	1.31	1.39	1.46
1.1	1.00	1.07	1.14	1.19	1.27	1.32	1.41	1.49
1.2	1.00	1.07	1.15	1.20	1.28	1.34	1.44	1.52
1.3	1.00	1.08	1.16	1.21	1.29	1.36	1.46	1.55
1.4	1.00	1.08	1.16	1.22	1.31	1.38	1.49	1.58
≥ 1.5	1.00	1.08	1.17	1.23	1.32	1.40	1.51	1.61

Table B-9   Spectral Shape Factors for Seismic Design Category E

T (sec.)	Period-Based Ductility, μ_T							
	1.0	1.1	1.5	2	3	4	6	≥ 8
≤ 0.5	1.00	1.03	1.06	1.09	1.12	1.14	1.18	1.21
0.6	1.00	1.04	1.07	1.10	1.13	1.16	1.20	1.23
0.7	1.00	1.04	1.08	1.11	1.14	1.17	1.22	1.26
0.8	1.00	1.04	1.09	1.12	1.16	1.19	1.24	1.28
0.9	1.00	1.05	1.09	1.12	1.17	1.21	1.26	1.31
1.0	1.00	1.05	1.10	1.13	1.18	1.22	1.28	1.33
1.1	1.00	1.05	1.11	1.14	1.20	1.24	1.30	1.36
1.2	1.00	1.06	1.11	1.15	1.21	1.25	1.32	1.38
1.3	1.00	1.06	1.12	1.16	1.22	1.27	1.35	1.41
1.4	1.00	1.06	1.13	1.17	1.24	1.29	1.37	1.44
≥ 1.5	1.00	1.07	1.13	1.18	1.25	1.31	1.39	1.46

B.5   Application to Site Specific Performance Assessment

Simplified spectral shape adjustment factors can be modified for site-specific and building-specific collapse performance assessment. To compute the *SSF* for a site-specific collapse performance assessment, use Equation B-4 with the following values:

- β_1 should be computed using Equation B-3.

- $\bar{\varepsilon}_0$ (T) should not be based on Section B.3.2. Instead, $\bar{\varepsilon}_0$ (T) should be based directly on the dissagregation of the probabilistic seismic hazard analysis for the site of interest and for the ground motion hazard level of the median collapse capacity.

- $\varepsilon(T)_{.records}$ will differ depending on the ground motion set utilized in the performance assessment. If the site is within 10 km of an active fault capable of producing an event larger than magnitude M6.5, then the Near-Field ground motion record set should be used (Appendix A); otherwise, the Far-Field record set should be used. If the Far-Field record set is used, then $\varepsilon(T)_{.records}$ should be taken from Figure B-3 and Equation B-1. If the Near-Field record set is used, then $\varepsilon(T)_{.records}$ should be taken to equal $\varepsilon(T)_{.records, NF}$ from Figure B-9 and Equation B-6.

B: Adjustment of Collapse Capacity Considering
Effects of Spectral Shape

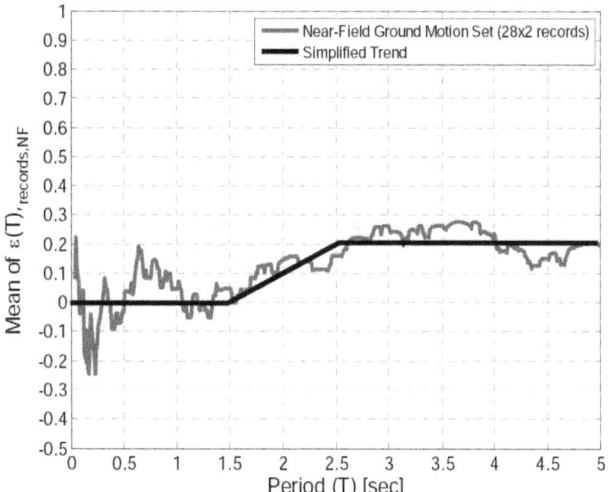

Figure B-9   Mean ε values for the Near-Field ground motion set,
$\overline{\varepsilon}(\mathrm{T})_{,records,NF}$.

The Near-Field set is nearly ε-neutral at all periods, but a slight simplified trend from Figure B-9 can be approximated using Equation B-6:

$$\overline{\varepsilon}(\mathrm{T})_{,records,NF} = (0.2)(T-1.5) \qquad (B-6)$$

where $0.0 \le \overline{\varepsilon}(\mathrm{T})_{,records,NF} \le 0.2$.

Appendix C
Development of Index Archetype Configurations

This appendix illustrates how index archetype configurations are developed for reinforced concrete moment frame systems and wood light-frame shear wall systems.

Development of structural system archetypes considers both structural configuration issues and seismic behavioral effects described in Chapter 4. Consideration of structural configuration issues is discussed in this appendix. Examination of seismic behavioral effects is described in Appendix D. Index archetype configurations are subsequently used to develop index archetype designs, which are then used to develop index archetype models. Development and calibration of nonlinear index archetype models is described in Appendix E.

C.1 Development of Index Archetype Configurations for a Reinforced Concrete Moment Frame System

In this section, index archetype configurations are developed for a reinforced concrete special moment frame system conforming to design requirements contained in ASCE/SEI 7-05 *Minimum Design Loads for Buildings and Other Structures* (ASCE, 2006a), and ACI 318-05 *Building Code Requirements for Structural Concrete* (ACI, 2005). The index archetype configurations described here are used in the reinforced concrete moment frame system example of Section 9.2. When the Methodology is applied to a proposed new seismic-force-resisting system, index archetype configurations will be based on existing code requirements, as appropriate, and new design requirements developed specifically for the proposed system.

C.1.1 Establishing the Archetype Design Space

To establish the archetype design space, design parameters that significantly affect seismic performance are first identified, and then the bounds on each design parameter are established. The overall range of permissible configurations, structural design parameters, and other features that define the application limits for a seismic-force-resisting system are specified in the system design requirements and in existing code requirements, as applicable.

Parameters that were identified as having a critical impact on the collapse performance of reinforced concrete moment frame systems, along with related physical properties and associated design variables, are listed in Table C-1. This organizational approach is useful for deciding which of the many design variables should be the focus of further investigation in index archetype configurations.

Table C-1   Important Parameters, Related Physical Properties, and Design Variables for Reinforced Concrete Moment Frame Systems

Important Parameter	Related Physical Properties	Design Variables
Column and beam plastic rotation capacity	Axial load ratio	Building height, bay width, ratio of tributary areas for gravity and lateral loads
	Column aspect ratio	Building height, bay width, story heights, allowable reinforcement ratio
	Confinement ratio	Confinement ratio used in design
	Stirrup spacing	Stirrup spacing used in design
	Longitudinal bar diameter	Longitudinal bar diameter used in design
	Reinforcement ratios	Reinforcement ratio allowed in design
	Concrete strength	Concrete strength used in design
Element Strengths	All element strengths	Conservatism of engineer, dead and live loads used in design
	Beam strengths	Slab width (steel) assumed effective
	Column strengths	Ratio of factored to expected axial loads, level of conservatism in applying strong-column weak-beam provision
Number of stories in collapse mechanism	Strength/stiffness irregularities	Presence of strength or stiffness irregularity, ratio of first to upper story heights, how column heights are stepped down over height
Lateral stiffness of frame	Member sizes in frame	Member/joint/footing stiffness used in design
Gravity system strength/stiffness	Gravity system	Not considered in this assessment

Key design variables are those that are likely to have a significant impact on the collapse performance of the proposed system. Key design variables identified for reinforced concrete moment frame systems, along with their applicable ranges, are listed in Table C-2.

Table C-2   Key Design Variables and Ranges Considered in the Design Space for Reinforced Concrete Moment Frame Systems

Key Design Variable	Range Considered in Archetype Design Space
Structural System	
Special Reinforced Concrete Moment Frame (as per ASCE/SEI 7-05, ACI 318-05)	All designs meet code requirements
Seismic framing system	Perimeter and space frames
Configuration	
Building Height	1 to 20 stories
Bay Width	20 to 30 feet
First and upper story heights	15 and 13 feet
Element Design	
Confinement ratio and stirrup spacing	Conforming to ACI 318-05
Concrete compressive strength	5 to 7 ksi
Longitudinal bar diameter	#8 and #9 are commonly used
Strength/stiffness irregularities	As permitted by existing code
Loading	
Ratio of tributary areas for gravity and lateral loads	0.1 (perimeter frame) to 1.0 (space frame)
Design floor loads	175 psf
Lower and upper bounds on design floor load	150 to 200 psf
Design floor live load	Constant: 50 psf

The design variables and ranges identified in Table C-2 provide the basis for identifying a finite number of design variations for use in developing index archetype configurations.

Figure C-1 illustrates one example of how index archetype configurations might change for a design variable associated with the ratio of tributary areas for gravity and lateral loads (A_{grav}/A_{lat}). In the case of moment frame systems, this variable is primarily affected by whether the building is designed as a space frame or a perimeter frame system. Table C-2 identifies a range of 0.1 to 1.0 for this design variable, which is schematically shown in the figure.

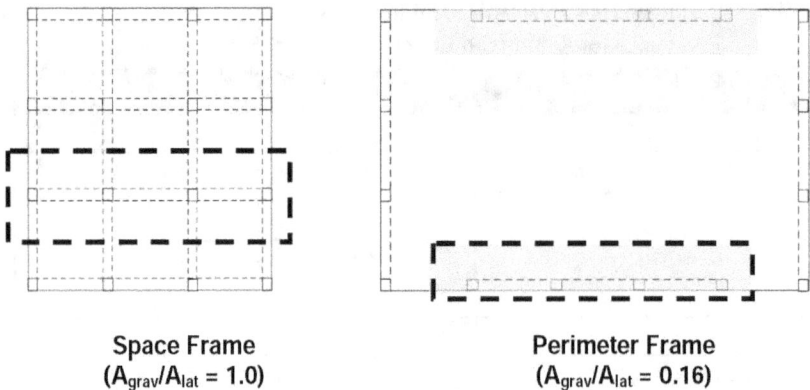

<table>
<tr><td>**Space Frame**
($A_{grav}/A_{lat} = 1.0$)</td><td>**Perimeter Frame**
($A_{grav}/A_{lat} = 0.16$)</td></tr>
</table>

Figure C-1   Different index archetype configurations for varying ratios of tributary areas for gravity A_{grav} and lateral loads A_{lat}.

C.1.2 Identifying Index Archetype Configurations and Populating Performance Groups

Identification of a set of index archetype configurations requires consideration of performance group binning in accordance with Section 4.3.1. As a minimum, performance groups should consider the design ground motion intensities for the governing Seismic Design Category (maximum and minimum SDC), two fundamental period domains (long-period and short-period), variations in gravity load intensity (high and low gravity loads), and building configuration. In the case of reinforced concrete moment frames, "high gravity" systems are space frames and "low gravity" systems are perimeter frames, as shown in Figure C-1. This results in the need for index archetype configurations to populate at least eight performance groups, as described in Table 4-3:

- High gravity loads, maximum SDC, short-period archetypes
- High gravity loads, maximum SDC, long-period archetypes
- High gravity loads, minimum SDC, short-period archetypes
- High gravity loads, minimum SDC, long-period archetypes
- Low gravity loads, maximum SDC, short-period archetypes
- Low gravity loads, maximum SDC, long-period archetypes
- Low gravity loads, minimum SDC, short-period archetypes
- Low gravity loads, minimum SDC, long-period archetypes

Additional performance groups need to be defined where other design variables are important. For example, performance groups for reinforced concrete moment frame also include variation in bay width (column spacing) of 20 feet and 30 feet, respectively, which may be an important parameter for system performance (Table 9-1). Each performance group must be robustly populated by buildings that have representative configurations and heights across the entire archetype design space. Each performance group is expected to contain at least three buildings, except in cases where that performance group represents an unusual rather than typical configuration of the structural system of interest.

Index archetype configurations must be conceived in such a way that index archetype models will capture important system behavior. For moment frames in general, three framing bays are considered to be the minimum number feasible for capturing variations in behavior related to interior and exterior columns and beam-column joints, strong-column weak-beam design provisions, and induced column axial loads due to overturning effects. As a result, the three-bay, variable story-height configuration shown in Figure C-2 was selected as the simplest model still capable of capturing important collapse performance behaviors for reinforced concrete moment frame systems.

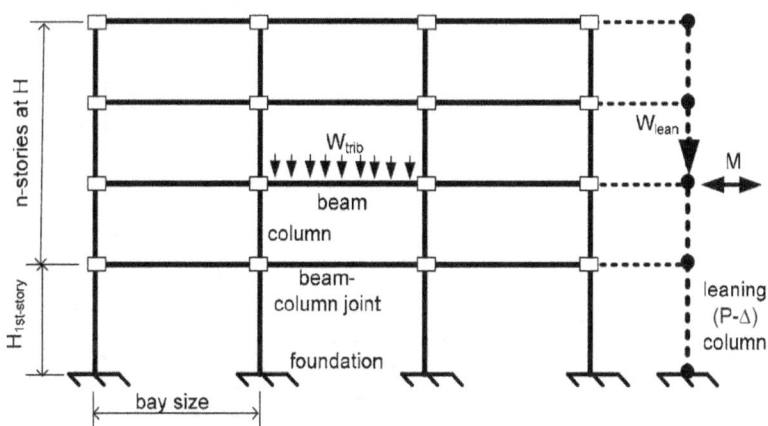

Figure C-2   Index archetype model for reinforced concrete moment frame systems.

To develop a set of index archetype configurations, the range of parameters that will be commonly utilized in design and construction must be understood. As identified in Table C-2, building heights between one and twenty stories are expected, since most buildings taller than twenty stories would include a core wall in addition to moment frames. Considering typical office occupancies, story heights are expected to be relatively consistent, taken as 15 feet for the first story and 13 feet for the upper stories. Plan dimensions of 120 feet by 120 feet and 120 feet by 180 feet, along with bay widths ranging between 20 feet and 30 feet, are also expected to be typical for such buildings.

Figure C-3 illustrates two set of index archetype configurations representing the range of expected building heights for 20-foot and 30-foot configurations, respectively, that are used to populate reinforced-concrete moment frame performance groups. In this case, performance groups with short-period archetypes include 1-story and 2-story (and 3-story) buildings and performance groups with long-period archetypes include 4-story and taller buildings.

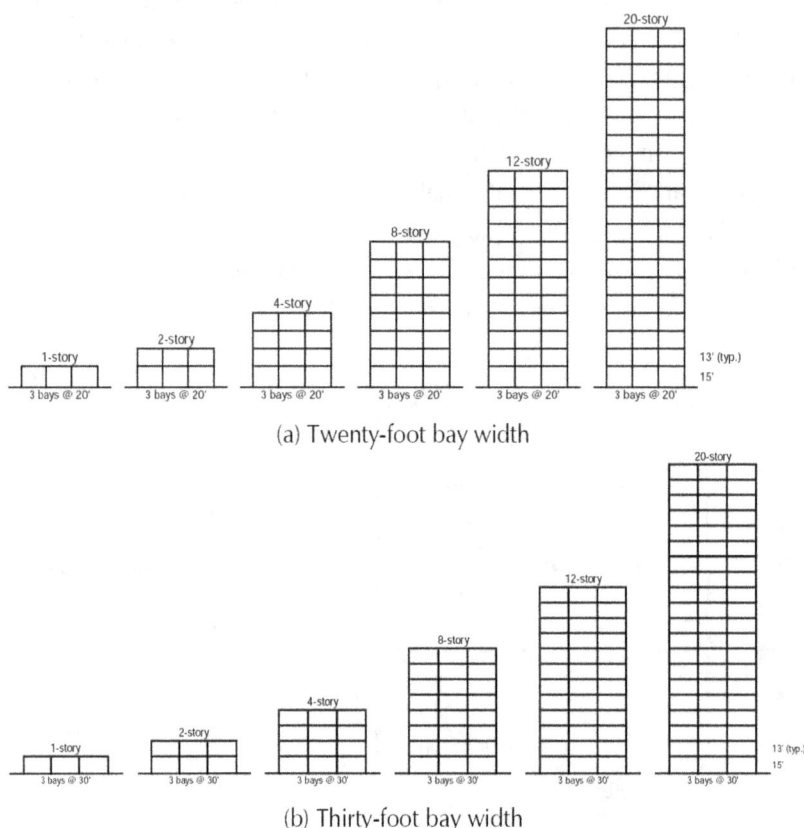

(a) Twenty-foot bay width

(b) Thirty-foot bay width

Figure C-3   Index archetype configurations for a reinforced concrete moment frame system.

Maximum and minimum seismic loads are defined by the range of possible design loads for Seismic Design Category D (SDC D_{max} and SDC D_{min}). High and low gravity load intensities are represented by the space frame or perimeter frame configurations.

Based on the information summarized in Table C-1 and Table C-2, and consideration of performance group binning and nonlinear analysis modeling, a matrix of index archetype configurations, such as the one described in Table C-3, can be developed. This matrix of index archetype configurations completely populates the performance groups expected to be most typical: 20-foot bay spacing of different heights, space and perimeter frames, designed for the maximum seismic load intensity (SDC D_{max}). Taller buildings (4, 8, 12, and 20 stories) respond in the long-period range and are classified in separate performance groups from shorter short-period buildings (1, 2, and 3 stories). In order to verify that these performance groups control the assessment, a small number of additional archetype configurations are considered in other performance groups, including archetypes with a 30-foot bay width and archetypes designed for minimum seismic load intensity (SDC D_{min}).

Table C-3   Matrix of Index Archetype Configurations for a Reinforced Concrete Moment Frame System

Perform. Group No.	Archetype ID (Ch. 9)	No. of Stories	Period Domain	Bay Width	Gravity Loads	SDC
PG-1	2061	1	Short	20'	High (Space)	D_{max}
	1001	2				
	--	3				
PG-2	1008	4	Long	20'	High (Space)	D_{max}
	1012	8				
	5014	12				
	5021	20				
PG-5	2069	1	Short	20'	Low (Perimeter)	D_{max}
	2064	2				
	--	3				
PG-6	1003	4	Long	20'	Low (Perimeter)	D_{max}
	1011	8				
	5013	12				
	5020	20				
PG-4	6021	20	Long	20'	High (Space)	D_{min}
PG-8	6011	8	Long	20'	Low (Perimeter)	D_{min}
	6013	12				
	6020	20				
PG-10	1009	4	Long	30'	High (Space)	D_{max}
PG-14	1010	4	Long	30'	Low (Perimeter)	D_{max}

C.1.3 Preparing Index Archetype Designs and Index Archetype Models

Each of the index archetype configurations in Table C-3 are used to prepare index archetype designs, which are represented with index archetype models. In the examples of Chapter 9, index archetype designs for reinforced concrete moment frames were prepared in accordance with the requirements of ASCE/SEI 7-05 and ACI 318-05. Complete adherence to design requirements is essential for adequately evaluating the performance of a class of buildings designed using the proposed system. Designs should address minimum design requirements, utilize nominal material properties, and not be overly conservative. Assumptions used in the development of index archetype designs for reinforced concrete moment frames are listed in Table C-4.

While index archetype designs are prepared using nominal material properties and other standard assumptions, index archetype models should be defined based on the expected behavior of the building. For modeling and assessment of the mean performance of reinforced concrete moment frame systems, this included the use of expected material properties, element stiffness assumptions based on test data, tributary slab contributions for beam strength and stiffness, and expected gravity loads. More details of modeling for reinforced concrete moment frame systems are described in Appendix D and Appendix E.

Table C-4   Index Archetype Design Assumptions for a Reinforced Concrete Moment Frame System

Design Parameter	Design Assumption
Assumed Stiffness	
Member stiffness assumed in design: Beams	$0.5EI_g$ (ASCE/SEI 41)
Member stiffness assumed in design: Columns	$0.7EI_g$ for all axial load levels (based on practitioner recommendation)
Slab consideration	Slab not included in stiffness/strength design of beams
Footing rotational stiffness assumed in design: 2-4 story	Effective stiffness of grade beam
Footing rotational stiffness assumed in design: 8-20 story	Basement assumed; exterior columns fixed at basement wall, interior columns consider stiffness of first floor beam and basement column
Joint stiffness assumed in design	Elastic joint stiffness
Expected Design Conservatisms	
Conservatism applied in element flexural and shear (capacity) strength design	1.15 times required strength
Conservatism applied in joint strength design	1.0 times required strength
Conservatism applied in strong-column weak-beam design	Use expected ratio of 1.3 instead of 1.2

C.2 Development of Index Archetype Configurations for a Wood Light-Frame Shear Wall System

In this section, index archetype configurations are developed for a wood light-frame shear wall system conforming to design requirements contained in ASCE/SEI 7-05. The system consists of wood light-frame bearing wall structures braced with wood structural panel shear walls. These archetypes are used in the wood light-frame bearing wall example of Section 9.4.

C.2.1 Establishing the Archetype Design Space

The following design variables were considered in the development of index archetype configurations for this system: (1) number of stories; (2) Maximum and minimum seismic intensity for the governing Seismic Design Category (SDC D in this case); (3) building occupancy and use; and (4) shear wall aspect ratio.

Wood light-frame buildings of upto three stories are common across most of the United States. Wood light-frame multi-family residential buildings of four to five stories represent a growing trend along the West Coast. The number of stories considered in the archetype design space ranges from one to five stories.

Minimum and maximum values of design spectral response acceleration, S_{DS}, used in this example are 0.375g and 1.00g, respectively. The minimum value of 0.375g is lower than 0.50g specified for SDC D in Table 5-1A, but is used in this illustrative example to make use of available designs. An S_{DS} of 1.00g represents the Maximum seismic intensity for regular, short-period SDC D structures. Design for wind loads is not considered.

The range of building occupancies considered includes residential and commercial occupancies, the latter including educational and institutional uses. The primary difference between residential and commercial buildings is the spacing between shear wall lines, which affects the tributary seismic mass. Residential buildings generally have more walls and closer spacing. A typical spacing of 25 feet between shear wall lines and a tributary width of 12.5 feet for seismic mass was used for residential occupancies. Commercial buildings are more likely to have open configurations with widely spaced shear walls at the perimeter. A typical spacing of 80 feet between shear wall lines and a tributary width of 40 feet for seismic mass was used for commercial occupancies.

Residential occupancies were further split into one- and two-family detached dwellings and multi-family dwellings, respectively. One- and two-family detached dwellings were assumed to be one- and two-stories tall, with typical wood-frame floor weights (without topping slabs). Multi-family dwellings were assumed to be three- to five-stories tall, with floor weights including gypcrete topping slabs.

Shear wall aspect ratios included high aspect ratio shear walls (height/width ratios of 2.7 to 3.3) and low aspect ratio shear walls (height/width ratios of 1.5 or less). In accordance with ASCE/SEI 7-05, a capacity reduction was considered for high aspect ratio shear walls, resulting in the need for configurations with more dense nailing patterns.

C.2.2 Identifying Index Archetype Configurations and Populating Performance Groups

A set of 16 index archetype configurations are developed to investigate the effect of wall aspect ratio, height (one- to five-stories), and lateral loading intensity, as shown in Table C-5. These configurations are chosen recognizing the predominant use of high aspect ratio shear walls in residential buildings and low aspect ratio shear walls in commercial buildings. These structures are grouped into performance groups according to the design lateral load intensity (SDC D_{max} or D_{min}), wall aspect ratio (high or low) and period domain dominating the response (short- or long-period) (see Table 9-23). Other configurations were eliminated because they are not representative of wood light-frame construction. In particular, nominal gravity loads are considered in all the designs because light wood-frame archetype design and performance is not influenced by gravity loads. Likewise, there are few archetype configurations falling in the long-period domain because these are not representative.

C.2.3 Preparing Index Archetype Designs and Index Archetype Models

Each of the index archetype configurations in Table C-5 are used to prepare index archetype designs, which are represented with index archetype models. In this example, index archetype designs were prepared in accordance with the requirements of ASCE/SEI 7-05, considering only the contribution of wood structural panel sheathing, and ignoring the possible beneficial effects of gypsum wallboard that is likely to be present.

Designs were developed using a response modification coefficient, R, equal to 6. Index archetype designs, including number of piers, sheathing, and fasteners for R=6 are provided in Table C-6.

Table C-5   Index Archetype Configurations for Wood Light-Frame Shear
           Wall Systems

Model No.	No. of Stories	Seismic Design Coef. (S_{DS})	Tributary Width for Seismic Weight (ft)	Floor/Roof Tributary Seismic Weight (kips)	Shear Wall Aspect Ratio	Occupancy
1	1	1.0	40	41/0	Low	Commercial
2	1	1.0	12.5	13.65/0	High	1&2 Family
3	1	0.375	40	41/0	High	Commercial
4	1	0.375	12.5	13.65/0	High	1&2 Family
5	2	1.0	40	82/41	Low	Commercial
6	2	1.0	12.5	17.3/13.65	High	1&2 Family
7	2	0.375	40	82/41	High	Commercial
8	2	0.375	12.5	17.3/13.65	High	1&2 Family
9	3	1.0	40	82/41	Low	Commercial
10	3	1.0	12.5	27.3/13.65	High	Multi-Family
11	3	0.375	40	82/41	Low	Commercial
12	3	0.375	12.5	27.3/13.65	High	Multi-Family
13	4	1.0	12.5	27.3/13.65	High	Multi-Family
14	4	0.375	12.5	27.3/13.65	High	Multi-Family
15	5	1.0	12.5	27.3/13.65	High	Multi-Family
16	5	0.375	12.5	27.3/13.65	High	Multi-Family

Table C-6   Index archetype designs for wood light-frame shear wall
           systems ($R = 6$)

Model No.	No. of Stories	No. of Piers	Pier Length (ft)	Sheathing	Shear Wall Nailing
1	1	2	9.0	7/16" OSB	8d at 6"
2	1	4	3.0	7/16"OSB	8d at 6"
3	1	4	3.0	7/16" OSB	8d at 6"
4	1	2	3.0	7/16" OSB	8d at 6"
5	2	3	8.0	7/16" OSB	8d at 4"
	1	3	8.0	7/16" OSB	8d at 2"

Table C-6   Index archetype designs for wood light-frame shear wall systems ($R = 6$) continued

Model No.	No. of Stories	No. of Piers	Pier Length (ft)	Sheathing	Shear Wall Nailing
6	2	5	3.0	7/16" OSB	8d at 6"
	1	5	3.0	7/16" OSB	8d at 3"
7	2	5	3.0	7/16" OSB	8d at 4"
	1	5	3.0	7/16" OSB	8d at 2"
8	2	2	3.0	7/16" OSB	8d at 6"
	1	2	3.0	7/16" OSB	8d at 4"
9	3	3	10.0	7/16" OSB	8d at 6"
	2	3	10.0	7/16" OSB	8d at 2"
	1	3	10.0	19/32" PLWD	10d at 2"
10	3	6	3.0	7/16" OSB	8d at 6"
	2	6	3.0	7/16" OSB	8d at 2"
	1	6	3.0	19/32" PLWD	10d at 2"
11	3	2	7.0	7/16" OSB	8d at 6"
	2	2	7.0	7/16" OSB	8d at 3"
	1	2	7.0	7/16" OSB	8d at 2"
12	3	3	3.0	7/16" OSB	8d at 6"
	2	3	3.0	7/16" OSB	8d at 3"
	1	3	3.0	7/16" OSB	8d at 2"
13	4	6	3.3	7/16" OSB	8d at 6"
	3	6	3.3	7/16" OSB	8d at 2"
	2	6	3.3	19/32" PLWD	10 at 2"
	1	6	3.3	19/32" PLWD	10 at 2"
14	4	4	3.0	7/16" OSB	8d at 6"
	3	4	3.0	7/16" OSB	8d at 4"
	2	4	3.0	7/16" OSB	8d at 3"
	1	4	3.0	7/16" OSB	8d at 2"
15	5	6	3.7	7/16" OSB	8d at 6"
	4	6	3.7	7/16" OSB	8d at 3"
	3	6	3.7	7/16" OSB	8d at 2"
	2	6	3.7	19/32" PLWD	10d at 2"
	1	6	3.7	19/32" PLWD	10d at 2"
16	5	4	3.3	7/16" OSB	8d at 6"
	4	4	3.3	7/16" OSB	8d at 4"
	3	4	3.3	7/16" OSB	8d at 3"
	2	4	3.3	7/16" OSB	8d at 2"
	1	4	3.3	7/16" OSB	8d at 2"

C.2.4 Other Considerations for Wood Light-Frame Shear Wall Systems

The following considerations were not addressed in this example, but are provided for the purpose of further illustrating the development of index archetype configurations.

A mix of shear wall aspect ratios could very likely occur within a given wood light-frame system. Careful thought should be given as to whether performance groups consisting of all high, all low, or mixed aspect ratio walls will be most representative or most critical. This could change with the system being considered.

It is becoming common practice to mix alternative bracing elements with conventional shear wall bracing in this type of construction. These elements could be designed to carry a small or large portion of the seismic story shear forces. Consideration should be given as to whether the seismic story shear carried by alternative bracing elements needs to be added as another variable in the index archetype configurations.

For wood light-frame systems in particular, wall finish materials are known to have a very significant impact on the overall collapse behavior of the system. The effects of wall finish materials should be considered in the context of the system being proposed. It may be appropriate to include a range of wall finish materials as another variable in the index archetype configurations.

Appendix D
Consideration of Behavioral Effects

This appendix illustrates how seismic behavioral effects are considered in the development of index archetype configurations for reinforced concrete moment frame systems conforming to design requirements contained in ASCE/SEI 7-05, *Minimum Design Loads for Buildings and Other Structures* (ASCE, 2006a), and ACI 318-05, *Building Code Requirements for Structural Concrete* (ACI, 2005).

Development of structural system archetypes considers both structural configuration issues and seismic behavioral effects described in Chapter 4. Consideration of structural configuration issues is discussed in Appendix C. Index archetype configurations are subsequently used to develop index archetype designs, which are then used to develop index archetype models. Development and calibration of nonlinear index archetype models for reinforced concrete moment frames is described in Appendix E.

D.1 Identification of Structural Failure Modes

Consideration of seismic behavioral effects includes identifying critical limit states, dominant deterioration modes and collapse mechanisms that are possible, and assessing the likelihood that they will occur. How a component or system behaves under seismic loading is often influenced by configuration decisions, so behavioral effects and configuration issues should be considered concurrently in the development of index archetype configurations. This process is illustrated in Figure D-1.

Once all possible failure modes have been identified for a given system, the list is narrowed to a subset of likely collapse mechanisms by ruling out failure modes that are unlikely to occur based on system design and detailing requirements, experimental data, engineering judgment, analytical models, or observations from past earthquakes. Remaining failure modes will be assessed through explicit simulation of failure modes through nonlinear analyses, or through evaluation of non-simulated failure modes using alternative limit state checks on demand quantities from nonlinear analyses. Index archetype configurations must address failure modes that will be explicitly simulated in nonlinear analysis models in Chapter 5.

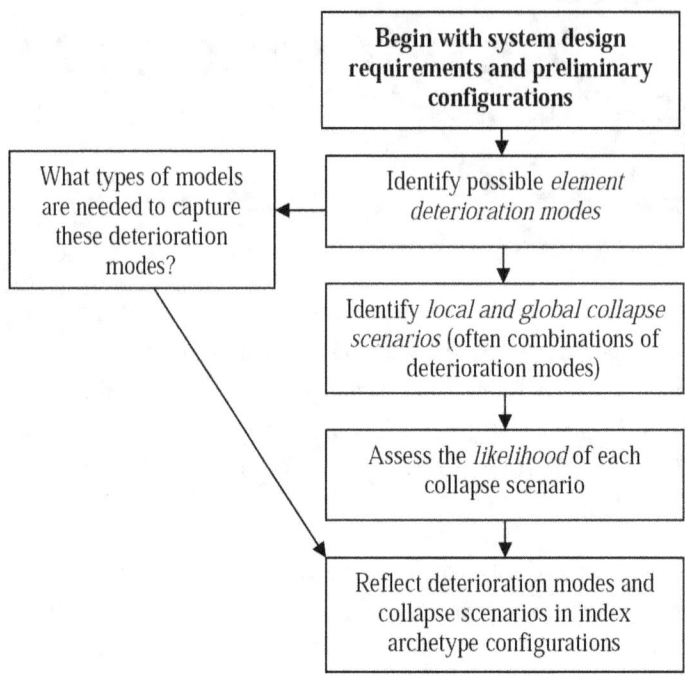

Figure D-1   Consideration of behavioral effects in developing index archetype configurations.

D.2   System Definition

Possible element deterioration and failure modes are influenced by the material properties, mechanical properties, and design requirements of the proposed seismic-force-resisting system.

Reinforced concrete moment frame systems in this example are assumed to meet design requirements for reinforced concrete moment frames specified in ASCE/SEI 7-05 and ACI 318-05. For *special* moment frames, these standards provide rigorous and specific design and detailing provisions intended to prevent certain failure modes from occurring, and to promote ductile failure modes in selected members. Special reinforced concrete moment frame requirements include, for example, an upper limit on longitudinal steel reinforcement ratio, seismic hoop detailing, confinement reinforcing in columns and beam-column joints, exclusion of lap splices from hinge regions, strong-column-weak-beam provisions, and other capacity design requirements. *Ordinary* moment frames are not subject to these provisions.

For a proposed seismic-force-resisting system, design and detailing requirements must be well-defined and specific before element deterioration modes and system collapse scenarios can be identified.

D.3   Element Deterioration Modes

In this example, failure modes have been identified for reinforced concrete moment frame components based on a review of experimental tests, available published information, and observations from past earthquakes. Potential deterioration modes for reinforced concrete moment frame components are listed in Table D-1 and shown in Figure D-2. They are classified into six groups (A to F) depending on the type of structural element and the physical behavior associated with deterioration, as shown in Figure D-3. For each mode, currently available nonlinear element models are rated for their ability to simulate the deterioration behavior.

Table D-1   Possible Deterioration Modes for Reinforced Concrete Moment Frame Components.

Deteri-oration Mode	Element	Behavior	Model Availability[1]		Description
			Simulation	Fragility	
A	Beam-column	Flexural	4	NR	Concrete spalling Reinforcing bar yielding Concrete core Reinforcing bar buckling (incl. Stirrup fracture) Reinforcing bar fracture
B	Beam-column	Axial compression	2	4	Concrete crushing, longitudinal bar yielding Stirrup-rupture, longitudinal bar buckling
C	Beam-column	Shear Shear +Axial	1	4	Concrete Shear Transverse tie pull-out of rupture Concrete Loss of aggregate interlock Possible loss of axial load-carrying capacity in columns
D	Joint	Shear	3	2	Panel shear failure
E	Reinforcing bar	Pull-out or Bond-slip	2	2	Reinforcing bar bond-slip or anchorage failure at joint Reinforcing bar lap-splice failure Reinforcing bar pull-out (in beams or at footing)
F	Slab connection	Shear	2	3	Punching shear or large shear eccentricity at slab column connection, in shear wall structures possible loss of connection between slab and wall Possible vertical collapse of slab

[1]Model availability ratings: 0 - nonexistent; 1 – low confidence; 5 – high confidence; NR – not required because behavior can be simulated.

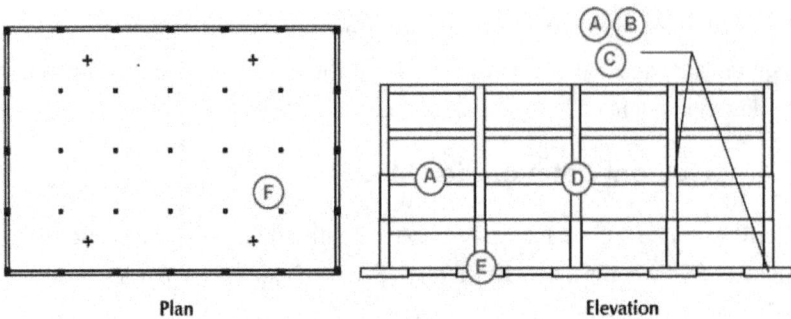

Plan	**Elevation**

Figure D-2　　Reinforced concrete frame plan and elevation views, showing location of possible deterioration modes.

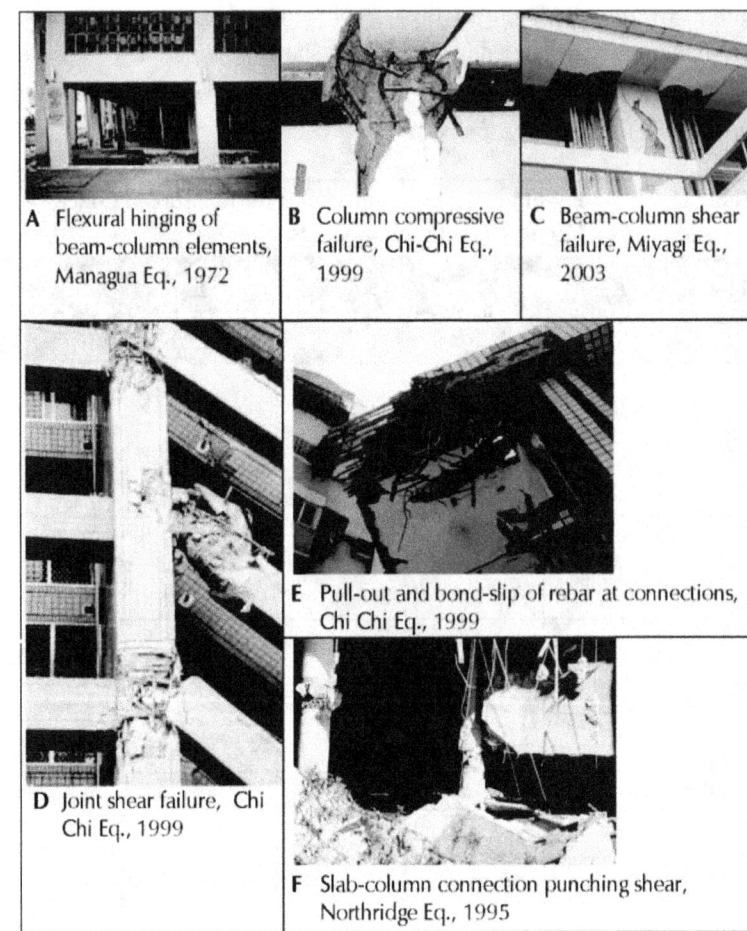

Figure D-3　　Illustration and classification of possible deterioration modes for reinforced concrete moment frame components.[1]

---

[1] Photo sources: (A) nisee.berkeley.edu/thumbnail/6257_3021_0662/IMG0071.jpg (B) www.structures.ucsd.edu/Taiwaneq/buildi25.jpg (C) www.disaster.archi. tohoku.ac.jp/eng/topicse/030726htm/5-28.jpg (D) www.structures. ucsd.edu/ Taiwaneq/ buildi21.jpg (E) www.structures.ucsd.edu/Taiwaneq/buildil9.jpg (F) www.nbmg.unr.edu/nesc/bobcox/ndx2.php

In this example, foundation failure modes have not been included because they are judged not critical for this system. Foundation failure modes should be considered if they are judged to have a potentially significant effect on the collapse performance of a proposed system.

D.3.1 Flexural Hinging of Beam and Columns

Deterioration mode 'A' consists of flexural hinging in beams and columns and associated concrete spalling, concrete core crushing, stirrup fracture, and reinforcing bar yielding, buckling, and fracture. Reinforcing bar yielding, concrete core crushing, and the associated strength and stiffness degradation can be simulated fairly accurately. In contrast, modeling of buckling and fracture of longitudinal reinforcement, or stirrup fracture is less accurate. These behaviors are important contributors to the deterioration of strength and stiffness in reinforced concrete frame elements at large deformations near collapse.

Fiber-type models capture the spread of plasticity along the length of the element and the constituent concrete models can be calibrated to adequately model the behavior associated with concrete deterioration from cracking to crushing (Haselton et al., 2007). However, currently available steel material models are not able to replicate the behavior of rebar as it buckles and fractures. Due to this limitation, current fiber models were judged inadequate for simulating collapse. While lumped plasticity models do not have the precision of fiber models, they can be calibrated to capture the deterioration associated with rebar buckling and stirrup fracture leading to loss of confinement (Haselton, 2006; Ibarra, 2003).

D.3.2 Compressive Failure of Columns

Deterioration mode 'B' corresponds to compressive failure of columns and is characterized by concrete crushing, buckling of longitudinal reinforcement, and yielding or fracture of transverse reinforcement. This deterioration mode occurs when compressive forces in the column exceed the compressive capacity. In earthquakes, this may occur due to overturning, which increases axial loads in some of the columns of a moment-resisting frame.

D.3.3 Shear Failure of Beam and Columns

Deterioration mode 'C' corresponds to shear failure of beam-columns, characterized by shear cracking in concrete and yielding and/or pull-out of transverse stirrups. This mode of deterioration is particularly dangerous for columns with significant axial load. Shear deterioration and increased displacement demands can lead to subsequent vertical collapse of the column (Elwood and Moehle, 2005), and possibly progressive collapse of a structure.

Modeling the cyclic response of a reinforced concrete element experiencing shear deterioration is complex due to interactions between shear, moment, and axial force, as well as the overall brittle nature of the deterioration mode. Elwood (2004) and others have simulated deterioration in the lateral strength and stiffness of a column by adding a shear spring. To date, models for vertical collapse of columns following shear failure have been challenged by a lack of experimental data for columns experiencing vertical collapse.

Due to the brittle nature of this failure mode, special and intermediate moment frames in seismically active areas are required to utilize capacity design principles that ensure flexural hinging prior to shear failure (ACI, 2005). Capacity design requirements, however, do not apply to ordinary concrete moment frames. Shear failure, and the possible subsequent loss of gravity-load carrying capacity, is treated as a non-simulated collapse mode in evaluating ordinary moment frame systems, due to limitations in the ability to directly simulate these failure modes.

D.3.4 Joint Panel Shear Behavior

Deterioration Mode 'D' is associated with deterioration in shear strength and stiffness of the joint panel region. In reinforced concrete frame models, shear panels are modeled with an inelastic rotational spring inserted at the joint (Lowes et al., 2004; Altoontash, 2004). This joint modeling capability is available in the Open System for Earthquake Engineering Simulation (OpenSees, 2006) software framework and in most structural analysis software.

Several researchers have used a detailed approach that employs the modified compression field theory (Vecchio and Collins,1986; Stevens et al., 1991) to develop a monotonic backbone that relates the panel shear force to the shear deformation angle (Altoontash, 2004; Lowes et al., 2004). Modified compression field theory has been shown to work well for conforming joints, but is not as reliable for non-conforming joints with less confinement. Modeling of joint shear behavior is especially important for joints that are not protected by capacity design requirements (e.g., ordinary concrete moment frame systems).

D.3.5 Bond-Slip of Reinforcing Bars

Deterioration Mode 'E' corresponds to slip of the reinforcing bar relative to the surrounding concrete, including anchorage pull-out, bond slip, and splice failure. The extent of slip, and the possible occurrence of bar pull-out, depends on the embedded length, bar diameter, concrete strength, and the number and magnitude of load cycles.

Slip without pull-out occurs when the embedment or splice length is sufficient to prevent the cut end of the rebar from moving relative to the surrounding concrete, even under cyclic loading. The effects of slip without pull-out include: (1) decreased pre-yield stiffness; (2) decreased post-yield stiffness; and (3) increased element plastic deformation capacity. These effects are included in the calibration of the plastic hinges for reinforced concrete moment frame models.

Slip with pull-out occurs when the rebar slips relative to the concrete over the full length of embedment. This occurs when the embedment or lap splice lengths are relatively short, as is typical of ordinary concrete moment frames. Slip with pull-out can lead to severe reduction in strength and stiffness, as well as a significant reduction in plastic rotation capacity. Models for pull-out are moderately well-developed and range from continuum finite element models (which attempt to model the interface between the rebar and concrete) to simple rotational spring models.

D.3.6 Punching Shear in Slab-Column Connections

Deterioration Mode 'F' corresponds to punching shear in slab-column connections or in wall-slab connections. Following punching shear failure, a slab may experience local collapse depending on the gravity load intensity and continuity of reinforcement in the area of the slab-column connection (Aslani 2005).

Punching shear deterioration can be modeled with a standard nonlinear lumped plasticity spring. The resulting vertical collapse is difficult to directly simulate, but this failure mode can be treated as a non-simulated collapse mode (Aslani and Miranda 2005).

D.4   Local and Global Collapse Scenarios

Collapse occurs when seismic loading causes element deterioration modes to combine in a way that forms a sidesway collapse mechanism or a vertical (local) collapse mechanism. For reinforced concrete frame systems, possible collapse scenarios and contributing element deterioration modes are identified and organized as shown in Tables D-2a and D2-b for sidesway and vertical collapse modes, respectively. These scenarios were established using engineering judgment based on examination of collapses in previous earthquakes, experimental test data, and analytical studies.

Table D-2a Collapse scenarios for Reinforced Concrete Moment Frames – Sidesway Collapse

Scenario	Element Deterioration Mode						Description
	A	B	C	D	E	F	
FS1	x						Beam and column flexural hinging, forming sidesway mechanism
FS2	x						Column hinging, forming soft-story mechanism
FS3	x		x				Beam or column flexural-shear failure, forming sidesway mechanism
FS4	x			x			Joint-shear failure, possibly with beam and/or column hinging
FS5	x				x		Reinforcing bar pull-out or splice failure in columns or beams, leading to sidesway mechanism

Table D-2b Collapse Scenarios For Reinforced Concrete Moment Frames – Vertical Collapse

Scenario	Element Deterioration Mode						Description
	A	B	C	D	E	F	
FV1			x				Column shear failure, leading to column axial collapse
FV2	x		x				Column flexure-shear failure, leading to column axial collapse
FV3						x	Punching shear failure, leading to slab collapse
FV4							Failure of floor diaphragm, leading to column instability
FV5		x					Crushing of column, leading to column axial collapse; possibly from overturning effects

D.5 Likelihood of Collapse Scenarios

Table D-3 includes the range of possible collapse scenarios for reinforced concrete frame structures. From this list, the scenarios which are most likely to occur for each type of moment frame defined in ACI 318-05 (ordinary, intermediate, and special) are identified.

Changes in detailing requirements for different reinforced concrete frame systems govern which collapse modes are likely to occur for that system. Design requirements for special concrete moment frames (ACI, 2005) are designed to promote ductile collapse modes and to prevent the formation of brittle collapse modes. These requirements serve to limit the likelihood of several possible collapse modes.

Ordinary moment frames are vulnerable to a wider range of possible collapse modes, due to less stringent design requirements. Several researchers (e.g. Aycardi et al., 1994; Kunnath et al., 1995; Filiatrault et al., 1998) have noted the tendency for soft story or column-hinging mechanisms to form in ordinary concrete moment frames. In addition, ordinary concrete moment frames might experience lap-splice failure, pull-out of reinforcing bars at beam-column joints, and column shear failures. The behavior of intermediate moment frames is anticipated to be somewhere between special and ordinary moment frames.

The likelihood of each collapse scenario for special, intermediate, and ordinary moment frames is shown in Table D-3. Depending on the proposed system and associated design requirements, there may be one or several collapse modes that are likely to occur.

Table D-3   Likelihood of Column Collapse Scenarios by Frame Type (H: High, M: Medium, L: Low)

System	Sidesway Collapse					Vertical Collapse				
	FS1	FS2	FS3	FS4	FS5	FV1	FV2	FV3	FV4	FV5
SMF	H	M	L	L	L	L	L	L	L	L-M
IMF	H	M-H	L	M	M	L	L	L	M	L-M
OMF	H	H	H	H	H	M	H	M	M	M-H

D.6   Collapse Simulation

For reinforced concrete moment frame systems, possible collapse modes are considered in the context of the three-bay variable-height archetype frame model described in Appendix C and Chapter 9. Index archetype configurations must account for all collapse scenarios that have been identified as likely to occur, and will be explicitly simulated in the nonlinear index archetype models. Critical limit states or failure modes that cannot be explicitly simulated in index archetype models are evaluated using the procedure for non-simulated collapse modes.

Appendix E
Nonlinear Modeling of Reinforced Concrete Moment Frame Systems

E.1 Purpose

This appendix describes the development of nonlinear index archetype models used for collapse assessment of reinforced concrete special moment frame and reinforced concrete ordinary moment frame example structures presented in Chapter 9. It includes an overview of critical modeling decisions and a description of the calibration procedures used for beam-column element models. Index archetype models represent index archetype configurations, which are developed with careful consideration of structural configuration issues (Appendix C) and seismic behavioral effects (Appendix D).

Although the information is specific to reinforced concrete moment frame systems, the example is illustrative of the issues and considerations typical of nonlinear collapse simulation for any structural system type. Similar procedures were used to create nonlinear models for the wood light-frame system presented in Chapter 9, and the steel special moment frame system presented in Chapter 10.

E.2 Structural Modeling Overview

Structural models should be capable of simulating the accrual of structural damage and the resulting sidesway collapse when a structure is subjected to severe ground shaking. Identification of key deterioration and collapse modes, as described in Appendix D, is an important precursor to the choice of the nonlinear analysis model.

The seismic-force-resisting system for reinforced concrete moment frames is represented by a two-dimensional, three-bay frame. The destabilizing P-delta effects are modeled using a leaning column. At the element level, frames are modeled with the following features illustrated in Figure E-1: beam-column elements with concentrated inelastic rotational hinges at each end and finite size beam-column joints that employ five concentrated inelastic springs to model joint panel shear distortion and bond slip at each face of the joint. The collapse behavior is simulated using the Open System for Earthquake Engineering Simulation (OpenSees, 2006) software.

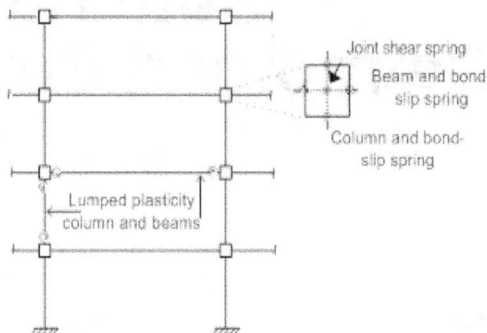

Figure E-1   Schematic diagram illustrating key elements of nonlinear frame model for reinforced concrete frame systems.

Although not shown in Figure E-1, the effect of foundation flexibility on the archetype models is incorporated using elastic, semi-rigid rotational springs at the base of the column. For shorter frame structures (fewer than 4 stories), the stiffness of the rotational springs is determined from assumed grade beam and soil stiffness. For taller buildings, the structure is assumed to have a basement, exterior columns are assumed to be fixed (connected to the basement wall), and the properties of the rotational spring attached to interior columns are computed based on the basement column and beam stiffnesses.

E.3   Beam-Column Element Model

Since the damage is likely to concentrate in the beams and columns of the reinforced moment frame structures, accurate modeling of the inelastic effects in beam and column elements is an essential component of collapse modeling of these structures. As shown in Figure E-1, the beam-column elements of the lateral resisting frame are modeled with lumped plasticity elements in the plastic hinge locations. Lumped plasticity elements are frequently used in structural analysis models, and their use here follows the precedent established in ASCE/SEI 41-06 *Seismic Rehabilitation of Existing Buildings* (ASCE, 2006b) and other guideline documents. When calibrated properly, they are capable of capturing degradation of strength and stiffness that is essential to collapse modeling. Their properties can also be easily modified in a sensitivity analysis to determine the effects of uncertainties in material modeling.

Researchers have also used a variety of other methods to simulate cyclic response of reinforced concrete beam-columns, including creating fiber models which can capture cracking behavior and the spread of plasticity throughout the element (see e.g., Filippou, 1999) The decision to use a lumped-plasticity approach here was based on simplicity and an inherent limitation in the fiber element formulation which makes simulation of the strain softening associated with rebar buckling difficult. The choice of element model should be carefully evaluated for any given structural system, and with careful consideration of available simulation technologies.

In this section, the properties of the lumped plasticity elements used to model the example reinforced concrete special moment frame structures are illustrated, including descriptions of both the element model used and the process through which the key modeling parameters were calibrated. The calibration process described here demonstrates the level of detail and sophistication that is possible in studies of this type. Depending on the structural system and type of analytical model, a greater or lesser degree of detail may be required.

E.3.1 Element and Hysteretic Model

Lumped plasticity element models for the reinforced concrete special moment frame structures utilize a material model developed by Ibarra, Medina, and Krawinkler (2005), and implemented in OpenSees. This model is chosen because it is capable of capturing the important modes of deterioration that precipitate sidesway collapse of reinforced concrete frames, but one could imagine using another material model or software platform that also met these requirements.

Figure E-2 shows the tri-linear monotonic backbone curve and associated hysteretic rules of the model, which permit versatile modeling of cyclic behavior. For simulating structural collapse, the most important aspect of this model is the post-peak response, which enables modeling the strain softening behavior associated with concrete crushing, rebar buckling and fracture, and/or bond failure. The model also captures four basic modes of cyclic deterioration: (1) strength deterioration of the inelastic strain hardening branch; (2) strength deterioration of the post-peak strain softening branch; (3) accelerated reloading stiffness deterioration; and (4) unloading stiffness deterioration. Cyclic deterioration is based on an energy index that includes normalized energy dissipation capacity (λ) and an exponent term to describe how the rate of cyclic deterioration changes with accumulation of damage (c).

This element model requires the specification of seven parameters to control the monotonic and cyclic behavior of the model: M_y, θ_y, M_c/M_y, $\theta_{cap,pl}$, θ_{pc}, λ, and c. The post-yield and post-capping stiffnesses are quantified by M_c/M_y and θ_{pc}; K_s and K_c can be easily computed as:

$$K_s = K_e\left(\theta_y/\theta_{cap,pl}\right)\left(\left(M_c - M_y\right)/M_y\right) \qquad \text{(E-1)}$$

$$K_c = -K_e\left(\theta_y/\theta_{pc}\right)\left(M_c/M_y\right). \qquad \text{(E-2)}$$

Symbols are defined in Figure E-2 and in the list of symbols at the end of the document.

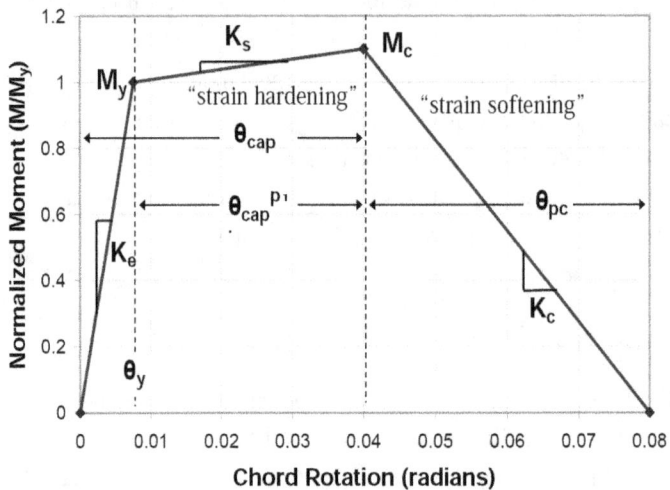

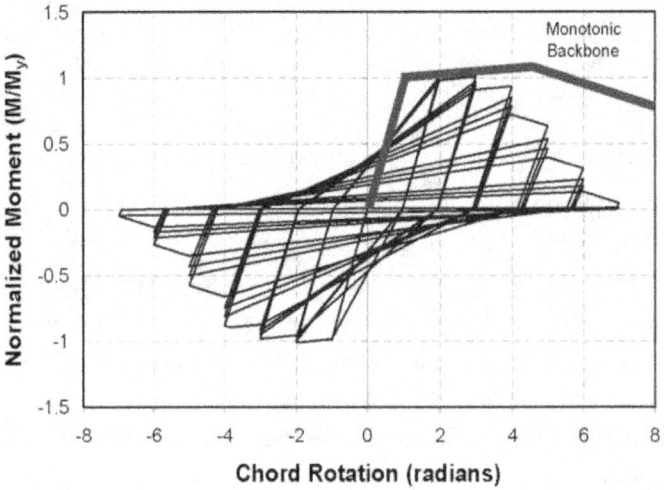

Figure E-2   Monotonic and cyclic behavior of component model used to model reinforced concrete beam-column elements.

In order to determine the appropriate values of each of these parameters, model parameters were carefully calibrated to 255 experimental tests of reinforced concrete columns. These calibrations were then used to develop empirical equations relating the design parameters of a beam-column to the modeling parameters needed for input in the lumped plasticity model. Calibrations are based on mean values. This type of calibration procedure can be used in conjunction with any type of material model, provided that there is sufficient test data available.

E.3.2 Calibration of Parameters for the Reinforced Concrete Beam-Column Element Model

Experimental Database

Parameters of the material model are calibrated with data from the PEER Structural Performance Database (PEER, 2006a; Berry et al., 2004), assembled by Berry and Eberhard. This database includes the results of cyclic and monotonic tests of 306 rectangular columns and 177 circular columns. For each column test, the database reports the force-displacement time history, the column geometry and reinforcement information, the failure mode, and other relevant information. All data is converted to that of an equivalent cantilever, regardless of experimental setup.

From this database, rectangular columns failing in a flexural mode (220 tests) or in a combined flexure-shear mode (35 tests) were selected, for a total of 255 tests. These tests cover the typical range of column design parameters: $0.0 < v < 0.7$, $0.0 < P/P_b < 2.0$, $1.5 < L_s/H < 6.0$, $20 < f'_c \ (MPa) < 120$, $340 < f_y \ (MPa) < 520$, $0.015 < \rho < 0.043$, $0.1 < s/d < 0.6$, $0.002 < \rho_{sh} < 0.02$. These symbols are defined in the list of symbols at the end of this document.

Calibration Procedure

In order to calibrate the element model parameters, each column test in the database was modeled as a cantilever column in OpenSees, idealized using an elastic element and a zero-length Ibarra model plastic hinge at the column base. The properties of the plastic hinge are the subject of this calibration effort. This appendix provides only a brief description of the calibration process and the interested reader is referred to Haselton et al. (2007) for more details.

The calibration of the beam-column element model to the data from each experimental test was done systematically and, referring to Figure E-3, each test was calibrated according to the following standardized procedure. First, the yield shear force (1) was estimated visually from the experimental results. Then, the "yield" displacement (2) was calibrated as the point of a significant observed change in lateral stiffness, i.e., where rebar yielding or significant concrete crushing occurs. Calibration of this point often required judgment, since concrete response is nonlinear well before rebar yielding. In addition, displacement at 40% of the yield force (3) was calibrated to provide an estimate of initial stiffness. In step (4), the post-yield hardening stiffness was visually calibrated.

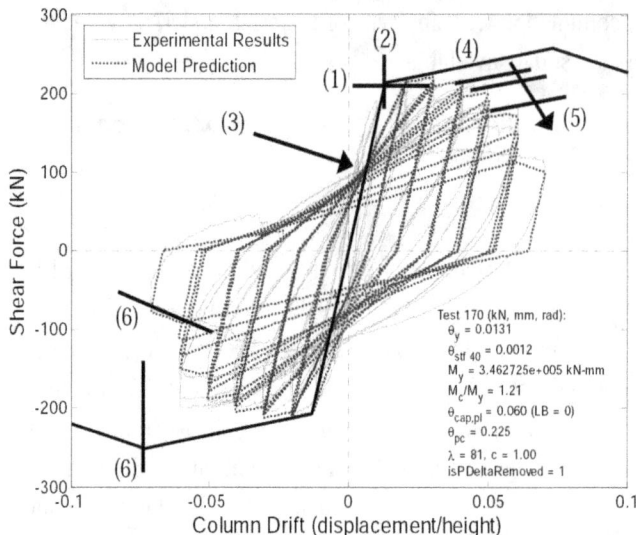

Figure E-3   Example of calibration procedure; calibration of reinforced
concrete beam-column element to experimental test by
Saatcioglu and Grira, specimen BG-6.

The next step (5) was to calibrate the normalized cyclic energy dissipation capacity, λ. This parameter is defined such that the total energy dissipated in a cycle is given by $E_t = \lambda M_y \theta_y$. The parameter c describes how energy dissipation changes in subsequent cycles (Ibarra et al., 2005). The element model allows cyclic deterioration coefficients λ and c to be calibrated independently for each of the four cyclic deterioration modes. However, based on a pilot study of 20 columns, $c = 1.0$ was found to be acceptable for columns failing in flexure and flexure-shear modes. The deterioration rate, λ, was set to be equal for the basic strength and post-capping strength deterioration modes, following recommendations by Ibarra (2003). Based on observations of the hysteretic response of the reinforced concrete columns, no accelerated stiffness deterioration was used. Unloading stiffness deterioration was neglected to avoid an error in the OpenSees implementation of the element model; in future, unloading stiffness deterioration should be employed, as it leads to better modeling of cyclic response. These simplifications reduce the calibration of cyclic energy dissipation capacity to one parameter (λ). When calibrating λ, the aim was to match the average deterioration for the full displacement history, but with a slightly higher emphasis on matching the deterioration rate of the later, more damaging, cycles.

The final step of the calibration process (6) involved quantification of the capping point and the post-capping deformation capacity. It has been shown (e.g., Haselton et al., 2007) that incorrect calibration of strength deterioration can have a significant impact on structural response prediction, and the calibration procedure carefully distinguished between in-cycle and cyclic strength deterioration. The capping point and post-capping stiffness were only calibrated when a negative post-peak stiffness was clearly observed in the data, such that strength loss occurs within a single cycle (sometimes called "in-cycle strength deterioration"). Often the test specimen did not undergo sufficient deformations for a capping point to be observed. In such cases where the data do not reveal the capping point, a lower-bound value of the capping point was determined.

It should be noted that the hinge model is based on the definition of a monotonic backbone and cyclic deterioration rules. This calibration process relied primarily on cyclic tests with many cycles to calibrate both the monotonic backbone parameters and the cyclic deterioration rules. As a result, the monotonic backbone and the cyclic deterioration rules are interdependent, and the approximation of the monotonic backbone depends on cyclic deterioration rules assumed and, to some extent, the displacement pattern used in experimental tests. This approximation of the monotonic backbone from cyclic data is not ideal, but is necessary because sufficient data are not available to calibrate the monotonic and cyclic behavior separately.

A full table of calibrated model parameters for each of the 255 experimental tests used can be found in the extended report on this study (Haselton et al., 2007).

Interpretation of Calibration Results and Regression Analysis

Calibrated model parameters from the 255 column tests are used to create empirical equations that predict model parameters based on the column design parameters. The functional form used in regression analysis was carefully determined based on trends in the data and isolated effects of individual variables, previous research and existing equations, and judgment based on mechanics and expected behavior. The regression analysis was performed using the natural logarithm of the model parameter, and the logarithmic standard deviation quantifies the uncertainty. More details on how the data were dissected to create empirical regression equations to predict each model parameter is available in Haselton et al. (2007).

As noted previously, those test specimens in which post-capping behavior was not observed were calibrated with a lower-bound value for deformation capacity. In the regression analysis, this lower-bound calibration data were given special consideration. In order to take advantage of lower-bound data, without unnecessarily biasing the results, the deformation capacity equations presented here are based on all data, and the prediction uncertainties are reported based on only the data with an observed capping point in order to avoid reporting the artificially high uncertainty associated with the lower-bound data. Note that this approach is still appropriate for elements with high deformation capacity because the lower-bound data underestimates the true deformation capacity.

Proposed Equations

Predictive equations were developed for each of the parameters of the element model for reinforced concrete columns. Each equation includes all statistically significant parameters, unless otherwise noted. The extended report on this study also includes more simplified equations (Haselton et al., 2007).

Effective Initial Stiffness (EI_y/EI_g and EI_{stf}/EI_g)

Due to nonlinearities in reinforced concrete behavior associated with cracking, the definition of the stiffness of a reinforced concrete element depends on the load and deformation level. In this work, two values of effective stiffness are defined: (1) the secant stiffness to the yield point of the component (termed EI_y or K_e) and (2) the secant stiffness to 40% of the yield force of the component (termed EI_{stf}). EI_{stf} is used as the initial stiffness in the creation of the structural models for reinforced concrete frames in Chapter 9, based on a study by Haselton et al. (2007). The component stiffness includes all modes of deformation including flexure, shear, and bond-slip.

The equation for secant stiffness to yield depends on both axial load ratio and shear span ratio of the column, and is given as follows:

$$\frac{EI_y}{EI_g} = -0.07 + 0.59\left[\frac{P}{A_g f_c'}\right] + 0.07\left[\frac{L_s}{H}\right] \tag{E-3}$$

where $\quad 0.2 \leq \dfrac{EI_y}{EI_g} \leq 0.6$

This equation represents the mean value of effective stiffness to yield. The prediction uncertainty, assuming the residuals are lognormally distributed, is given by the logarithmic standard deviation: $\sigma_{LN} = 0.28$. R^2, a measure of the extent to which the proposed equation explains the data, is 0.80. For this and all equations to follow, these values are reported below in Table E-1. The upper and lower limits on the stiffness were imposed because there is limited data for columns with very low axial loads and, at high levels of axial load, the positive trend diminishes and the scatter in the data is large. The limits were chosen based on a visual inspection of data.

The effective initial stiffness, defined as the secant stiffness to 40% of the yield force of a reinforced concrete column, can be predicted as follows:

$$\frac{EI_{stf}}{EI_g} = -0.02 + 0.98\left[\frac{P}{A_g f_c^i}\right] + 0.09\left[\frac{L_s}{H}\right] \qquad \text{(E-4)}$$

where $0.35 \le \dfrac{EI_{stf}}{EI_g} \le 0.8$

For a typical column, Equation (E-4) predicts the initial stiffness to be approximately 1.7 times stiffer than the secant stiffness to yield (Equation (E-3)).

The effective stiffness of reinforced concrete columns has been the subject of much research; for comparison, selected studies are presented here. The guidelines in FEMA 356 *Prestandard and Commentary for Seismic Rehabilitation of Buildings* (FEMA, 2000) permit the use of standard simplified values based on the level of axial load for linear analysis: $0.5E_cI_g$ when $P/(A_g f'_c) < 0.3$, and $0.7E_cI_g$ when $P/(A_g f'_c) > 0.5$. More recently, Elwood and Eberhard (2006) proposed an equation for effective stiffness that includes all components of deformation (flexure, shear, and bond-slip), where the effective stiffness is defined as the secant stiffness to the yield point of the component. Their equation proposes $0.2E_cI_g$ when $P/(A_g f'_c) < 0.2$ and $0.7E_cI_g$ when $P/(A_g f'_c) > 0.5$, with a linear transition between these two extremes.

The equations proposed here for EI_y are similar to those recently proposed by Elwood and Eberhard (2006), but also includes an L_s/H term. Due to this additional term, the proposed Equation (E-3) has a lower prediction uncertainty of $\sigma_{LN} = 0.28$, as compared to the coefficient of variation of 0.35 reported by Elwood et al. The stiffness predictions in FEMA 356 are much higher than the values of EI_y predicted in this study, and slightly larger than the predicted values of EI_{stf}. Elwood and Eberhard (2006) found that most of this difference can be explained if it is assumed that the FEMA 356 values only include flexural deformation, and do not account for the bond-slip deformations, which can account for a significant proportion of an element's flexibility.

Flexural Strength (M_y)

Panagiotakos and Fardis (2001) have published equations to predict flexural strength. Their equation works well, so for this study their method is used to predict M_y. When comparing the calibrated values to flexural strength for the 255 columns to predictions by Panagiotakos and Fardis (2001) for these columns, the median ratio of $M_y / M_{y,(Fardis)}$ is 0.97, and the coefficient of variation is 0.36.

Plastic rotation capacity ($\theta_{cap,pl}$)

The following equation is proposed for predicting plastic rotation capacity, including all parameters that are statistically significant:

$$\theta_{cap,pl} = 0.12\left(1+0.55a_{sl}\right)\left(0.16\right)^{\nu}\left(0.02+40\rho_{sh}\right)^{0.43}$$
$$\left(0.54\right)^{0.01c_{units}f_c'}\left(0.66\right)^{0.1s_n}\left(2.27\right)^{10.0\rho} \tag{E-5}$$

Possible correlations between ρ_{sh} and s_n were verified to be small to eliminate concerns regarding co-linearity in regression. This equation is based on all data, including the lower-bound data for plastic rotation capacity, as discussed previously. The shear-span ratio (L_s/H) is notably absent from the predictive equations for rotation capacity. The stepwise regression process consistently showed the shear-span ratio to be statistically insignificant. These findings differ from those of Panagiotakos and Fardis (2001).

The experimental data used in this study is limited to tests of columns with symmetrical arrangements of reinforcement and, as a result, Equation (E-5) applies only to columns with symmetric reinforcement. To eliminate the symmetric reinforcement limitation from the rotation capacity equation, it is proposed to multiply the rotation capacity obtained from these equations by the term proposed by Fardis and Biskini (2003), which accounts for the ratio between the areas of compressive and tensile steel.

It is also useful as verification to compare the predicted rotation capacity to the ultimate rotation capacity predicted by Fardis and Biskini based on their study of more than 700 reinforced concrete elements including beams, columns, and walls (Fardis and Biskini ,2003; Panagiotakos and Fardis, 2001). The equation based on these studies consistently predicts higher values, which is expected since Equation (E-5) predicts the capping point where the ultimate rotation capacity equation is based on the ultimate (20% strength loss) point. To make a more consistent comparison, the predictions of the ultimate rotation (at 20% strength loss) are combined with the calibrated values of post-capping slope (θ_{pc}) to back-calculate a prediction of

$\theta_{cap,pl}$ from the equation in Fardis and Biskini. This comparison shows that the proposed equation here predicts lower deformation capacity, with a mean ratio of 0.94 and a median ratio of 0.69. Based on this comparison, Equation (E-5) may still include some conservatism. This conservatism exists even though the predicted deformation capacities are already much higher than what is typically used; for example, values in ASCE/SEI 41 are typically less than one-half of those shown in Table E-2. Note: The recent supplement to ASCE/SEI 41 for concrete elements, particularly those with non-ductile detailing, has reduced some of the conservatism in ASCE/SEI provisions (Elwood et al., 2007).

Post-capping Rotation Capacity (θ_{pc})

Previous research on predicting post-capping rotation capacity has been limited, despite its important impact on predicting collapse capacity. The key parameters considered in the development of an equation for post-capping response are those that are known to most impact deformation capacity: axial load ratio (v), transverse steel ratio (ρ_{sh}), rebar buckling coefficient (s_n), stirrup spacing, and longitudinal steel ratio. The equation is based on only those tests where a post-capping slope was observed. The proposed equation for post-capping rotation capacity is as follows:

$$\theta_{pc} = (0.76)(0.031)^v (0.02 + 40\rho_{sh})^{1.02} \le 0.10 \qquad \text{(E-6)}$$

The upper bound imposed on Equation (E-6) is due to lack of reliable data for elements with shallow post-capping slopes. The upper limit of 0.10 may be conservative for well-confined, conforming elements, but existing test data does not justify using a larger value.

Post-Yield Hardening Stiffness

Regression analysis shows that axial load ratio and concrete strength statistically impact the prediction of hardening stiffness (M_c/M_y). Even so, inclusion of both of the parameters scarcely improved the regression analysis, so a constant value of M_c/M_y is recommended. Equation (E-7) led to an acceptably low prediction uncertainty of $\sigma_{LN} = 0.10$.

$$M_c/M_y = 1.13 \qquad \text{(E-7)}$$

Cyclic Energy Dissipation Capacity

Based on the observed trends in the data, the following equation is proposed for the mean energy dissipation capacity, including all statistically significant predictors:

$$\lambda = (131.0)(0.18)^{\nu}(0.26)^{s/d}(0.57)^{V_p/V_n}(61.4)^{\rho_{sh.eff}} \qquad \text{(E-8)}$$

As discussed previously, this value is calibrated for both types of strength deterioration and should be used with c = 1.0.

Discussion of Proposed Equations

The proposed equations can be evaluated to determine (a) how well they reflect the mean tendencies in the data, and (b) whether they provide suitably small prediction errors. Table E-1 reports the median ratio of predicted to observed values (from model calibrations) for each proposed equation. The regression analyses (e.g., Equations (E-3) – (E-8)) assumed the prediction uncertainty is lognormally distributed, so this median ratio should be close to 1.0; this ratio varies between 1.00 and 1.06 for the proposed equations, showing that the predictive equations have little bias.

Table E-1   Prediction Uncertainties and Bias in Proposed Equations

Equation	Median (predicted/ observed)	σ_{LN}	R^2
Effective Stiffness to Yield (E-1)	1.03	0.28	0.80
Effective Stiffness to 40% Yield (E-2)	1.02	0.33	0.59
Plastic Rotation Capacity (E-3)	1.02	0.54	0.60
Post-Capping Rotation Capacity (E-4)	1.00	0.72	0.51
Post-Yield Hardening Stiffness (E-5)	1.01	0.10	n/a
Cyclic Energy Dissipation Capacity (E-6)	1.06	0.47	0.51

Table E-1 also shows that the prediction uncertainty is large for many of the important parameters. For example, the prediction uncertainty (σ_{LN}) for plastic deformation capacity is 0.54. Previous research has shown that these large uncertainties in element deformation capacity cause similarly large uncertainties in collapse capacity (Ibarra, 2003; Goulet et al., 2006; Haselton et al., 2007). These large uncertainties are associated with wide variability in the physical phenomena and limitations in the available test data. With the availability of more test data it may be possible to reduce these uncertainties.

Due to its particularly large effect on collapse assessment, it is useful to also examine the prediction bias for selected subsets of the data for $\theta_{cap,pl}$ as given in Equation (E-5). For non-conforming elements (i.e., those with $\rho_{sh} <$ 0.003) the prediction is unbiased, with a median ratio of predicted to observed values of 1.02. The plastic rotation capacity equation tends to overpredict the plastic rotation capacity by approximately 12% for conforming elements ($\rho_{sh} > 0.006$), and underpredicts the plastic rotation capacity by approximately 10% for elements with extremely high axial load (i.e., axial load ratios above 0.65). Considering the large uncertainty in the prediction of plastic rotation capacity and the small number of datapoints in some of the subsets, these computed biases seem reasonable. With more data, the prediction error and biases could be further reduced.

To illustrate the impact of column design variables on the model parameters predicted by Equation (E-3) through (E-8), modeling parameters obtained for an 8-story reinforced concrete special moment frame perimeter system are shown in Table E-2. When the proposed equations are used, a typical column at the base should be modeled with an effective stiffness to yield of 35% of EI_g, a post-yield hardening ratio of 1.13, a plastic rotation capacity of 0.085, a post-capping rotation capacity of 0.10 and λ equal to 154.

Table E-2   Predicted Model Parameters for an 8-Story Reinforced Concrete Special Moment Frame Perimeter System (Interior Column, 1st-Story Location)

Design Parameter	Value	EI_{stf}/EI_g	M_c/M_y	$\theta_{cap,pl}$	θ_{pc}	λ
All	Baseline[1]	0.35	1.13	0.085	0.100	154
v	0.0	0.35	1.13	0.095	0.100	170
	0.3	0.52		0.055	0.100	104
ρ_{sh}	0.002	0.35	1.13	0.042	0.059	154
	0.010			0.078	0.100	
	0.020			0.105	0.100	
L_s/h	2	0.35	1.13	0.085	0.100	154
	6	0.58				
s/d	0.1	0.35	1.13	0.085	0.100	163
	0.4					106
	0.6					80
ρ	0.01	0.35	1.13	0.080	0.100	154
	0.03			0.094		

[1]Baseline values: $v = 0.06$, $\rho_{sh} = 0.0121$, $L_s/h = 2.7$, $s/d = 0.14$, $\rho = 0.018$, $f'_c = 35$ MPa, $\alpha_{sl} = 1$, $s_n = 7.5$.

Table E-2 also illustrates how changes in each design parameter impact predicted model parameters. For example, suppose various design parameters were changed such that the column under consideration in the first row of Table E-2 has a different axial load ratio, amount of transverse reinforcement, or column reinforcement ratio; these changes would lead to modifications in the parameters used to model that column. Within the range of column parameters considered in Table E-2, the predicted plastic rotation capacity varies from 0.042 to 0.105 radians. Axial load ratio (v) and the lateral confinement ratio (ρ_{sh}) have the largest effect on the predicted value of $\theta_{cap.pl}$, while concrete strength (f'_c), rebar buckling coefficient (s_n), and longitudinal reinforcement ratio (ρ) have less dominant effects (f'_c and s_n are not shown here). Table E-2 also shows the effects of the design parameters on initial stiffness, hardening stiffness ratio, post-capping stiffness, and cyclic deterioration parameters.

Summary and Limitations

The calibration process provided a comprehensive set of equations capable of predicting the parameters of a lumped plasticity element model for a reinforced concrete beam-column, with the empirical predictive equations having the ability to capture the effects of design aspects such as detailing and level of axial load. These equations are proposed for use with the element model developed by Ibarra et al. (2003, 2005), and can be used to model cyclic and in-cycle strength and stiffness degradation to capture element behavior up to the point of structural collapse. Even so, these equations are general and can be used with slight modifications with most lumped plasticity models currently used in research.

The predictive empirical equations proposed here provide a critical link between column design parameters and element modeling parameters, facilitating the creation of nonlinear structural models of reinforced concrete frame elements. The empirical equations predict the mean value of each parameter, so the prediction uncertainty associated with each equation is also quantified and reported. This provides an indication of the uncertainty in the prediction of model parameters, and can be used in sensitivity analyses and propagation of structural modeling uncertainties.

As with any study, there are limitations in terms of the applicability of these equations, which are discussed briefly here.

The equations developed here are based on a comprehensive database assembled by Berry et al. (Berry et al., 2004; PEER, 2006a). Even so, the range of column parameters included in the database is limited, and the derived equations may not be applicable outside the range of column parameters considered.

The equations are also limited more generally by the number of test specimens available that have an observed capping point. There are only a small number of experimental tests with clearly observable negative post-capping stiffness. For model calibration and understanding of element behavior, it is important that future testing continue to deformation levels large enough to clearly show the negative post-capping stiffness. In addition, there is very limited data that show post-peak cyclic deterioration behavior. With additional data, it may be possible to further remove some of the conservatism still in the proposed equations and reduce the prediction uncertainties. A further limitation is associated with the testing protocols; virtually all of the available test data are based on a cyclic loading protocol with many cycles and 2-3 cycles per deformation level. This type of loading may not be representative of the type of earthquake loading that typically causes structural collapse, which would generally only contain a few large displacement cycles before collapse occurs. Tests conducted under a variety of loading histories will lead to a better understanding of how load history impacts cyclic behavior, and provide a basis for better development and calibration of the element model cyclic rules.

In addition to the above mentioned issues, it is also important to remember that the empirical equations developed in this work are all based on laboratory test data where the test specimen was constructed in a controlled environment, and thus indicative of a high quality of construction. Actual buildings are constructed in a less controlled environment, so we expect the elements of actual buildings to have a lower level of performance than that predicted using the equations presented here. This work does not attempt to quantify this difference in performance coming from construction quality.

E.4   Joint Modeling

Nonlinear models of reinforced concrete structures employ a two-dimensional joint model developed and implemented by Lowes et al. (2004). This model accounts for the finite joint size, and includes rotational springs and systems of constraints for direct modeling of the shear panel and bond-slip behavior. Figure E-4 shows a schematic diagram of this model.

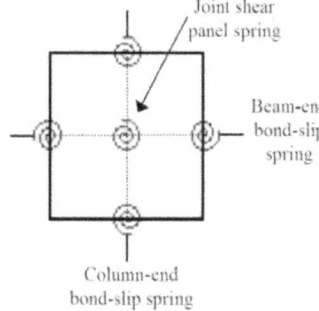

Figure E-4   Schematic diagram of joint model.

The joint model and parameters are based on a careful review of previous research. This approach is different than the detailed calibration study presented earlier to determine the parameters for beam-column elements.

E.4.1   Shear Panel Spring

Current building code provisions require that the joint shear capacity be based on capacity design principles for special moment frames, so if properly implemented, the joint shear demand should never exceed the capacity. A review of available research finds that the joint shear capacity designs provisions are conservative, and should be able to prohibit joint shear failure in all properly designed reinforced concrete special moment frame structures. For example, Brown and Lowes (2006) reviewed results of 45 experimental tests of conforming joints, finding that not one of these conforming joints exhibited damage requiring joint replacement. As a result of observations like these, the joint shear modeling is not judged to be a critical part of the overall behavior of the frame; damage in the joint shear panel will increase the flexibility of the frame, but the joint shear panel is not a dominant damage/failure mode (see also Appendix D).

For reinforced concrete special moment frame structures, joint models are developed to accurately reflect system stiffness. For simplicity, joints are modeled as elastic elements with the cracked stiffness based on simple mechanics (Umemura and Aoyama, 1969). Reinforced concrete ordinary moment frame structures are modeled to account for the deterioration that may occur in the joints of those structures. For readers interested in a more detailed approach, the modified compression field theory (MCFT) (Vecchio and Collins, 1986; Stevens et al., 1991; Altoontash, 2004, Lowes et al., 2004; and Lowes and Altoontash, 2003) is often used to develop panel shear models.

E.4.2   Bond-Slip Spring Model

The effect of bond-slip in joints and column footings is to decrease the pre-yield and post-yield stiffnesses and increase the plastic rotation capacity of reinforced concrete elements. In this study, bond-slip deformations are directly included in the plastic hinge calibration and thus do not need to be separately defined.

By lumping the effects of bond-slip in the plastic hinge model, it is inherently assumed that the bond-slip component of deformation is linear from zero to the yield load. This is a simplification, and a quadrilinear model could instead be used to capture the nonlinear bond-slip behavior prior to rebar yielding.

Appendix F

Collapse Evaluation of Individual Buildings

F.1   Introduction

Although developed as a tool to establish seismic performance factors for generic seismic-force-resisting systems, the Methodology could be readily adapted for collapse assessment of an individual building system. As such, it could be used to demonstrate adequate collapse performance for a new building designed using performance-based design methods (as permitted by ASCE/SEI 7-05), or for collapse safety evaluation of an existing building. This appendix describes an adaptation of the Methodology for use in collapse evaluation of an individual building system.

F.2   Feasibility

Buildings designed or evaluated using performance-based methods are often large or important structures. Such projects typically utilize detailed models for analysis of the building, and peer review is often required. In this way, they are already set up to utilize many aspects of the Methodology.

In contrast to index archetype models, analytical models for individual buildings might have more elements based on the actual configuration and specific geometry of the building. Collapse evaluation can be performed using two-dimensional or three-dimensional models of the building. Depending on the needs of the project, individual building models may, or may not include a high level of sophistication in modeling nonlinear behavior.

F.3   Approach

The Methodology is based on the concept of collapse level ground motions, defined as the level of ground motions that cause median collapse (i.e., one-half of the records in the set cause collapse). For a building to meet the collapse performance objectives of this Methodology, the median collapse capacity must be an acceptable ratio above the Maximum Considered Earthquake (MCE) ground motion demand level (i.e., the adjusted collapse margin ratio, $ACMR$, must exceed acceptable values).

By starting with an acceptable collapse probability for MCE ground motions, and working backwards through the Methodology, values of the spectral shape factor, *SSF*, and collapse margin ratio, *CMR*, can be calculated to determine the ground motion intensity corresponding to median collapse. The Methodology can be "reverse engineered" to determine the level of ground motions for which not more than one-half of the records should cause collapse.

By scaling the record set to this level, trial designs for a subject new building can be evaluated. If the analytical model of the trial design survives one-half or more of the records without collapse, then the building has a collapse probability that is equal to (or less than) the acceptable collapse probability for MCE ground motions, and meets the collapse performance objectives of the Methodology.

Similarly, an existing building can be evaluated for records scaled to the intensity corresponding to median collapse. If the analytical model of the existing building survives one-half or more of the records, then the existing building meets the collapse performance objectives of the Methodology. If different seismic criteria are needed, as is often the case with existing buildings, the process can begin with any probability of collapse deemed acceptable for the project.

F.4 Collapse Evaluation of Individual Building Systems

The process for collapse evaluation of an individual building system is summarized in the following steps.

F.4.1 Step One: Develop Nonlinear Model(s)

Development of a representative model (or models) of the building must incorporate the nonlinear behavioral characteristics of building-specific components.

- **Two-Dimensional Versus Three-Dimensional Models.** It is likely that the project has developed a three-dimensional (3-D) model of the building (for design), which could be used for collapse evaluation. Two-dimensional (2-D) models could be used, but such models would need to address response in both horizontal directions (e.g., two 2-D models) and account for torsion and potential coupling of bi-directional response.

- **Gravity System.** Since the construction of the building is known, elements of the gravity system can be incorporated into the building-specific model. Whether or not they are directly simulated, potential collapse modes of the gravity system must be included in the collapse assessment process for an individual building. This is necessary because displacement compatibility requirements do not prohibit gravity system collapse from becoming a controlling behavior mode; they only address compatibility of displacements at the MCE ground motion level, not at collapse level ground motions. P-delta effects must consider the full weight of the building.

Inclusion of gravity system collapse modes is a departure from the formal application of the Methodology, which does not account for gravity system collapse. Accounting for collapse modes of the gravity system is typically accomplished by using limit state criteria for non-simulated collapse modes. Practically speaking, there is little experimental data on which to base limit state criteria for many gravity system collapse modes. In these cases, a peer review team could assist in determining appropriate criteria for use in the collapse assessment.

- **Nonlinear Elements.** Model(s) should incorporate nonlinear properties for all elements of a seismic force-resisting system (and gravity system) that cannot be shown to remain fully elastic in a near-collapse condition.

F.4.2 Step Two: Define Limit States and Acceptance Criteria

- **Limit States.** Accurately simulating sidesway collapse requires sophisticated modeling of element degradation and strain softening. Depending on the availability of component test data, and ability to develop high-fidelity representations of individual building components, use of non-simulated collapse limit states (Chapter 5) can be considered in lieu of direct simulation of collapse.

- **Acceptable Collapse Probability.** The maximum acceptable probability of collapse should be determined. An acceptable collapse probability of 10% is consistent with the collapse performance objectives of this Methodology.

F.4.3 Step Three: Determine Total System Uncertainty and Acceptable Collapse Margin Ratio

Total System Uncertainty (β_{TOT}). The value of total system collapse uncertainty, β_{TOT}, is determined in accordance with Section 7.3, based on quality of the design requirements, quality of the test data used to develop nonlinear properties, and quality of the nonlinear model.

The quality of design requirements (Chapter 3) could be "Superior" for buildings generally conforming to materials and detailing requirements of ASCE/SEI 7-05. In the case of an existing building, a judgment would need to be rendered on the quality of the code under which the building was originally constructed. The quality of test data (Chapter 3) could range from "Good" to "Fair" depending on the extent to which available test data accurately capture degrading behavior of existing building components. For an individual building, the extent to which a model captures the full range of the "design space" would be considered "High," because the building configuration is a known quantity. In this case, when determining model quality in accordance with Chapter 5, the first row in Table 5-3 should be used.

- **Acceptable Value of Adjusted Collapse Margin Ratio (e.g., $ACMR_{10\%}$).** The acceptable value of the adjusted collapse margin ratio should be determined. Based on an acceptable collapse probability of 10%, $ACMR_{10\%}$ is consistent with the collapse performance objectives of this Methodology.

F.4.4 Step Four: Perform Nonlinear Static Analysis (NSA)

- **Nonlinear Static Analysis.** A nonlinear static (pushover) analysis is performed to check nonlinear behavior of the model, and to verify that all elements assumed to be essentially elastic have not yielded at the point that a collapse mechanism develops in the structure. Pushover analyses should be completed in both horizontal directions.

- **Structure Period-Based Ductility, μ_T.** The period-based ductility, μ_T, is determined from pushover analysis results, for both horizontal directions, in accordance with Chapter 6 and including non-simulated limit states as appropriate.

- **Spectral Shape Factor, SSF.** The spectral shape factor, SSF, for both horizontal directions, is determined based on the period-based ductility, μ_T, the building Seismic Design Category, and the fundamental period, T, in the direction of interest, in accordance with Section 7.2.2.

F.4.5 Step Five: Select Record Set and Scale Records

- **Record Set Selection.** If the building is located within 10 km of an active fault, then the Near-Field record set should be selected for collapse evaluation, otherwise the Far-Field record set should be used.

- **Record Set Scaling.** Both components of each record in the record set should be scaled up to the required collapse level intensity at which one-half or more of the records must not cause collapse.

  To scale the records to this level, all normalized records are multiplied by the same scale factor, SF. The scaling factor, SF, should be computed for both horizontal directions and the average value used to scale both components of each record.

$$SF = \frac{ACMR_{10\%}}{C_{3D}\,SSF}\left(\frac{S_{MT}}{S_{NRT}}\right)$$ (G-1)

where:

SF = record (and component) scale factor in the direction of interest, required for collapse evaluation of an individual building system.

$ACMR_{10\%}$ = acceptable value of the adjusted collapse margin ratio from Section 7.4, corresponding to an acceptable collapse probability of 10%.

SSF = spectral shape factor in the direction of interest, as defined in Chapter 7.

C_{3D} = three-dimensional analysis coefficient, taken as 1.2 for three-dimensional analysis, and 1.0 for two-dimensional analysis.

S_{MT} = MCE, 5% damped, spectral response acceleration at the fundamental period, T, of the building in the direction of interest, as defined in Section 11.4.3 of ASCE/SEI 7-05, including Site Class.

S_{NRT} = median value of normalized record set, 5% damped, spectral response acceleration at the fundamental period, T, of the building in the direction of interest.

T = the fundamental period of the building in the direction of interest, based on the limits of Section 12.8.2 of ASCE/SEI 7-05, computed in accordance with Equation 5-5 in Chapter 5.

F.4.6   Step Six: Perform Nonlinear Dynamic Analysis (NDA) and Evaluate Performance

- **Nonlinear Dynamic Analysis.** A nonlinear dynamic analysis of the model(s) is performed separately for each scaled record of the record set. The response is classified as either "collapse" or "non-collapse," based on sidesway collapse of the analytical model or through the use of non-simulated collapse component acceptance criteria. Ground motion record pairs should be applied to two-dimensional and three-dimensional models in accordance with Chapter 6.

- **Collapse Evaluation.** If less than one-half of the records cause collapse, then the trial design (or the existing building) meets the collapse performance objective, and the building has an acceptably low probability of collapse for MCE ground motions. If one-half or more of the records cause collapse, then the design does not meet the collapse performance objective, and re-design and re-evaluation are required.

Symbols

A = force normalized by effective seismic weight, W, corresponding to arbitrary post-yield displacement, D, of the isolation system in the horizontal direction under consideration, used to define bi-linear spring properties of isolators

A_g = gross cross-sectional area of an element

a_{sl} = indicator variable (0 or 1) to signify possibility of longitudinal rebar slip past the column end, where $a_{sl} = 1$ if slip is possible (Panagiotakos, 2001)

A_y = force normalized by effective seismic weight, W, corresponding to idealized yield displacement, D_y, of the isolation system in the horizontal direction under consideration, used to define bi-linear spring properties of isolators

$ACMR$ = adjusted collapse margin ratio

$ACMR_{10\%}$ = acceptable value of the adjusted collapse margin ratio ($ACMR$), on average, for the performance group of interest

$ACMR_{20\%}$ = acceptable value of the adjusted collapse margin ratio ($ACMR$) for an individual archetype of the performance group of interest

b = element width

B_{1E} = numerical coefficient as set forth in Table 18.6-1 of ASCE/SEI 7-05 for the effective damping equal to $\beta_I + \beta_{VI}$ and period equal to T

c = cyclic deterioration calibration term (exponent); describes the change in the *rate* of cyclic deterioration as the energy dissipation capacity is exhausted

C_{3D} = 3-dimensional analysis coefficient, 1.2 for 3-dimensional analysis (and 1.0 for 2-dimensional analysis)

C_d = deflection amplification factor (current values given in Table 12.2-1 of ASCE/SEI 7-05)

C_s = seismic response coefficient as determined in Section 12.8.1.1 of ASCE/SEI 7-05)

C_t	=	approximate period coefficient as determined in Table 12.8-2 of ASCE/SEI 7-05)
C_u	=	upper-limit period coefficient as determined in Table 12.8-1 of ASCE/SEI 7-05)
C_{units}	=	a units conversion variable that equals 1.0 when f'_c is in MPa units and 6.9 for ksi units
CI	=	confidence interval
CMR	=	collapse margin ratio
C_0	=	coefficient relating fundamental-mode (SDOF) displacement to roof (MDOF) displacement of an index archetype model, as defined in Section 6.3
d	=	column depth
D	=	effect of dead load for use in load combinations of Section 12.4 of ASCE/SEI 7-05
D	=	arbitrary post-yield displacement, in inches, of the isolation system in the horizontal direction under consideration used to define bi-linear spring properties of isolators (Chapter 10)
D_M	=	maximum displacement, in inches, at the center of rigidity of the isolation system in the direction under consideration, as prescribed by Eq. 17.5-3 of ASCE/SEI 7-05
D_{TM}	=	total maximum displacement, in inches, of the isolation system in the direction under consideration considering both translational and torsional displacement, as prescribed by Eq. 17.5-6 of ASCE/SEI 7-05
D_y	=	idealized yield displacement, in inches, of the isolation system in the horizontal direction under consideration used to define bi-linear spring properties of isolators
EI_g	=	gross cross-sectional moment of inertia
EI_{stf}	=	effective cross-sectional moment of inertia such that the secant stiffness is defined to 40% of the yield moment/force of the component
EI_y	=	effective cross-sectional moment of inertia that provides a secant stiffness through the yield point
E_t	=	total energy dissipation capacity
EXP	=	exponential

F_a = short-period site coefficient (at 0.2-second period) as given in Section 11.4.3 of ASCE/SEI 7-05

F_c = maximum strength (at capping point)

F_r = residual strength

F_v = long-period site coefficient (at 1.0-second period) as given in Section 11.4.3 of ASCE/SEI 7-05

F_x = nonlinear static (pushover) analysis force at level x

F_y = yield strength of material

f_c = compressive strength of unconfined concrete, based on standard cylinder test

f_y = yield stress of longitudinal reinforcement (Appendix E).

g = constant acceleration due to gravity

H = foundation loads due to lateral earth pressure, ground water pressure, or pressure of bulk materials, Chapter 2 of ASCE/SEI 7-05

h = element height

h_n = height in feet above the base to the highest level of the structure (ASCE/SEI 7-05)

h_r = height above the base to the roof level of the archetype building

I = the importance factor in Section 11.5.1 of ASCE/SEI 7-05

K = residual strength present in material model, defined as a ratio of M_y

K_c = post-capping stiffness, i.e., stiffness beyond $\theta_{cap,pl}$

K_e = effective elastic secant stiffness to the yield point

K_{pc} = post-capping tangent stiffness

k_{eff} = effective stiffness of the isolation system, in kips/in., at displacement, D, in the horizontal direction under consideration

k_{Mmin} = minimum effective stiffness, in kips/in., of the isolation system at the maximum displacement in the horizontal direction under consideration, as prescribed by Eq. 17.8-6 of ASCE/SEI 7-05

K_s = hardening stiffness, i.e., stiffness between θ_y and $\theta_{cap,pl}$

L = effect of live load for use in load combinations of Section 12.4 of ASCE/SEI 7-05

L_s = shear span, distance between column end and point of inflection

LN = natural logarithm

M_c = moment capacity at the capping point; used for prediction of hardening stiffness

M_y = yield moment for Ibarra material model (nominal moment capacity of the column)

$M_{y\ (Fardis)}$ = yield moment as calculated based on predictive equations (Panagiotakos and Fardis, 2001)

M_p = plastic moment capacity of element

m_x = mass of index archetype model at level x

N = number of levels (stories) in an index archetype model

NMi = normalization factor of the ith record of the set of interest

$NTH_{1,i}$ = normalized i^{th} record, horizontal component 1

$NTH_{2,i}$ = normalized i^{th} record, horizontal component 2

P = perimeter frame system (Chapter 9)

P = axial load

P_b = axial load at the balanced condition

PGV_{PEER} = peak ground velocity of the ith record based on the geometric mean of the two horizontal components considering different record orientations, based on the PEER-NGA database

Q_E = the effect of horizontal seismic force from total design base shear, V, for use in load combinations of Section 12.4 of ASCE/SEI 7-05

R = response modification coefficient (current values given in Table 12.2-1 of ASCE/SEI 7-05)

R_{eff} = effective R factor which accounts for the additional strength caused by the minimum base shear requirement

R_I = numerical coefficient related to the type of seismic force-resisting system above the isolation system (i.e., three-eights of the R value given in Table 12.2.-1 of ASCE/SEI 7-05, not to exceed 2.0, need not be taken as less than 1.0)

RC = reinforced concrete

s = spacing of transverse reinforcement in the column hinge region

S = snow load, Chapter 2 of ASCE/SEI 7-05

S = space frame system (Chapter 9)

S_I = mapped MCE, 5-percent damped, spectral response acceleration parameter at a period of 1 second as defined in Section 11.4.1 of ASCE/SEI 7-05

S_{CT} = random variable representing collapse level earthquake, 5-percent damped, spectral response acceleration at the fundamental period, T, of the building (Site Class D)

$S_{CT(SC)}$ = collapse level earthquake, 5-percent damped, spectral response acceleration at the fundamental period, T, of the building (Site Class D) obtained from simulated collapse failure modes.

$S_{CT(NSC)}$ = collapse level earthquake, 5-percent damped, spectral response acceleration at the fundamental period, T, of the building (Site Class D) obtained from non-simulated collapse failure modes.

$\hat{S}_{CT}$ = median value of collapse level earthquake, 5-percent damped, spectral response acceleration at the fundamental period, T, of the building (Site Class D)

S_{CTI} = collapse level earthquake, 5-percent damped, spectral response acceleration at the fundamental period, T_I, of the building (Site Class D)

S_{DS} = design, 5-percent damped, spectral response acceleration parameter at short periods as defined in Section 11.4.4 of ASCE/SEI 7-05

S_{DI} = design, 5-percent damped, spectral response acceleration parameter at a period of 1 second as defined in Section 11.4.4 of ASCE/SEI 7-05

S_{max} = maximum lateral force of the fully-yielded seismic force-resisting system normalized by the effective seismic weight of the building, W

S_{MS} = the MCE, 5-percent damped, spectral response acceleration parameter at short periods adjusted for site class effects as defined in Section 11.4.3 of ASCE/SEI 7-05

S_{MT} = MCE, 5-percent damped, spectral response acceleration at the fundamental period, T, of the building, as defined in Section 11.4.3 of ASCE/SEI 7-05 (Site Class D)

S_{M1} = the MCE, 5-percent damped, spectral response acceleration parameter at a period of 1 second adjusted for site class effects as defined in Section 11.4.3 of ASCE/SEI 7-05 (Site Class D)

s_n = rebar buckling coefficient, $(s/d_b)(f_y/100)^{0.5}$, where f_y is in MPa (similar to a term proposed by Dhakal and Maekawa (2002))

$\hat{S}_{NRT}$ = median value of normalized record set, 5-percent damped, spectral response acceleration at the fundamental period, T, of the building

S_S = mapped MCE, 5-percent damped, spectral response acceleration parameter at short periods as defined in Section 11.4.1 of ASCE/SEI 7-05

S_T = median value of normalized record set, 5-percent damped, spectral response acceleration at the fundamental period, T, for record set scaled to an arbitrary intensity. This parameter is also the anchor point for individual ground motion records of the record set at this intensity

S_a = spectra acceleration, g

$S_a(T)$ = the spectral acceleration at the period, T

$SCWB$ = ratio of flexure strengths of column and beams framing into a joint, computed as per ACI 318-05 (ACI 2005)

SD_{CT} = median value of collapse level earthquake, 5-percent damped, spectral response displacement at the fundamental period, T, of the building corresponding to spectral response acceleration, $\hat{S}_{CT}$

SD_{MT} = MCE, 5-percent damped, spectral response displacement at the fundamental period, T, of the building, corresponding to spectral response acceleration, S_{MT}

SF = record (and component) scale factor required for collapse evaluation of an individual building

SMF = special moment frame

SSF = spectral shape factor

SSF_i = spectral shape factor of ith index archetype analysis model

T = the fundamental period of the building, in seconds, based on the limits of Section 12.8.2 of the ASCE/SEI 7-05 and the approximate fundamental period, T_a

T_1 = the fundamental period of the building, as determined by eigenvalue analysis of the structural model (seconds)

T_a = the approximate fundamental period of the building, in seconds, as determined in Section 12.8.2.1 of ASCE/SEI 7-05

T_{eff} = effective period of isolation system, in seconds, at displacement, D, in the horizontal direction under consideration.

$TH_{1,i}$ = record i, horizontal component 1, PEER database

$TH_{2,i}$ = record i, horizontal component 2, PEER database

T_L = long-period transition period of the building, in seconds, as defined in Section 11.4.5 of ASCE/SEI 7-05

T_M = effective period, in seconds, of the seismically isolated structure at the maximum displacement in the direction under consideration, as prescribed by Eq. 17.5-4 of ASCE/SEI 7-05

T_s = short-period transition period of the building, in seconds, equal to S_{D1}/S_{DS}, as defined by Section 11.3 of ASCE/SEI 7-05

V = total design lateral force or shear at base (ASCE/SEI 7-05)

V_c = shear capacity of concrete, as per ACI 318-05

V_E = lateral force that would be developed in the seismic force-resisting system if the system remained entirely elastic for design earthquake ground motions (FEMA, 2004b)

V_{max} = maximum lateral force of the fully-yielded seismic force-resisting system (same as V_Y parameter, FEMA, 2004b)

V_{mpr} = capacity shear demand, caused by fully plastic moments at each end of the element, and including 25% overstrength, as per ACI 318-05 (ACI 2005)

V_n = nominal shear capacity including contributions from concrete and steel, as per the ACI 318-05

V_s = total lateral seismic design force or shear on elements above the isolation system, as prescribed by Eq. 17.5-8 of ASCE/SEI 7-05

V_u = shear demand

W = effective seismic weight of the structure above the isolation interface, as defined in Section 17.5.3.4 of ASCE/SEI 7-05

W = effective seismic weight of the building as defined in Section 12.7.2 of ASCE/SEI 7-05

w_x = portion of W that is located at or assigned to Level x

x	=	parameter of Equation (12.8-7) given in Table 12.8-2 of ASCE/SEI 7-05
β_0	=	regression coefficient from the equation $\text{LN}[S_{CT1}] = \beta_0 + \beta_1 \varepsilon$
β_1	=	regression coefficient from the equation $\text{LN}[S_{CT1}] = \beta_0 + \beta_1 \varepsilon$
β_{DR}	=	design requirements-related collapse uncertainty
β_{eff}	=	effective damping of the isolation system, percent of critical damping, at displacement, D, in the horizontal direction under consideration.
β_F	=	uncertainty associated with ductile fracture of steel reduced beam sections, logarithmic standard deviation
β_I	=	component of effective damping of the structure due to the inherent dissipation of energy by elements of the structure, at or just below the effective yield displacement of the seismic force-resisting system, Section 18.6.2.1 of ASCE/SEI 7-05
β_M	=	effective damping, percent of critical, of the isolation system at the maximum displacement in the direction under consideration, as prescribed by Eq. 17.8-8 of ASCE/SEI 7-05
β_{MDL}	=	modeling-related collapse uncertainty
β_{RTR}	=	record-to-record collapse uncertainty
β_{TD}	=	test data-related collapse uncertainty
β_{TOT}	=	total system collapse uncertainty
β_{V1}	=	component of effective damping of the fundamental mode of vibration of the structure in the direction of interest due to viscous dissipation of energy by the damping system, at or just below the effective yield displacement of the seismic force-resisting system, Section 16.6.2.3 of ASCE/SEI 7-05
δ	=	roof drift of the seismic force-resisting system for design earthquake ground motions (FEMA, 2004b)
δ_c	=	displacement associated with maximum strength, F_c (at the capping point)
δ_E	=	roof drift of the seismic force-resisting system if the system remained entirely elastic for design earthquake ground motions (FEMA, 2004b)

δ_u = roof displacement used to approximate the ultimate displacement capacity of the seismic force-resisting system, as derived from pushover analysis

δ_p = pre-capping plastic deformation capacity

$\delta_{y,eff}$ = effective roof displacement used to approximate full yield of the seismic force-resisting system, as derived from pushover analysis

ε = number of standard deviations between observed spectral value and the median prediction from an attenuation function

$\varepsilon(T)$ = the epsilon value of a ground motion, evaluated at period, T

$\varepsilon(T)_{,records}$ = the mean epsilon value of the Far-Field ground motion set, evaluated at period, T

$\bar{\varepsilon}_0(T)$ = the expected ε value for the site and hazard-level of interest

ϕ = strength reduction factor

$\phi_{1,x}$ = ordinate of the fundamental mode in the direction of interest at level x

$\phi_{1,r}$ = ordinate of the fundamental mode in the direction of interest at the roof

λ = normalized energy dissipation capacity; defined such that $E_t = \lambda M_y \theta_y$ (Ibarra et al. 2005)

λ_{DR} = random variable representing design requirements-related collapse uncertainty

λ_{MDL} = random variable representing modeling-related collapse uncertainty

λ_{RTR} = random variable representing record-to-record collapse uncertainty

λ_{TD} = random variable representing test data-related collapse uncertainty

μ_T = period-based ductility of an index archetype model

ν = axial load ratio ($P/A_g f'_c$)

θ_{cap} = total chord rotation at capping, sum of elastic and plastic deformations (radians)

$\theta_{cap,pl}$ (or θ_p) = plastic chord rotation from yield to cap (radians)

θ_p = plastic hinge rotation (radians)

θ_{pc} = post-capping plastic rotation capacity, from the cap to point of zero strength (radians)

$\hat{\theta}_p$ = median value of plastic hinge rotation at which ductile fracture in steel RBS is initiated (radians)

θ_y = chord rotation at yielding, taken as the sum of flexural, shear and bond-slip components; yielding is defined as the point of significant stiffness change, i.e., steel yielding or concrete crushing (radians)

ρ = redundancy factor based on the extent of structural redundancy present in a building as defined in Section 12.3.4 of ASCE/SEI 7-05

ρ (or ρ_{tot}) = ratio of total area of longitudinal reinforcement (for columns) or ratio of tensile longitudinal reinforcement (for beams)

ρ' = ratio compressive longitudinal reinforcement (for beams)

ρ_{sh} = area ratio of transverse reinforcement in column hinge region

$\rho_{sh,eff}$ = effective ratio of transverse reinforcement in column hinge region ($\rho_{sh}f_{y,w}/f'_c$)

σ_{LN} = logarithmic standard deviation for the prediction uncertainty

Ω = calculated overstrength of an index archetype analysis model

Ω_O = overstrength factor appropriate for use in the load combinations of Section 12.4 of ASCE/SEI 7-05 (current values of Ω_O are given in Table 12.2-1 of ASCE/SEI 7-05)

Glossary

Definitions

Archetype: A prototypical representation of a seismic-force-resisting system.

Archetype Design Space: The overall range of permissible configurations, structural design parameters, and other features that define the application limits for a seismic-force-resisting system.

Base: The level at which the horizontal seismic ground motions are considered to be imparted to the structure (ASCE/SEI 7-05).

Base Shear: Total design lateral force or shear at the base (ASCE/SEI 7-05).

Building: Any structure whose intended use includes shelter of human occupants (ASCE/SEI 7-05).

Collapse Level Earthquake Ground Motions: The level of earthquake ground motions that cause collapse of the seismic force-resisting system of interest.

Component: A part or element of an architectural, electrical, mechanical or structural system (ASCE/SEI 7-05).

Damping Device: A flexible structural element of the damping system that dissipates energy due to relative motion of each end of the device (ASCE/SEI 7-05).

Damping System: The collection of structural elements that includes all individual damping devices, all structural elements or bracing required to transfer forces from damping devices to the base of the structure, and the structural elements required to transfer forces from damping devices to the seismic force-resisting system (ASCE/SEI 7-05).

Design Earthquake Ground Motions: The earthquake ground motions that are two-thirds of the corresponding MCE ground motions (ASCE/SEI 7-05).

Design Requirements-Related Uncertainty: Collapse uncertainty associated with the quality of the design requirements of the system of interest.

Displacement Restraint System: A collection of structural elements that limits lateral displacement of seismically isolated structures due to the maximum considered earthquake (Chapter 17, ASCE/SEI 7-05).

Effective Damping: The value of equivalent viscous damping corresponding to energy dissipated during cyclic response of the isolation system (Chapter 17, ASCE/SEI 7-05).

Effective Stiffness: The value of the lateral force in the isolation system, or element thereof, divided by the corresponding lateral displacement (Chapter 17, ASCE/SEI 7-05).

Importance Factor: A factor assigned to each structure according to its Occupancy Category, as prescribed in Section 11.5.1 of ASCE/SEI 7-05.

Incremental Dynamic Analysis: Series of nonlinear response history analyses using an input ground motion that is incrementally scaled to increasing intensities until collapse is detected in the analysis.

Index Archetype Configuration: A prototypical representation of a seismic-force-resisting system configuration that embodies key features and behaviors related to collapse performance when subjected to earthquake ground motions.

Index Archetype Design: An index archetype configuration that has been proportioned and detailed using the design requirements of the system of interest.

Index Archetype Model: An idealized mathematical representation of an index archetype design used to simulate collapse using nonlinear static and dynamic analyses.

Isolation Interface: The boundary between the upper portion of the structure, which is isolated, and the lower portion of the structure, which moves rigidly with the ground (Chapter 17, ASCE/SEI 7-05).

Isolation System: The collection of structural elements that includes all individual isolator units, all structural elements that transfer force between elements of the isolation system, and all connections to other structural elements. The isolation system also includes the wind-restraint system, energy-dissipation devices, and/or the displacement restraint system, if such systems and devices are used to meet the design requirements of Chapter 17 (ASCE/SEI 7-05).

Isolator Unit: A horizontally flexible and vertically stiff structural element of the isolation system that permits large lateral deformations under design seismic loads. An isolator unit is permitted to be used either as part of, or in addition to, the weight-supporting system of the structure (Chapter 17, ASCE/SEI 7-05).

Maximum Considered Earthquake (MCE) Ground Motions: The most severe earthquake effects considered, as defined by Section 11.4 of ASCE/SEI 7-05.

Modeling Uncertainty: Collapse uncertainty associated with the quality of the index archetype models.

Nonbuilding Structure: A structure, other than a building, constructed of a type included in Chapter 15 of ASCE/SEI 7-05.

Non-Simulated Collapse: Structural collapse caused by collapse modes that are not represented in the analytical model. Non-simulated collapse occurs when a component limit state is exceeded, as defined by component fragility functions.

Occupancy: The purpose for which a building or other structure, or part thereof, is used or intended to be used (ASCE/SEI 7-05).

Occupancy Category: A classification assigned to a structure based on occupancy as defined in Table 1-1 of ASCE/SEI 7-05.

Performance Group: A subset of the archetype design space containing a group of index archetype configurations that share a set of common features or behavioral characteristics, binned for statistical evaluation of collapse performance.

Record-to-Record Uncertainty: Collapse uncertainty due to variability in response to different ground motions.

Record-to-Record Variability: Variation in the response of a structure under multiple input ground motions that are scaled to a consistent ground motion intensity.

Seismic Design Category: A classification assigned to a structure based on Occupancy Category and the severity of design earthquake ground motions at the site, as defined in Section 11.4 of ASCE/SEI 7-05.

Seismic Force-Resisting System: That part of the structural system that is considered to provide the required resistance to seismic forces prescribed in ASCE/SEI 7-05.

Sidesway Collapse: Structural collapse due to excessive story drift associated with loss of lateral strength and stiffness due to material and geometric nonlinearities.

Simulated Collapse: Structural collapse caused by collapse modes that are directly represented in the analytical model.

Site Class: A classification assigned to a site based on the types of soils present and their engineering properties, as defined in Chapter 20 of ASCE/SEI 7-05.

Structure: That which is built or constructed and limited to buildings and nonbuilding structures, as defined in ASCE/SEI 7-05.

Test Data-Related Uncertainty: Collapse uncertainty associated with the quality of the test data for the system of interest.

Total Maximum Displacement: The maximum considered earthquake lateral displacement, including additional displacement due to actual and accidental torsion, required for verification of the stability of the isolation system or elements thereof, design of structure separations, and vertical load testing of isolator unit prototypes (Chapter 17, ASCE/SEI 7-05).

Vertical Collapse: Structural collapse due to the loss of vertical-load-carrying capacity of a critical component.

References

Abrahamson, N.A., and Silva, W.J., 1997, "Empirical spectral response attenuation relations for shallow crustal earthquakes," *Seismological Research Letters*, 68 (1), pp. 94-126.

ACI, 2005, *Building Code Requirements for Structural Concrete* (ACI 318-05) *and Commentary*, (ACI 318R-05), American Concrete Institute, Farmington Hills, Michigan.

ACI, 2002b, *Building Code Requirements for Masonry Structures* (ACI 530/ASCE 5/TMS 402), Masonry Standards Joint Committee of the American Concrete Institute, Farmington Hills, Michigan; Structural Engineering Institute of the American Society of Civil Engineers, Reston, Virginia; and The Masonry Society, Boulder, Colorado.

ACI, 2002a, *Building Code Requirements for Structural Concrete* (ACI 318-02) *and Commentary* (ACI 318R-02), American Concrete Institute, Farmington Hills, Michigan.

ACI, 2001, *Acceptance Criteria for Moment Frames Based on Structural Testing* (ACI T1.1-01) and *Commentary* (ACI T1.1R-01), Innovation Task Group 1 and collaborators, American Concrete Institute, Farmington Hills, Michigan.

AISC, 2005, *Seismic Provisions for Structural Steel Buildings*, ANSI/AISC 341-05, American Institute for Steel Construction, Chicago, Illinois.

Altoontash, A., 2004, *Simulation and Damage Models for Performance Assessment of Reinforced Concrete Beam-Column Joints*, Ph.D. Dissertation, Department of Civil and Environmental Engineering, Stanford University, Stanford, California.

ANSI/AF&PA, 2005, *National Design Specification for Wood Construction* (ANSI/AF&PA NDS-2005), American National Standards Institute and American Forest and Paper Association, Washington, D.C.

ASCE, 2006b, *Seismic Rehabilitation of Existing Buildings*, ASCE Standard ASCE/SEI 41-06, American Society of Civil Engineers, Reston, Virginia.

ASCE, 2006a, *Minimum Design Loads for Buildings and Other Structures,* ASCE Standard ASCE/SEI 7-05, including Supplement No. 1, American Society of Civil Engineers, Reston, Virginia.

ASCE, 2005, *Minimum Design Loads for Buildings and Other Structures,* ASCE 7-05, American Society of Civil Engineers, Reston, Virginia.

ASCE, 2003, *Minimum Design Loads for Buildings and Other Structures.* ASCE Standard ASCE 7-02, American Society of Civil Engineers, Washington, D.C.

ASCE, 2002, *Minimum Design Loads for Buildings and Other Structures,* ASCE 7-02, American Society of Civil Engineers, Reston, Virginia.

ASTM, 2003, *Standard Test Method for Cyclic (Reversed) Load Test for Shear Resistance of Walls for Buildings,* ASTM ES 2126-02a, American Society for Testing and Materials, West Conshohocken, Pennsylvania.

ATC, 2009, *Guidelines for Seismic Performance Assessment of Buildings - 50% Draft,* Report No. ATC-58, prepared by the Applied Technology Council for the Federal Emergency Management Agency, Washington, D.C.

ATC, 2008, *Interim Guidelines on Modeling and Acceptance Criteria for Seismic Design and Analysis of Tall Buildings, 90% Draft,* Report No. ATC-72-1, Applied Technology Council, Redwood City, California.

ATC, 1992, *Guidelines for Cyclic Seismic Testing of Components of Steel Structures,* Report No. ATC-24, Applied Technology Council, Redwood City, California.

ATC, 1978, *Tentative Provisions for the Development of Seismic Regulations for Buildings,* Report No. ATC 3-06, Applied Technology Council, Redwood City, California; also NSF Publication 78-8 and NBS Special Publication 510.

Aslani, H., 2005, *Probabilistic Earthquake Loss Estimation and Loss Deaggregation in Buildings,* Ph.D. Dissertation, Department of Civil and Environmental Engineering, Stanford University, Stanford, California.

Aslani, H., and Miranda, E., 2005, "Fragility assessment of slab-column connections in existing non-ductile reinforced concrete buildings," *Journal of Earthquake Engineering,* 9 (6), pp. 777-804.

Aycardi, L.E., Mander, J., and Reinhorn, A., 1994, "Seismic resistance of reinforced concrete frame structures designed only for gravity loads: experimental performance of subassemblages." *ACI Structural Journal*, 91 (5), pp. 552-563.

Baker, J.W., 2007, "Quantitative classification of near-fault ground motions using wavelet analysis," *Bulletin of the Seismological Society of America*, 97 (5), pp. 1486-1501.

Baker, J.W., 2005, *Vector-Valued Ground Motion Intensity Measures for Probabilistic Seismic Demand Analysis*, Ph.D. Dissertation, Department of Civil and Environmental Engineering, Stanford University, Stanford, California.

Baker, J.W. and Cornell, C.A., 2006, "Spectral shape, epsilon and record selection," *Earthquake Engineering and Structural Dynamics*, 34 (10), pp. 1193-1217.

Benjamin, J.R., and Cornell, C.A, 1970, *Probability, Statistics, and Decision for civil engineers*, McGraw-Hill, New York, New York, 684 pp.

Berry, M., Parrish, M., and Eberhard, M., 2004, *PEER Structural Performance Database User's Manual*, Pacific Earthquake Engineering Research Center, University of California, Berkeley, California, 38 pp. Available at http://nisee.berkeley.edu/spd/ and http://maximus.ce.washington.edu/~peera1/.

Boore, D.M., Joyner, W.B., and Fumal, T.E., 1997, "Equations for estimating horizontal response spectra and peak accelerations from western North America earthquakes: a summary of recent work," *Seismological Research Letters*, 68 (1), pp. 128-153.

Brown, P.C., and Lowes, L.N., 2006, "Fragility functions for modern reinforced concrete beam-column joints," *Earthquake Spectra*, 23 (2), pp. 263-289.

Campbell, K.W., and Borzorgnia, Y., 2003, "Updated near-source ground-motion (attenuation) relations for the horizontal and vertical components of peak ground acceleration and acceleration response spectra," *Bulletin of the Seismological Society of America*, Vol. 93, No. 1, pp. 314-331.

Chatterjee, S., Hadi, A.S., and Price, B., 2000, *Regression Analysis by Example*, Third Edition, John Wiley and Sons Inc., New York, ISBN: 0-471-31946.

Chopra, A.K., Goel, R.K., and De la Llera, J.C., 1998, "Seismic code improvements based on recorded motions of buildings during earthquakes," *SMIP98 Seminar Proceedings*, California Geological Survey, Sacramento, California.

Clark, P., Frank, K., Krawinkler, H., and Shaw, R., 1997, *Protocol for Fabrication, Inspection, Testing, and Documentation of Beam-Column Connection Tests and Other Experimental Specimens*, SAC Steel Project Background Document, Report No. SAC/BD-97/02.

CoLA, 2001, *Report of a Testing Program of Light-Framed Walls with Wood-Sheathed Panels*, Final Report to the City of Los Angeles Department of Building and Safety, by Structural Engineers Association of Southern California and COLA-UCI Light Frame Test Committee, Department of Civil Engineering, University of California, Irvine, California, 93 pp.

Dhakal, R.P., and Maekawa, K., 2002, "Modeling of postyielding buckling of reinforcement," *Journal of Structural Engineering*, Vol. 128, No. 9, pp. 1139-1147.

Ekiert, C., and Hong, J., 2006, *Framing-to-Sheathing Connection Tests in Support of NEESWood Project*, Network of Earthquake Engineering Simulation; host institution: University at Buffalo, Buffalo, New York, 20 pp.

Elghadamsi, F.E., and Mohraz, B., 1987, "Inelastic earthquake spectra," *Earthquake Engineering and Structural Dynamics*, Vol. 15, pp. 91-104.

Ellingwood, B., Galambos, T.V., MacGregor, J.G., and Cornell, C.A., 1980, *Development of a Probability-Based Load Criterion for American National Standard A58*, National Bureau of Standards, Washington, DC, 222 pp.

Elwood, K.J., 2004, "Modeling failures in existing reinforced concrete columns," *Canadian Journal of Civil Engineering*, 31 (5), pp. 846-859.

Elwood, K.J., and Eberhard, M.O., 2006, "Effective stiffness of reinforced concrete columns," *PEER Research Digest* 2006-1, pp. 1- 5.

Elwood, K. J., and Moehle, J., 2005, "Axial capacity model for shear-damaged columns," *ACI Structural Journal*, 102 (4).

Engelhardt, M., Winneberger, T., Zekany, A.J., and Potyraj, T.J., 1998, "Experimental investigation of dogbone moment connections," *Engineering Journal of Steel Construction,* Vol. 4, pp. 128-139.

Fardis, M.N., and Biskini, D.E., 2003, "Deformation capacity of RC members, as controlled by flexure or shear," *Otani Symposium,* University of Tokyo, Japan, pp. 511- 530.

FEMA, 2009, *Effects of Strength and Stiffness Degradation on Seismic Response,* FEMA P440A, prepared by the Applied Technology Council for the Federal Emergency Management Agency, Washington, D.C.

FEMA, 2007, *Interim Testing Protocols for Determining the Seismic Performance Characteristics of Structural and Nonstructural Components,* FEMA 461, prepared by the Applied Technology Council for the Federal Emergency Management Agency, Washington, D.C.

FEMA, 2005, *Improvement of Nonlinear Static Seismic Analysis Procedures,* FEMA 440, Federal Emergency Management Agency, Washington, D.C.

FEMA, 2004b, *NEHRP Recommended Provisions for Seismic Regulations for New Buildings and Other Structures,* FEMA 450-2/2003 Edition, Part 2: Commentary, Federal Emergency Management Agency, Washington, D.C.

FEMA, 2004a, *NEHRP Recommended Provisions for Seismic Regulations for New Buildings and Other Structures,* FEMA 450-1/2003 Edition, Part 1: Provisions, Federal Emergency Management Agency, Washington, D.C.

FEMA, 2001, *NEHRP Recommended Provisions for Seismic Regulations for New Buildings and Other Structures,* FEMA 369, Federal Emergency Management Agency, Washington, D.C.

FEMA, 2000, *Prestandard and Commentary for Seismic Rehabilitation of Buildings,* FEMA 356, Prepared by the American Society of Civil Engineers for the Federal Emergency Management Agency, Washington, D.C.

Filiatrault, A., Wanitkorkul, A., and Constantinou, M.C., 2008, *Development and Appraisal of a Numerical Cyclic Loading Protocol for Quantifying Building System Performance,* Technical Report MCEER-08-0013, MCEER, University at Buffalo, Buffalo, New York.

Filiatrault, A., Lachapelle, E., and Lamontagne, P., 1998, "Seismic performance of ductile and nominally ductile reinforced concrete moment resisting frames I: experimental study," *Canadian Journal of Civil Engineering*, 25 (2), pp. 342-358.

Filippou, F.C., 1999, "Analysis platform and member models for performance-based earthquake engineering," *U.S.-Japan Workshop on Performance-Based Earthquake Engineering Methodology for Reinforced Concrete Building Structures*, PEER Report 1999/10, Pacific Earthquake Engineering Research Center, University of California, Berkeley, California, pp. 95-106.

Folz, B., and Filiatrault, A., 2004b, "Seismic analysis of woodframe structures II: model implementation and verification," *ASCE Journal of Structural Engineering*, Vol. 130, No. 8, pp. 1361-1370.

Folz, B., and Filiatrault, A., 2004a, "Seismic analysis of woodframe structures I: model formulation," *ASCE Journal of Structural Engineering*, Vol. 130, No. 8, pp. 1353-1360.

Folz, B., and Filiatrault, A., 2001, "Cyclic analysis of wood shear walls," *ASCE Journal of Structural Engineering*, 127 (4), pp. 433-441.

Fonseca, F., Rose, S., and Campbell, S., 2002, *Nail, Wood Screw, and Staple Fastener Connections*, CUREE Report No. W-16, Consortium of Universities for Earthquake Engineering Research, Richmond, California.

Gatto, K., and Uang, C.M., 2002, *Cyclic Response of Woodframe Shearwalls: Loading Protocol and Rate of Loading Effects*, CUREE Report No. W-16, Consortium of Universities for Earthquake Engineering Research, Richmond, California.

Goulet, C., Haselton, C.B., Mitrani-Reiser, J., Beck, J., Deierlein, G.G., Porter, K.A., and Stewart, J., 2006, "Evaluation of the seismic performance of a code-conforming reinforced-concrete frame building - from seismic hazard to collapse safety and economic losses," *Earthquake Engineering and Structural Dynamics*, 36 (13), pp. 1973-1997.

Goulet, C., Haselton, C.B., Mitrani-Reiser, J., Stewart, J., Taciroglu, E., and Deierlein, G.G., 2006, "Evaluation of seismic performance of a code-conforming reinforced-concrete frame buildings - part I, ground motion selection and structural collapse simulation," *Proceedings*, 8th National Conference on Earthquake Engineering, San Francisco, California.

Harmsen, S.C., 2001, "Mean and Modal ε in the Deaggregation of Probabilistic Ground Motion," *Bulletin of the Seismological Society of America*, 91 (6), pp. 1537-1552.

Harmsen, S.C., Frankel, A. D., and Petersen, M. D., 2003, "Deaggregation of U.S. Seismic Hazard Sources: The 2002 Update," *U.S. Geological Survey Open-File Report 03-440*, Available at http://pubs.usgs.gov/of/2003/ofr-03-440/.

Haselton, C.B., 2006, *Assessing Seismic Collapse Safety of Modern Reinforced Concrete Moment-Frame Buildings*, Ph.D. Dissertation, Department of Civil and Environmental Engineering, Stanford University, Stanford, California.

Haselton, C.B., Baker, J.W., Liel, A.B. and Deierlein, G.G., 2009, "Accounting for expected spectral shape (epsilon) in collapse performance assessment," *American Society of Civil Engineers Journal of Structural Engineering, Special Publication on Ground Motion Selection and Modification* (submitted).

Haselton, C.B., Mitrani-Reiser, J., Goulet, C., Deierlein, G.G., Beck, J., Porter, K.A., Stewart, J., and Taciroglu, E., 2008, *An Assessment to Benchmark the Seismic Performance of a Code-Conforming Reinforced-Concrete Moment-Frame Building*, PEER Report 2007/12, Pacific Earthquake Engineering Research Center, University of California, Berkeley, California.

Haselton, C.B., and Deierlein, G.G., 2007, *Assessing Seismic Collapse Safety of Modern Reinforced Concrete Frame Buildings*, John A. Blume Earthquake Engineering Center Technical Report No. 156, Stanford University, Stanford, California.

Haselton, C.B., Liel, A., Taylor Lange, S., and Deierlein, G.G., 2007, *Beam-Column Element Model Calibrated for Predicting Flexural Response Leading to Global Collapse of RC Frame Buildings*, PEER 2007/03, Pacific Earthquake Engineering Research Center, University of California, Berkeley, California.

Haselton, C.B., and Baker, J.W., 2006, "Ground motion intensity measures for collapse capacity prediction: Choice of optimal spectral period and effect of spectral shape," *Proceedings*, 8th National Conference on Earthquake Engineering, San Francisco, California.

Ibarra, L.F., and Krawinkler, H., 2005b, *Global Collapse of Frame Structures Under Seismic Excitations*, PEER Report 2005/06, and John A. Blume Earthquake Engineering Center Technical Report No.

152, Department of Civil Engineering, Stanford University, Stanford, California.

Ibarra, L., and Krawinkler, H., 2005a, "Effect of uncertainty in system deterioration parameters on the variance of collapse capacity," *Proceedings,* ICOSSAR'05, Rome, Italy, pp. 3583-3590, Millpress, Rotterdam, ISBN 90 5966 040 4.

Ibarra, L.F., Medina, R.A., and Krawinkler, H., 2005, "Hysteretic models that incorporate strength and stiffness deterioration," *International Journal for Earthquake Engineering and Structural Dynamics,* Vol. 34, No.12, pp. 1489-1511.

Ibarra, L., Medina, R., and Krawinkler, H., 2002, "Collapse assessment of deteriorating SDOF systems," *Proceedings,* 12th European Conference on Earthquake Engineering, London, Elsevier Science Ltd, paper #665.

ICBO, 1997, *Uniform Building Code,* 1997 Edition, International Conference of Building Officials, Whittier, California.

ICC, 2009, *Acceptance Criteria for Prefabricated Wood Shear Panels,* AC130, International Code Council, Available at www.icc-es.org.

ISO/IEC, 2005, *General Requirements for the Competence of Testing and Calibration Laboratories,* ISO/IEC 17025, International Organization for Standardization, Geneva, Switzerland.

Isoda, H., Furuya, O., Tatsuya, M., Hirano, S. and Minowa, C., 2007, "Collapse behavior of wood house designed by minimum requirement in law," *Journal of Japan Association for Earthquake Engineering* (under review).

Isoda, H., Folz, B., and Filiatrault, A., 2001, *Seismic Modeling of Index Woodframe Buildings,* CUREE Report No. W-12, Consortium of Universities for Research in Earthquake Engineering, Richmond, California, 144 pp.

Kircher, C.A., 2006, "Seismically isolated structures," *NEHRP Recommended Provisions: Design Examples,* Chapter 11 of FEMA 451, prepared by the Building Seismic Safety Council for the Federal Emergency Management Agency, Washington, D.C.

Krawinkler, H., 1996, "Cyclic loading histories for seismic experimentation on structural components," *Earthquake Spectra*, Vol. 12, Number 1, pp. 1-12.

Krawinkler, H., 1978, "Shear design of steel frame joints," *Engineering Journal*, AISC, Vol. 15, No. 3.

Krawinkler, H., and Zareian, F., 2007, "Prediction of collapse – how realistic and practical is it, and what can we learn from it?," *The Structural Design of Tall and Special Buildings,* Vol. 16, No. 5, pp. 633-653.

Krawinkler, H., Parisi, F., Ibarra, L., Ayoub, A., and Medina, R., 2000, *Development of a Testing Protocol for Woodframe Structures,* CUREE Publication No. W-02, Consortium of Universities for Earthquake Engineering Research, Richmond, California, 74 pp.

Krawinkler, H., Bertero, V.V., and Popov, E.P., 1971, *Inelastic behavior of steel beam-to-column sub-assemblages*, Report No. UCB/EERC-71/07, Earthquake Engineering Research Center, University of California, Berkeley, California.

Kunnath, S.K., Hoffmann, G., Reinhorn, A.M., and Mander, J.B., 1995, "Gravity-load-designed reinforced concrete buildings -- part I: seismic evaluation of existing construction," *ACI Structural Journal*, 92 (3).

Lai, S.P., and Biggs, J.M., 1980, "Inelastic response spectra for aseismic buildings," *Journal of the Structural Division*, ASCE, Vol. 106, No. ST6, pp. 1295-1310.

Liel, A. B., 2008, *Assessing the Collapse Risk of California's Existing Reinforced Concrete Frame Structure: Metrics for Seismic Safety Decisions,* Ph.D. Dissertation, Department of Civil and Environmental Engineering, Stanford University, Stanford, California.

Lignos, D.G., and Krawinkler, H., 2007, "A database in support of modeling of component deterioration for collapse prediction of steel frame structures," *ASCE Structures Congress*, Long Beach, California.

Line, P., Waltz, N., and Skaggs, T., 2008, "Seismic equivalence parameters for engineered wood frame wood structural panel shear walls," *Wood Design Focus*, 18 (2).

Lowes, L.N., Mitra, N., and Altoontash, A., 2004, *A Beam-Column Joint Model for Simulating the Earthquake Response of Reinforced Concrete Frames*, PEER 2003/10, Pacific Earthquake Engineering Research Center, University of California, Berkeley, California.

Lowes, L.N., and Altoontash, A., 2003, "Modeling of reinforced-concrete beam-column joints subjected to cyclic loading," *Journal of Structural Engineering*, 129 (12).

Miranda, E. and Bertero, V.V., 1994, "Evaluation of strength reduction factors for earthquake-resistant design," *Earthquake Spectra*, Vol. 10, No. 2, pp. 357-379.

Nassar, A.A. and Krawinkler, H., 1991, *Seismic Demands for SDOF and MDOF Systems*, The John A. Blume Earthquake Engineering Center Technical Report No. 95, Stanford University, Stanford, California.

Newmark, N.M. and Hall, W.J., 1973, "Seismic design criteria for nuclear reactor facilities," *Building Practices for Disaster Mitigation*, Report No. 46, National Bureau of Standards, U.S. Department of Commerce, pp. 209-236.

OpenSees, 2006, *Open System for Earthquake Engineering Simulation*, Pacific Earthquake Engineering Research Center, University of California, Berkeley, Available at http://opensees.berkeley.edu/.

Panagiotakos, T.B., and Fardis, M.N., 2001, "Deformations of reinforced concrete members at yielding and ultimate," *ACI Structural Journal*, 98 (2).

PEER, 2006a, *PEER NGA Database*, Pacific Earthquake Engineering Research Center, University of California, Berkeley, California, Available at http://peer.berkeley.edu/nga/.

PEER, 2006b, *PEER Structural Performance Database*, Pacific Earthquake Engineering Research Center, University of California, Berkeley, Available at http://nisee.berkeley.edu/spd/ and http://maximus.ce.washington.edu/~peera1/.

Ricles, J.M., Zhang, X., Lu, L.W., Fisher, J., 2004, *Development of seismic guidelines for deep-column steel moment connections*, ATLSS Report No. 04-13, Center for Advanced Technology for Large Structural Systems, Lehigh University, Bethlehem, Pennsylvania.

Riddell, R., Hidalgo, P., and Cruz, E., 1989, "Response modification factors for earthquake-resistant design of short-period structures," *Earthquake Spectra*, Vol. 5, No. 3, pp. 571-590.

Sivaselvan M., and Reinhorn, A.M., 2000, "Hysteretic models for deteriorating inelastic structures," *ASCE Journal of Engineering Mechanics*, Vol. 126 (6), pp. 633-640.

Stevens, N., Uzumeri, S., and Collins, M., 1991, "Reinforced concrete subjected to reversed cyclic shear," *ACI Structural Journal,* 88 (2), pp. 135-146.

Takeda, T., Hwang, H.H.M., and Shinozuka, M., 1988, "Response modification factors for multiple-degree-of-freedom systems," *Proceedings,* 9th World Conference on Earthquake Engineering, Tokyo-Kyoto, Japan, Vol. V, pp. 129-134.

Uang, C.M., Yu, K., and Gilton, C., 2000, *Cyclic Response of RBS Moment Connections: Loading Sequence and Lateral Bracing Effects,* Structural Steel Research Project Report No. SSRP-99/13.

Umemura, H., and Aoyama, H., 1969, "Empirical evaluation of the behavior of structural elements," *Proceedings,* 4th World Conference on Earthquake Engineering, Santiago, Chile.

Vamvatsikos, D., and Cornell, C.A., 2002, "Incremental Dynamic Analysis," *Earthquake Engineering and Structural Dynamics,* Vol. 31, Issue 3, pp. 491-514.

Vecchio, F.J., and Collins, M.P., 1986. "Modified compression-field theory for reinforced concrete elements subjected to shear," *ACI Journal,* 83 (2), pp. 219-231.

Vidic, T., Fajfar, P., and Fischinger, M., 1992, "A procedure for determining consistent inelastic design spectra," *Proceedings,* Workshop on Nonlinear Seismic Analysis of RC Structures, Bled, Slovenia.

White, T.W. and Ventura, C., 2007, "Seismic behavior of residential wood-frame construction in British Columbia: Part I – modeling and validation," *Proceedings,* Ninth Canadian Conference on Earthquake Engineering, Ottawa, Canada, 10 pp.

Zareian, F., 2006, *Simplified Performance-Based Earthquake Engineering,* Ph.D. Dissertation, Dept. of Civil and Environmental Engineering, Stanford University, Stanford, California.

Project Participants

ATC Management and Oversight

Christopher Rojahn (Project Executive)
Applied Technology Council
201 Redwood Shores Parkway, Suite 240
Redwood City, California 94065

Jon A. Heintz (Project Manager)
Applied Technology Council
201 Redwood Shores Parkway, Suite 240
Redwood City, California 94065

William T. Holmes (Project Technical Monitor)
Rutherford & Chekene
55 Second Street, Suite 600
San Francisco, California 94105

FEMA Project Officer

Michael Mahoney
Federal Emergency Management Agency
500 C Street, SW
Washington, DC 20472

FEMA Technical Monitor

Robert D. Hanson
Federal Emergency Management Agency
2926 Saklan Indian Drive
Walnut Creek, California 94595

Project Management Committee

Charles Kircher (Project Technical Director)
Kircher & Associates, Consulting Engineers
1121 San Antonio Road, Suite D-202
Palo Alto, California 94303

Michael Constantinou
University at Buffalo
Dept. of Civil, Structural & Environ. Engineering
132 Ketter Hall
Buffalo, New York 14260

Gregory Deierlein
Stanford University
Dept. of Civil & Environmental Engineering
240 Terman Engineering Center
Stanford, California 94305

James R. Harris
J.R. Harris & Company
1776 Lincoln Street, Suite 1100
Denver, Colorado 80203

John Hooper
Magnesson Klemencic Associates
1301 Fifth Avenue, Suite 3200
Seattle, Washington 98101

Allan R. Porush
URS Corporation
915 Wilshire Blvd., Suite 700
Los Angeles, California 90017

Christopher Rojahn (ex-officio)
William Holmes (ex-officio)
Jon A. Heintz (ex- officio)

Working Group on Nonlinear Static Analysis

Michael Constantinou, Group Leader
University at Buffalo
Dept. of Civil, Structural & Environ. Engineering
132 Ketter Hall
Buffalo, New York 14260

Assawin Wanitkorkul
The University at Buffalo Foundation
Suite 211, The UB Commons
520 Lee Entrance
Amherst, New York 14228

Working Group on Nonlinear Dynamic Analysis

Gregory Deierlein, Group Leader
Stanford University
Dept. of Civil & Environmental Engineering
Blume Earthquake Engineering Center
Stanford, California 94305

Curt Haselton
California State University, Chico
475 East 10th Avenue
Chico, California 95926

Abbie Liel
University of Colorado at Boulder
1793 Yellow Pine Avenue
Boulder, Colorado 80304

Jason Chou
University of California, Davis
11713 New Albion Drive
Gold River, California 95670

Stephen Cranford
Stanford University
119 Quillen Court, Apt. 417
Stanford, California 94305

Brian Dean
Stanford University
159 Melville Ave.
Palo Alto, California 94301

Kevin Haas
Stanford University
218 Ayrshire Farm Lane #108
Stanford, California 94305

Jiro Takagi
Stanford University
322 College Ave., #C
Palo Alto, California 94306

Working Group on Wood-frame Construction

Andre Filiatrault, Group Leader
University at Buffalo
Dept. of Civil, Structural & Environ. Engineering
134 Ketter Hall
Buffalo, New York 14260

Jiannis Christovasilis
The University at Buffalo Foundation
Suite 211, The UB Commons
520 Lee Entrance
Amherst, New York 14228

Kelly Cobeen
Wiss Janney Elstner Associates, Inc.
2200 Powell St., Ste 925
Emeryville, California 94608

Working Group on Autoclaved Aerated Concrete

Helmut Krawinkler, Group Leader
Stanford University
Dept. of Civil & Environmental Engineering
380 Panama Mall
Stanford, California 94305

Farzin Zareian
University of California, Irvine
Dept. of Civil & Environmental Engineering
E/4141 Engineering Gateway
Irvine, California 92697

Project Review Panel

Maryann T. Phipps (Chair)
Estructure
8331 Kent Court, Suite 100
El Cerrito, California 94530

Amr Elnashai
Mid-America Earthquake Center
University of Illinois at Urbana-Champaign
Department of Civil and Environmental
Engineering 1241 Newmark
Civil Engineering Lab
Urbana, Illinois 61801

S.K. Ghosh
S.K. Ghosh Associates Inc.
334 East Colfax Street, Unit E
Palatine, Illinois 60067

Ramon Gilsanz
Gilsanz Murray Steficek LLP
129 W. 27th Street, 5th Floor
New York, New York 10001

Ronald O. Hamburger
Simpson Gumpertz & Heger
The Landmark @ One Market, Suite 600
San Francisco, California 94105

Jack Hayes
National Institute of Standards and Technology
100 Bureau Drive, MS 8610
Gaithersburg, Maryland 20899

Richard E. Klingner
University of Texas at Austin
10100 Burnet Rd
Austin, Texas 78758

Philip Line
American Forest and Paper Association (AFPA)
1111 19th Street, NW, Suite 800
Washington, DC 20036

Bonnie E. Manley, P.E.
AISI Regional Director
4 Canvasback Way
Walpole, Maryland 02081

Andrei M. Reinhorn
University at Buffalo
Dept. of Civil, Structural & Envir. Engineering
231 Ketter Hall
Buffalo, New York 14260

Rafael Sabelli
Walter P. Moore
595 Market Street, Suite 950
San Francisco, California 94105

Workshop Participants – Chicago, Illinois

John Abruzzo
Thornton Tomasetti
1617 JFK Boulevard, Suite 545
Philadelphia, Pennsylvania 19103

Victor Azzi
Consulting Structural Engineer
1100 Old Ocean Boulevard
Rye, New Hampshire 03870

William Baker
Skidmore Owings & Merrill LLP
224 South Michigan Avenue, Suite 1000
Chicago, Illinois 60604

Charles Carter
American Institute of Steel Construction
One East Wacker Drive Suite 700
Chicago, Illinois 60601-1802

Finley Charney
Virginia Polytechnic Institute
Dept. of Civil & Environmental Engineering
200 Patton Hall, MS 0105
Blacksburg, Virginia 24061

Peter Cheever
LeMessurier Consultants, Inc.
675 Massachusetts Avenue
Cambridge, Massachusetts 02139

Helen Chen
American Institute of Steel Construction
One East Wacker Drive Suite 700
Chicago, Illinois 60601-1802

Ned Cleland
Blue Ridge Design, Inc.
19 West Cork Street, Suite 300
Winchester, Virginia 22601-4749

Bruce Ellingwood
Georgia Institute of Technology
790 Atlantic Drive
School of Civil and Environmental Engineering
Atlanta, Georgia 30332-0355

Perry Green
Steel Joint Institute (SJI)
3127 Mr. Joe White Avenue
Myrtle Beach, South Carolina 29577-6760

Greg Greenlee
United Steel Products Company
2150 Kitty Hawk Road
Livermore, California 94551-9522

Kirk Grundahl
QUALTIM, Inc.
6300 Enterprise Lane
Madison, WI 53719

Kurt Gustafson
AISC
One East Wacker Drive Suite 700
Chicago, Illinois 60601-1802

Jerome Hajjar
University of Illinois
2129b Newmark Civil Engineering Lab.
05 N. Mathews Ave., MC 250
Urbana, Illinois 61801

Bob Hannen
Wiss, Janney, Elstner Associates
330 Pfingsten Road
Northbrook, Illinois 60062

Joe Jun
Allied Tube & Conduit
16100 South Lathrop Avenue
Harvey, Illinois 60426

Mike Kempfert
Computerized Structural Design
8989 N. Port Washington Rd. Suite 101
Milwaukee, Wisconsin 53217

Bonghwan Kim
Skidmore Owings & Merrill LLP
14 Wall Street
New York, New York 10005

Vladimir Kochkin
NAHB Research Center
400 Prince George's Blvd.
Upper Marlboro, Maryland 20774

Jay Larson
American Iron and Steel Institute
3810 Sydna Street
Bethlehem, Pennsylvania 18107-1048

Judy Liu
Purdue University
School of Civil Engineering
550 Stadium Mall Drive
West Lafayette, Idiana 47907

Douglas Nyman
DJ Nyman & Associates
12337 Jones Road, Suite 232
Houston, Texas 77070-4844

Marjan Popovski
FPInnovations
2665 East Mall
Vancouver, British Columbia C V6T 2G1 0

Randy Poston
Whitlock Dalrymple Poston & Associates
10621 Gateway Boulevard Suite 200
Manassas, Virginia 20110

Lawrence Reaveley
University of Utah
160 S. Cventral Camus Dr., Rm 104
Salt Lake City, Utah 84112-6931

Brett Schneider
Guy Nordenson & Associates
225 Varick, 6th Floor
New York, New York 10014

Lee Shoemaker
Metal Building Manufacturers Association
1300 Sumner Avenue
Cleveland, Ohio 44115-2851

Mike Tong
Federal Emergency Management Agency
500 C Street SW
Washington, DC 20472

Workshop Participants – San Francisco, California

Daniel Abrams
University of Illinois
205 N. Mathews Ave., MC 250
Urbana, Illinois 61801

Don Allen
Steel Framing Alliance (SFA)
1201 15th Street, N.W., Suite 320
Washington, DC 20005-2842

Robert Bachman
RE Bachman Consulting Structural Engrs.
25152 La Estrada Drive
Laguna Niguel, California 92677

Yousef Bozorgnia
PEER Center
University of California, Berkeley
325 Davis Hall - MC1792
Berkeley, California 94720

Edwin T. Dean
Nishkian Dean
425 SW Start Street, Second Floor
Portland, Oregon 97204

Ronald DeVall
Read Jones Christoffersen
201-628 W. 13th Avenue
Vancouver, British Columbia V6T 1N9 Canada

Ken Elwood
University of British Columbia
Dept. of Civil Engineering
6250 Applied Science Lane
Vancouver, British Columbia V6T 1Z4 Canada

Johanna Fenten
FEMA Region IX
1111 Broadway, Suite 1200
Oakland, California 94607-4052

David Garza
Garza Structural Engineers
1134 Shady Mill Road
Corona, California 92882

Avik Ghosh
Borm
5161 California Suite 250
Irvine, California 92617

David S. Gromala
Weyerhaeuser; WTC-2B2
P.O. Box 9777
Federal Way, Washington

Bernadette Hadnagy
Applied Technology Council
201 Redwood Shores Pkwy., Suite 240
Redwood City, California 94065

Gary Hart
Weidlinger Associates
4551 Glencoe Ave., Suite 350
Marina del Rey, California 90292

Ayse Hortacsu
Applied Technology Council
201 Redwood Shores Pkwy., Suite 240
Redwood City, California 94065

Ronald Klemencic
Magnusson Klemencic Associates
1301 Fifth Avenue, Suite 3200
Seattle, Washington 98101-2699

Peter L. Lee
Skidmore, Owings, & Merrill LLP
1 Front Street
San Francisco, California 94111

Roy Lobo
State of California Facilities Development
    Division
1600 9th Street, Room 420
Sacramento, California 95814

Suikai Lu
Wienerberger AG
A-1100 Wien
Wienerberg City, Austria

Ray Lui
Department of Building Inspection
1660 Mission Street, Second Floor
San Francisco, California 94103

James A. Mahaney
Wiss, Janney, Elstner Associates, Inc.
104 El Dorado Street
Auburn, California 95603

Steven Mahin
University of California, Berkeley
Dept. of Civil Engineering
777 Davis Hall
Berkeley, California 94720

James O. Malley
Degenkolb Engineers
235 Montgomery Street, Suite 500
San Francisco, California 94104

Michael Mehrain
URS Corporation
915 Wilshire Blvd., Suite 700
Los Angeles, California 90017

Farzad Naeim
John A. Martin & Associates, Inc.
950 S. Grand Avenue, 4th Floor
Los Angeles, California 90015

Steven Pryor
Simpson Strong-Tie
5956 W. Las Positas Blvd.
Pleasanton, California 94588

Josh Richards
KPFF Consulting Engineers
111 SW Fifth Avenue, Suite 2500
Portland, Oregon 97204

Charles Roeder
Unversity of Washington
Dept. of Civil and Environmental Engineering
201 More Hall
Seattle, Washington 98195-2700

James E. Russell
Building Codes Consultant
3654 Sun View Way
Concord, California 94520-1346

Reynaud Serrette
Santa Clara University
Department of Civil Engineering
Santa Clara, California 96053-0563

Benson Shing
University of California, San Diego
409 University Center
9500 Gilman Drive, MC0085
La Jolla, California 92093

Thomas Skaggs
APA
7011 S. 19th Street
Tacoma, Washington 98466

William Staehlin
Division of the State Architect,
1102 Q Street, Suite 5200
Sacramento, California 95814

Kurt Stochlia
International Code Council
5360 Workman Mill Road
Whittier, California 90601-2298

Steve Tipping
Tipping Mar & Associates
1906 Shattuck Avenue
Berkeley, California 94704

Chia-Ming Uang
University of California, San Diego
Dept. of Structural Engineering
409 University Center
La Jolla, California 92093-0085

Williston L. Warren , IV
SESOL, Inc.
1001 West 17th St. Suite K
Costa Mesa, California 92627

Nabih Youssef
Nabih Youssef & Associates
800 Wilshire Blvd., Suite 200
Los Angeles, California 90017

Victor Zayas
Earthquake Protection Systems, Inc.
451 Azuar Drive, Building 759
Vallejo, California 94592

www.ingramcontent.com/pod-product-compliance
Lightning Source LLC
Chambersburg PA
CBHW081429170526
45166CB00008B/2143